Interface Oral Health Science 2007

M. Watanabe, O. Okuno
Editors
K. Sasaki, O. Suzuki, N. Takahashi
Associate Editors

Interface Oral Health Science 2007

Proceedings of the 2nd International Symposium for Interface Oral Health Science, Held in Sendai, Japan, Between 18 and 19 February, 2007

Springer

Editors:
Makoto Watanabe, DDS, Ph.D
Director, Tohoku University Graduate School of Dentistry
4-1 Seiryo-machi, Aoba-ku, Sendai 980-8575 Japan

Osamu Okuno, Ph.D
Division of Dental Biomaterials, Tohoku University Graduate School of Dentistry
4-1 Seiryo-machi, Aoba-ku, Sendai 980-8575 Japan

Associate Editors:
Keiichi Sasaki, DDS, Ph.D
Division of Advanced Prosthetic Dentistry, Tohoku University Graduate School of Dentistry
4-1 Seiryo-machi, Aoba-ku, Sendai 980-8575 Japan

Osamu Suzuki, Ph.D
Division of Craniofacial Function Engineering, Tohoku University Graduate School of Dentistry
4-1 Seiryo-machi, Aoba-ku, Sendai 980-8575 Japan

Nobuhiro Takahashi, DDS, Ph.D
Division of Oral Ecology and Biochemistry, Tohoku University Graduate School of Dentistry
4-1 Seiryo-machi, Aoba-ku, Sendai 980-8575 Japan

Editorial Liaison:
Haruhiko Takada, DDS, Ph.D
Department of Microbiology and Immunology, Tohoku University Graduate School of Dentistry
4-1 Seiryo-machi, Aoba-ku, Sendai 980-8575, Japan

ISBN 978-4-431-99840-2 e-ISBN 978-4-431-76690-2

Springer is a part of Springer Science+Business Media
springer.com

Printed in Japan

Printed on acid-free paper

Preface

Since 2002, the Tohoku University Graduate School of Dentistry has proposed "Interface Oral Health Science" as a major theme for next-generation dental research. That theme is based on the innovative concept that healthy oral function is provided by biological and biomechanical harmony among three systems: (1) oral tissues including the teeth, mucosa, bones, and muscles (host); (2) parasitic microorganisms of the oral cavity (parasites); and (3) biomaterials. The concept posits that oral diseases such as dental caries, periodontal disease, and temporomandibular disorders should be interpreted as interface diseases that result from disruption of the intact interfaces among these systems. The uniqueness of this concept rests on the fact that it not only encompasses the entire field of dentistry and dental care, but also expands the common ground shared with many other fields, including medicine, pharmaceutical science, agriculture, material science, and engineering. Our Graduate School of Dentistry aims to promote advances in dental research and to activate interdisciplinary research with related fields by putting interface oral health science into practice. On this basis we organized the First International Symposium for Interface Oral Health in February 2005, with productive discussions stimulated by two special lectures, three symposia, and poster presentations. A monograph titled *Interface Oral Health Science* that summarized the contents of the symposium was published in the autumn of 2005 (International Congress Series 1284, Elsevier, Amsterdam).

The Second International Symposium was the most recent. We invited researchers active in the global forefront of our field from Japan and abroad, and arranged a special lecture by President Akihisa Inoue of Tohoku University, an educational lecture by Vice-Dean Stephen Challacombe of King's College London Dental Institute, two symposia on biomaterials, and a symposium celebrating the agreement on international scientific exchanges with King's College London Dental Institute. There were more poster presentations than in the previous symposium, and we would like to take this opportunity to express our thanks to those researchers for their contributions. Because this year happens to be the 100th anniversary of the founding of Tohoku University, it was a special pleasure for us to be able to commemorate that anniversary with this symposium.

This book compiling the presentations at the symposium is published as a serial entitled *Interface Oral Health Science*. We hope that with the cooperation of our fellow researchers, this symposium and the book that grew out of it become forums for communication among dental researchers and point the directions for dental research for future generations. In closing, I would like to extend my best wishes for the health and success of those who participated in this symposium and presented such outstanding papers.

Makoto Watanabe
President, Second International Symposium for Interface Oral Health Science
Director, Tohoku University Graduate School of Dentistry
Sendai, Japan
February 2007

Acknowledgment

The Editors wishes to acknowledge the following members of Tohoku University Graduate School of Dentistry, who have contributed their expertise and time to the review of manuscripts submitted to *Interface Oral Health Science 2007*. These colleagues have provided the important assistance that made it possible for this monograph to publish critically reviewed papers in a timely manner.

Takahisa ANADA
Haruhide HAYASHI
Shiniji HATA
Yoshinori HATTORI
Motohide IKAWA
Shinji KAMAKURA
Hiroyasu KANETAKA
Shin KASAHARA
Masafumi KIKUCHI
Ken-ichiro KOMAKI
Takeyoshi KOSEKI
Kazuko NAKAJO
Eiji NEMOTO
Osamu OKUNO
Keiichi SASAKI
Takashi SASANO
Yasuyuki SASANO
Hideki SATO
Takuichi SATO
Hidetoshi SHIMAUCHI
Shunji SUGAWARA
Osamu SUZUKI
Haruhiko TAKADA
Yukyo TAKADA
Ichiro TAKAHASHI
Masatoshi TAKAHASHI
Nobuhiro TAKAHASHI
Masahiro TSUCHIYA
Akiko UEHARA
Jumpei WASHIO
Masanobu YODA

Contents

Plenary lectures

Interface Oral Health Science 2007: 3–20

Possibility of bulk glassy and nanogranular alloys as biomedical materials

Akihisa Inoue[1]* **and Xin Min Wang**[2]
[1]*Tohoku University, Sendai 980-8577;* [2]*Institute for Materials Research, Tohoku University, Sendai 980-8577, Japan*
*a.inoue@imr.tohoku.ac.jp

Abstract. This paper reviews our recent results on the formation, fundamental properties and application examples of nanogranular body-centered cubic (bcc) Ti-based alloys, bulk glassy Ti-based alloys and porous Pd- and Zr-based glassy alloys with the aim of clarifying the possibility of practical uses as biomedical materials. The bcc Ti-based alloys with low Young's modulus, large elastic elongation and high mechanical strength have already been used as eyeglass frame materials. New Ti-based bulk glassy alloys with a critical diameter of 7 mm in Ti–Zr–Cu–Pd system have been developed and tested as artificial dental root materials in various environmental conditions. The Pd- and Zr-based bulk glassy alloys can include spherical or polyhedral pores in a wide porosity range. The porous bulk glassy alloys have unique mechanical properties which are comparable to bones of human beings. These results indicate the possibility that the present nanogranular bcc Ti-based alloys as well as the bulk glassy alloys in Ti-, Pd- and Zr-based systems are used as biomedical materials in the near future.

Key words. bulk glassy alloy, titanium–based alloy, nanogranular phase, biomedical material, porous glassy alloy

1 Nanogranular body-centered cubic (bcc) Ti-based alloys

For the past decade, much attention has been paid to high-entropy alloys which are defined as multi-component crystalline solid solution alloys consisting of transition metals with nearly zero heat of mixing [1–4]. By use of solid solution strengthening, solid solution strain and atomic configuration complexity in their bcc base alloys, there is a high possibility of developing a bcc single-phase alloy with unique mechanical properties. Based on this concept, we examined the compositional dependence of mechanical properties of Ti–Zr–Nb ternary alloys subjected to cold rolling, followed by appropriate annealing treatments. As shown in Fig. 1, a very low Young's modulus below 60 GPa was found to be obtained for $Ti_{65}Zr_{17}Nb_{18}$ (atomic percentage). This ternary alloy also exhibited a large elastic elongation limit of about 2% in conjunction with rather high yield strength of 800–900 MPa. The structure of the Ti-based alloy was identified to consist of nanogranular bcc grains with an average grain size of about 5 nm. The nanogranular bcc Ti-based alloy has been commercialized as "Bio-Soft Titanium" [5] and used as frame wire materials for eyeglasses. Table 1

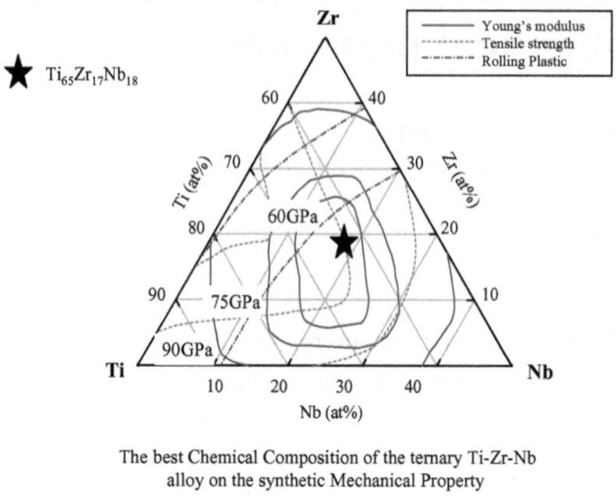

Fig. 1. Compositional dependence of Young's modulus of nanogranular bcc Ti-Zr-Nb ternary alloys subjected to cold rolling to 50% to 80% and appropriate annealing treatment

Table 1. Mechanical properties of $Ti_{65}Zr_{17}Nb_{18}$ and $Ti_{50}Zr_{30}Nb_{10}Ta_{10}$ alloys consisting of nanogranular bcc grains

Alloy	Tensile strength, Mpa	Yield strength, Mpa	Young's modulus, Gpa	Vickers hardness, Hv	Elastic elongation, %	Elongation, %
Pure Ti	345	275	102	100	~0.3	20
Ti-6Al-4V	910	840	110	260	~0.5	18
$Ti_{65}Zr_{17}Nb_{18}$	690~1,200	650~1,190	52~65	200~330	~1.8	<20
$Ti_{50}Zr_{30}Nb_{10}Ta_{10}$	750~1,250	680~1,190	47~58	270~350	~2.0	<18

shows that the addition of 10at%Ta to the alloy series of $Ti_{50}Zr_{40-x}Ta_xNb_{10}$ causes a further decrease in Young's modulus from 58 GPa for $Ti_{65}Zr_{17}Nb_{18}$ to 53 GPa for $Ti_{50}Zr_{30}Nb_{10}Ta_{10}$, in addition to the increases of yield strength and ultimate tensile strength. It was also found that these nanogranular bcc alloys in Ti–Zr–Nb and Ti–Zr–Nb–Ta systems exhibited much better corrosion resistance than those for pure titanium and SUS316L in 1N HCl solution open to air at 298 K.

We have also noticed that stent materials made of the nanogranular bcc $Ti_{50}Zr_{30}Nb_{10}Ta_{10}$ alloy can have much improved radio-opacity characteristics in comparison with conventional SUS316L stent material in both states before and after extension of their stents [6]. In addition, the nanogranular bcc Ti–Zr–Nb–Ta alloy did not exhibit any harmful influence in the histo-pathological examination of alloy-implanted rabbit skeletal muscle [6], suggesting that the bcc Ti-based quaternary alloy is acceptable as an implantable material. Figure 2 also shows that the bcc Ti-based alloy has good cold workability and can be cold drawn to a long pipe form with an outer diameter of 1.22 mm and a thickness of 0.07 mm [6]. It was also rather easy for the alloy pipe to be worked to a stent form for heart blood vessel extension, as shown in Fig. 3.

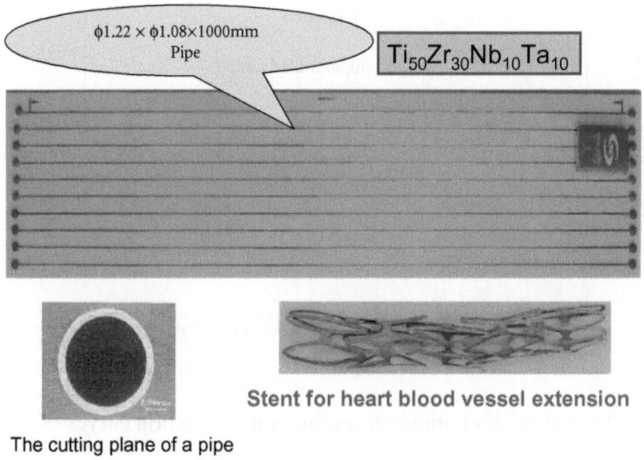

Fig. 2. Outer shape and cross section of nanogranular bcc $Ti_{50}Zr_{30}Nb_{10}Ta_{10}$ alloy pipes produced by heavy cold drawing. The stent made of the pipe is also shown for reference

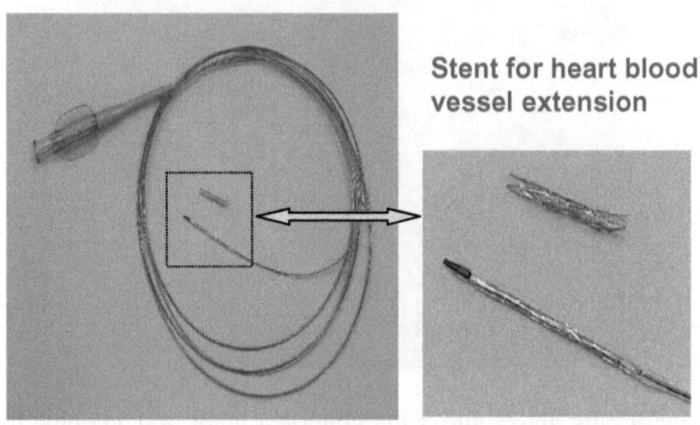

Fig. 3. Outside view of the stent for heart blood vessel extension made of nanogranular bcc $Ti_{50}Zr_{30}Nb_{10}Ta_{10}$ alloy pipe

2 Formation of bulk glassy Ti-based alloys

All metallic alloys in a bulk form with critical dimensions of over several millimeters which had been used by human beings for several thousand years before 1990 consisted of only a crystalline structure [7, 8]. The strict limitation to the crystalline structure originates from very short incubation times for the transition from supercooled liquid to crystalline phase because of an extreme instability of supercooled liquid for metals and alloys. Such an extreme instability is due to very high atomic mobility for metallic-bond type alloys at high temperatures just below melting temperature. Figure 4 shows a schematic illustration of continuous cooling transformation (CCT) curves for some metallic alloys [7, 8]. When we pay attention to the incubation time at the nose temperature for the transition from supercooled liquid to crystalline phase, its time scale is as short as less than 10^{-5} s

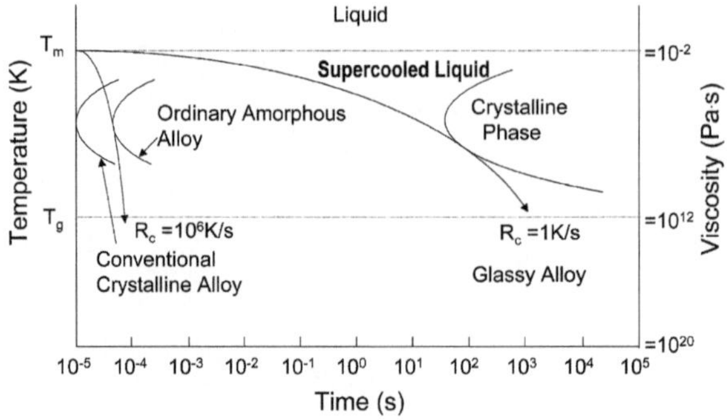

Fig. 4. Schematic illustration of continuous cooling transformation curves of glassy, amorphous, and crystalline alloys

Fig. 5. Outside view of bulk glassy alloys in cylindrical ingot, rod and pipe forms

for conventional crystalline alloys and of the order of 10^{-4} s for ordinary amorphous alloys. These incubation times are too short to fabricate engineering metallic materials through the control of supercooled liquid. Such an uncontrolled situation was maintained for many years just before around 1990. However, since 1988, the incubation time increased significantly. The longest incubation time reaches about 4,000 s at present, and the lowest critical cooling rate for glass formation is as low as 0.033 K/s [9, 10]. Considering that the critical cooling rate for formation of amorphous phase is usually 10^6 K/s, it is concluded that the glass-forming ability has been enhanced 8- to 9-fold during the past 19 years. The dramatic increase in the stability of supercooled liquid against crystallization for specific metallic alloys has enabled us to fabricate various kinds of nonequilibrium alloys exhibiting unique functional properties even in a bulk form. As a result, we have rather easily produced bulk glassy alloys with large critical dimensions and various outer shapes such as cylindrical ingots 75 mm in diameter and 85 mm in height, rod specimens 25 mm in diameter and 300 mm in length, and pipe specimens 10 mm in outer diameter, 1 mm in thickness, and 1,500–2000 mm in length for Zr-, Pd- and La-based alloys, as shown in Fig. 5 [8]. Very recently, glassy alloy sheets with uniform thicknesses ranging from 0.3 to 2 mm and large surface areas have also been produced, and the maximum size of the glassy alloy sheets reaches as large as an A-4 paper size (XM Wang and A Inoue, 2007, unpublished).

3 Feature of alloy components for bulk glassy alloys with centimeter critical size

Table 2 summarizes typical bulk glassy alloy systems reported to date together with the calendar years when their first papers or patents were published. The alloy systems can be classified into nonferrous and ferrous alloy groups. The former alloy group consists of Mg-, Ln (lanthanide metal)-, Zr-, Ti-, Pd-, Ca-, Cu-, Pt-, and Au-based systems, whereas the latter alloy group consists of Fe-, Co-, and Ni-based systems. Considering that the Ln metals can utilize at least 15 kinds of elements, the total number of glassy alloy systems reaches about 500 kinds. As recognized in Table 1, the 5–6 years between 1988 and 1993 can be regarded as an incunabula period in the research field of bulk glassy alloys. In addition, more than 50% of the total numbers of these alloy systems have been developed for the past 12 years, since 1995, indicating that the research field of bulk glassy alloys is still growing significantly even at present. For the past several years, the glass-forming ability

Table 2. Typical bulk glassy alloy systems reported to date and the calendar years when their first papers or patents were published

1. Nonferrous alloy systems	Year	2. Ferrous alloy systems	Year
Mg-Ln-M (Ln = Lanthanide Metal, M = Ni, Cu, Zn)	1988	*Fe-(Al,Ga)-(P,C,B,Si,Ge)*	1995
Ln-Al-TM (TM = Fe, Co, Ni, Cu)	1989	*Fe-(Nb,Mo)-(Al,Ga)-(P,B,Si)*	1995
Ln-Ga-TM	1989	*Co-(Al,Ga)-(P,B,Si)*	1996
Zr-Al-TM	1990	*Fe-(Zr,Hf,Nb)-B*	1996
Zr-Ln-Al-TM (double T_g)	1992	*Co-(Zr,Hf,Nb)-B*	1996
Ti-Zr-TM	1993	*Ni-(Zr,Hf,Nb)-B*	1996
Zr-Ti-TM-Be	1993	*Fe-Co-Ln-B*	1998
Zr-(Ti,Nb,Pd)-Al-TM	1995	Fe-Ga-(Cr,Mo)-(P,C,B)	1999
Pd-Cu-Ni-P	1996	*Fe-(Nb,Cr,Mo)-(C,B)*	1999
Pd-Ni-Fe-P	1996	*Ni-(Nb,Cr,Mo)-(P,B)*	1999
Ti-Ni-Cu-Sn	1998	*Co-Ta-B*	1999
Ca-Cu-Ag-Mg	2000	*Fe-Ga-(P,B)*	2000
Cu-Zr (nanophase)	2001	Ni-Zr-Ti-Sn-Si	2001
Cu-(Zr,Hf)-Ti (glass, nanophase)	2001	*Ni-(Nb,Ta)-Zr-Ti*	2002
Cu-(Zr,Hf)-Ti-(Y,Be)	2001	*Fe-Si-B-Nb*	2002
Cu-(Zr,Hf)-Ti-(Fe,Co,Ni)	2002	*Co-Fe-Si-B-Nb*	2002
Ti-Cu-(Zr,Hf)-(Co,Ni)	2004	*Co-Fe-Ta-B-Si*	2003
Ca-Mg-Zn	2004	Fe-(Cr,Mo)-(C,B)-Ln	2004
Pt-Cu-P	2004		

Alloy systems in italics were found by the Sendai Group

Table 3. Alloy system, maximum diameter and preparation method for bulk glassy alloys with critical diameters above 10 mm

Alloy system	Maximum diameter (mm)	Preparation method
Zr-Al-Cu	10	Tilt-casting
Pd-Ni-P	25	Water quenching
Ni-Pd-P	12	Water quenching
Zr-Al-Ni-Cu	30	Suction casting
Pd-Cu-Ni-P	<80	Water quenching
Pd-Pt-Cu-P	>30	Water quenching
Pt-Cu-Ni-P	20	Water quenching
Cu-Zr-Al-Y	10	Copper mold casting
Cu-Zr-Al-Ag	15	Copper mold casting
Y-Sc-Al-Co	20	Copper mold casting
Mg-Cu-Ag-Gd	25	Copper mold casting
Zr-Ti-Ni-Cu-Be	40	Copper mold casting
Ce-Cu-Al-Si-Fe	20	Copper mold casting
Fe-(Cr,Mo)-C-B-Y	12	Copper mold casting
Fe-(Cr,Mo)-C-B-Tm	12	Copper mold casting
Co-(Cr,Mo)-C-B-Y	10	Copper mold casting
Co-(Cr,Mo)-C-B-Tm	10	Copper mold casting
Fe-Co-(Cr,Mo)-C-B-Tm	16	Copper mold casting

has been further improved, and the large critical diameters of more than 10 mm have been achieved in various alloy systems such as Zr-, Pd-, Ni-, Pt-, Cu-, Y-, Mg-, Ce-, Fe-, and Co-based alloys, as summarized in Table 3. It is notable that bulk glassy alloys with maximum diameters of more than 20 mm can be formed in ten kinds of alloy systems including conventional metallic components such as Mg, Zr, and Cu bases [11–14].

4 Fundamental properties of bulk glassy alloys

Figure 6 shows the relation between tensile strength and Young's modulus for bulk glassy alloys, together with the data of conventional crystalline alloys [8]. It is notable that these fundamental characteristics of bulk glassy alloys are significantly different from those of crystalline alloys. For instance, when the comparison of tensile strength is made at the same Young's modulus level, the strength of bulk glassy alloys is about three times higher than that for crystalline alloys. On the other hand, the Young's moduli of bulk glassy alloys are about one-third when the comparison is made at the same tensile strength level. One can also recognize a good linear relation between both properties, indicating the satisfaction of Hook's law. Consequently, the slope of the linear relation corresponds to an elastic elongation limit. The elastic elongation limit of bulk glassy alloys is 1.9% which is about three times larger than that (0.65%) for crystalline alloys.

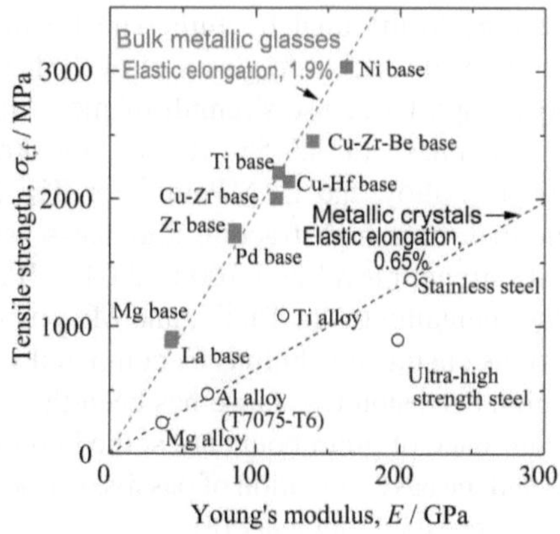

Fig. 6. Relation between tensile strength and Young's modulus for bulk glassy alloys. The data of conventional crystalline alloys are also shown for comparison

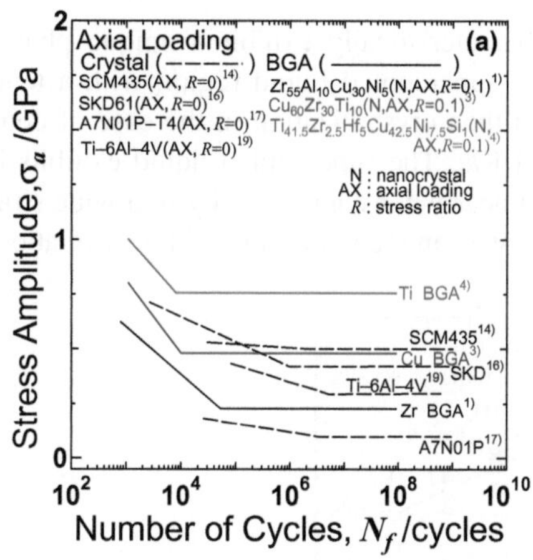

Fig. 7. Fatigue stress amplitude as a function of cycle for bulk glassy alloys. The data of conventional crystalline alloys are also shown for comparison

The endurance strength limit and fracture behavior under cyclic loading conditions were also examined for various bulk glassy alloys. It was reported that the fatigue endurance limits (σ_a/σ_B) of Zr-, Pd-, Ti-, and Cu-based bulk glassy alloys are in the range of 0.25 to 0.45 after the cyclic cumber of 10^7 [15, 16]. Here, σ_a is the applied stress amplitude and σ_B is the yield strength. Reflecting the high yield strength values of their bulk glassy alloys, their stress amplitude values lie in the high strength range from 500 to 750 MPa after the cyclic number of 10^7, being comparable to or higher than those for conventional SCM435 and SKD alloys, as summarized in Fig. 7. The fatigue crack usually generated at the defect site on the outer surface of the rod specimen and propagated to the inner region accompanying the striation pattern, leading to the final fracture. The fracture surface consisted mainly of a vein pattern typical for ductile glassy alloys. These bulk glassy alloys also exhibit high Sharpy impact fracture strength of more than 100 kJ/m^2 [8] as well as high fracture toughness of 45–55 $MPam^{1/2}$ for $Zr_{55}Al_{10}Ni_5Cu_{30}$ alloy, 68 $MPam^{1/2}$ for $Cu_{60}Zr_{30}Ti_{10}$ alloy and 22 $MPam^{1/2}$ for $Ti_{41.5}Zr_{2.5}Hf_5Cu_{42.5}Ni_{7.5}Si_1$ alloy [17]. It is notable that these high fracture toughness values are obtained in spite of very high tensile strength level of 1,700 to 2,040 MPa.

The bulk glassy alloys containing Nb, Ta, Cr, and Mo possess much better corrosion resistance in various chemical solutions as compared with the highest class of stainless steel. The high corrosion resistance has been thought to originate from the combination of the absence of grain boundaries, the homogeneous distribution of constituent elements and the easy formation of passive surface layer with enriched compositions of corrosion resistant elements [8].

5 Viscous flow working of bulk glassy alloys

Bulk glassy alloys characteristically exhibit a unique phase transition of glass transition, followed by supercooled liquid region with a temperature interval of about 50 to 130 K and then crystallization in the case of continuous heating at a heating rate of about 1 K/s. The supercooled liquid exhibits Newtonian viscosity as is evidenced from a nearly constant viscosity in a wide strain rate range shown in Fig. 8 [18]. Besides, we can recognize a good linear relation between true stress

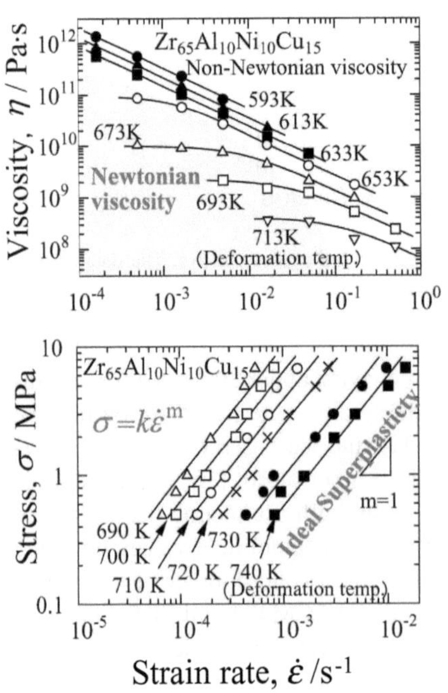

Fig. 8. Relation between viscosity or true flow stress and strain rate for bulk glassy $Zr_{65}Al_{10}Ni_{10}Cu_{15}$ and $La_{55}Al_{20}Ni_{25}$ alloys.

(σ) and strain rate (ε) which can be expressed by $\sigma = k\varepsilon^m$. The slope of the linear relation corresponds to the strain rate sensitivity exponent (m-value). The m-value is as high as 1.0, indicating the achievement of an ideal superplasticity which cannot be obtained for any kind of crystalline alloys.

By use of the ideal superplasticity, the nano-pyramid pattern with a size of 100 nm has been fabricated by pressing Pt-based bulk glassy alloy into concave nano-pyramid silicon die in the supercooled liquid region [19]. The minimum size of the nano-pyramid pattern is as small as 22 nm [20], indicating the possibility of forming imprinted nano-data-pit patterns for next-generation digital video devices.

6 Net-shape castability of bulk glassy alloys

Figure 9 shows a schematic illustration of specific volume of glassy alloys as a function of temperature, together with the data of conventional crystalline alloys. It is notable that the specific volume of the supercooled liquid in glassy type alloys decreases continuously with decreasing temperature, in good contrast to abrupt shrinkage due to solidification at the specific temperature for conventional crystalline alloys. This unique solidification mode enables us to form a mirror-like pattern on a nanometer scale between cast copper mold surface and cast glassy alloy surface [21], when an appropriate stress is applied to the supercooled liquid of glassy type alloys. These results indicate clearly that the bulk glassy alloys have good imprint-ability on a nanometer scale which cannot be obtained for any crystalline alloys including super-plastic alloys.

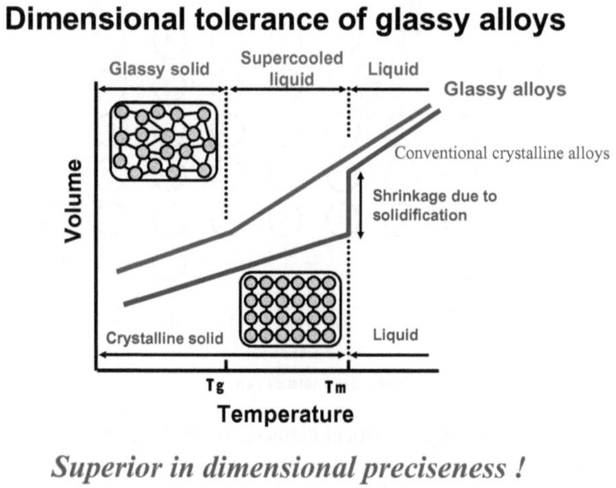

Fig. 9. Changes in the specific volumes of glassy and crystalline alloys with temperature

7 Ti-based bulk glassy alloys as biomedical materials

Based on the above-described advantages of bulk glassy alloys, great effort has been devoted to developing useful biomaterials using Ti-based bulk glassy alloys. Although a number of data on the formation of bulk gassy alloys in Ti-based systems have been presented for the past decade after the first synthesis of Ti-based bulk glassy alloys in 1993 [22], little is known about the synthesis of Ti-based bulk glassy alloys without toxic elements. For the past several years, we have searched for a new Ti-based bulk glassy alloy exhibiting high glass-forming ability and useful characteristics in the absence of toxic element.

Figure 10 shows the compositional dependence of maximum diameter of cast glassy alloy rods in Ti–Zr–Cu–Pd system without toxic elements [23]. It is notable that bulk glassy alloy rods of 6 and 7 mm in diameter are formed in a rather wide composition range of 10%–16% Zr, 34%–40% Cu and 10%–16% Pd. Considering that the critical diameters for various biomedical materials in practical are usually smaller than 5 mm, the maximum diameters of 6 to 7 mm are large enough to be applied to biomedical materials. The Ti–Zr–Cu–Pd bulk glassy alloys with maximum diameters of 6 to 7 mm exhibit relatively high yield strength of 2,000 MPa in conjunction with distinct plastic strains, as shown in Fig. 11 [23]. The fracture takes place along the maximum shear stress plane which is declined by about 45° to the direction of an applied load, and the fracture surface consists mainly of well-

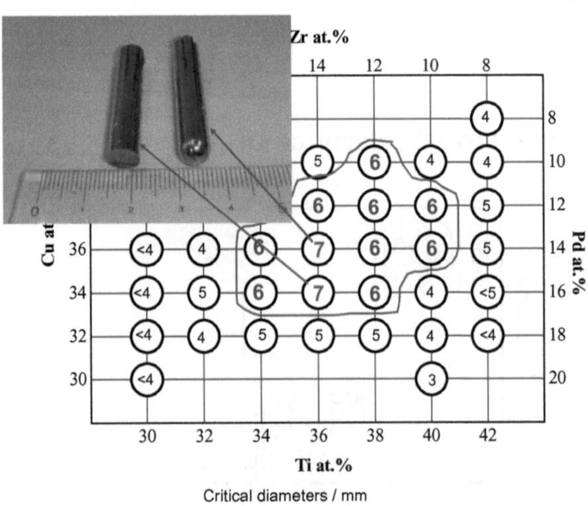

Fig. 10. Compositional dependence of critical diameter of Ti-Zr-Cu-Pd quaternary glassy alloys produced by conventional copper mold casting. The outside view of bulk glassy Ti-Zr-Cu-Pd quaternary alloy rods with a diameter of 7 mm is also shown for reference

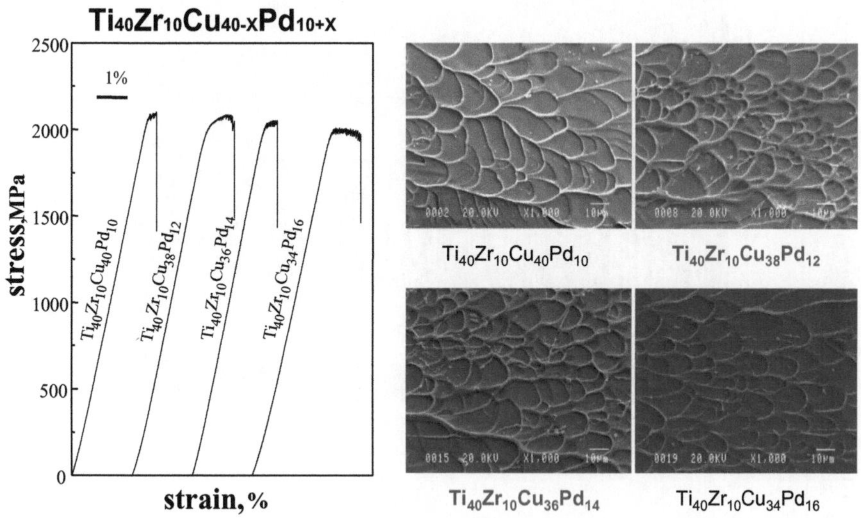

Fig. 11. Compressive stress–strain curves and fracture surface appearance of bulk glassy $Ti_{40}Zr_{10}Cu_{40-x}Pd_{10+x}$ (x = 0, 2, 4 and 6 at%) alloys rods with a diameter of 2 mm

developed vein pattern. The $Ti_{40}Zr_{10}Cu_{36}Pd_{14}$ bulk glassy alloy also shows considerably higher anodic corrosive potential and lower current density than those for pure Ti metal and conventional Ti-6Al-4V alloy in Hank's solution at 310 K [24]. However, the anodic potential at the breaking point of the passive state for the Ti-based bulk glassy alloy is considerably lower than those for pure Ti metal and Ti-6Al-4V alloy, and hence the improvement of the stabilization of the passive state has been desired. Based on the results that the Ti–Zr–Cu–Pd bulk glassy alloys had high glass-forming ability, good mechanical properties, good machinability, and high corrosion resistance; artificial dental roots were made of the Ti–Zr–Cu–Pd bulk glassy alloy, as shown in Fig. 12. The corrosion tests at different conditions in Hank's solution have been carried out at present.

We have further tried to produce glassy alloy composites in which hydroxylapatite (HA) particles are dispersed homogeneously in a glassy matrix [25]. Figure 13 shows the microstructure of the Ti–Zr–Cu–Pd glassy alloy composite including homogeneously dispersed HA particles. In addition to the HA phase, the other crystalline phase is also observed because of the decrease in glass-forming ability of the Ti–Zr–Cu–Pd alloy resulting from the dissolution of Ca, O, H, and P elements which are components in the HA particles. Besides, it has been confirmed that the HA dispersed particles in the glassy matrix can act as subsequent growth sites for HA phase, based on the data (Fig. 14) that the HA phase precipitates preferentially at the HA/glassy interface during immersion in Hank's solution containing saturated HA phase.

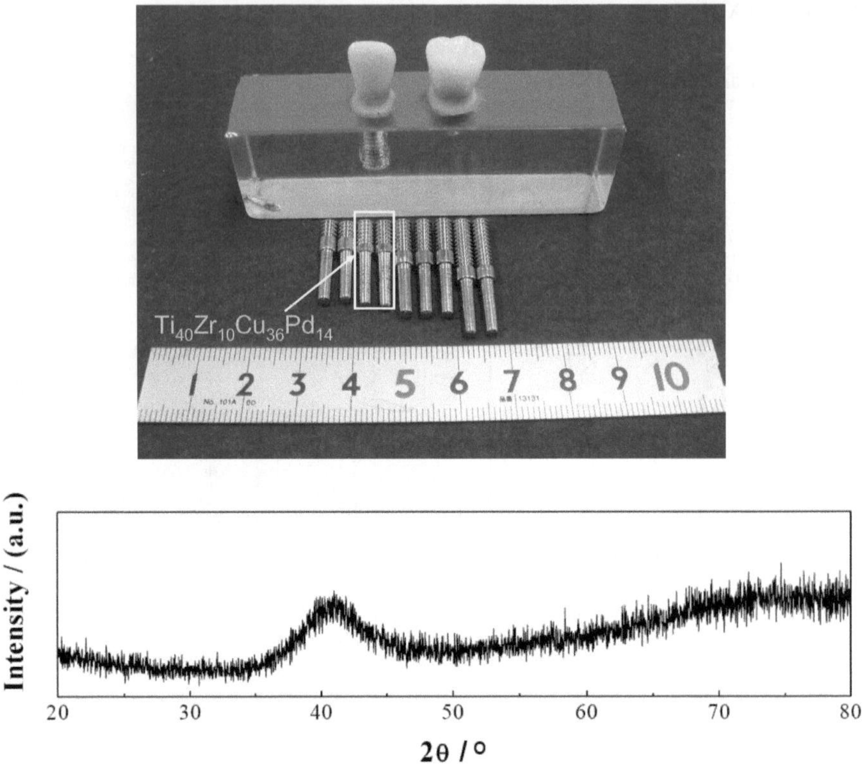

Fig. 12. Outer shape of artificial tooth roots made of $Ti_{40}Zr_{10}Cu_{36}Pd_{14}$ glassy alloy

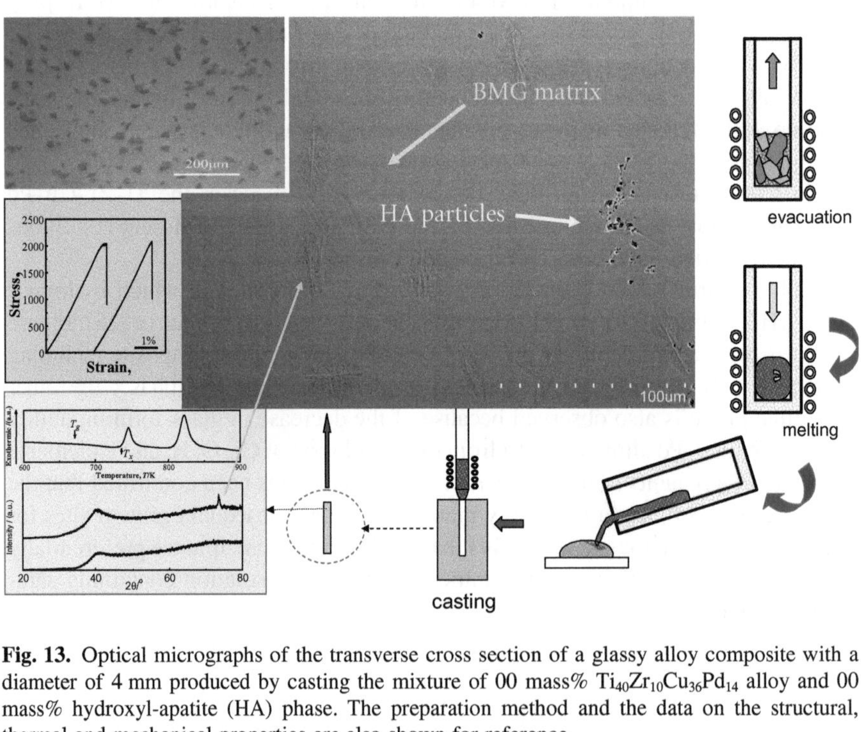

Fig. 13. Optical micrographs of the transverse cross section of a glassy alloy composite with a diameter of 4 mm produced by casting the mixture of 00 mass% $Ti_{40}Zr_{10}Cu_{36}Pd_{14}$ alloy and 00 mass% hydroxyl-apatite (HA) phase. The preparation method and the data on the structural, thermal and mechanical properties are also shown for reference

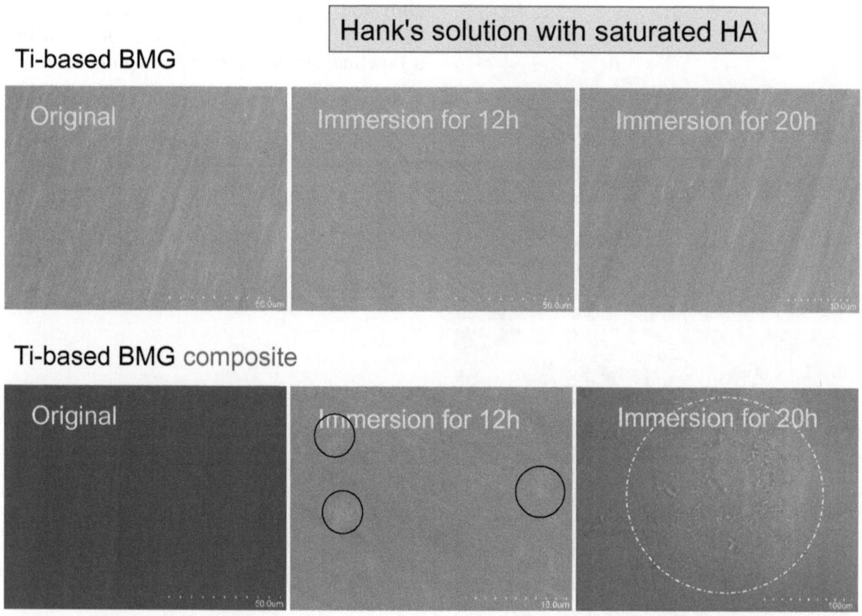

Fig. 14. Optical micrographs of the transverse cross sections of cast $Ti_{40}Zr_{10}Cu_{36}Pd_{14}$ alloy rods consisting of glassy single phase and glassy + HA composite phases immersed for 12 h and 20 h in Hank's solution containing saturated HA phase

8 Porous bulk glassy alloys

It is known that bulk glassy alloys containing transition metals, such as Ti, Zr, Pd, and Ni, having high affinity with hydrogen can dissolve large amounts of hydrogen in liquid and solid states. By use of the difference in solution amounts of hydrogen in bulk glassy alloys melted under different high hydrogen pressure atmospheres (by use of Dievert's law), porous bulk glassy alloys with a wide porosity range from 2 to 70% have been produced for $Pd_{42.5}Cu_{30}Ni_{7.5}P_{20}$ alloy [26, 27]. As shown in Fig. 15, each pore has a spherical form, and the pore diameter can be controlled in the range from 10 to 30 μm. In addition, porous bulk glassy alloys containing polyhedral pores have also been produced by dissolving the less-noble second phase embedded in the more noble glassy matrix with chemical etching solutions [28, 29].

The 0.2% proof stress ($\sigma_{0.2}$) and Young's modulus (E) as a function of porosity (P) are expressed by the relations of $\sigma = \sigma_0(1P)^n$ and $E = E_0(1P)^n$, respectively, where σ_0 and E_0 are the 0.2% proof stress and Young's modulus, respectively, for the porous free bulk glassy alloy and n is 2.4 for spherical pore and 3.2 for polyhedral pore, as shown in Fig. 16 [26, 27]. This result indicates that the fundamental characteristics for the porous bulk glassy alloys can be estimated from the porosity data. The porous bulk Pd–Cu–Ni–P glassy alloys with porosities of 50 to 60% exhibit tensile yield strength of 300 MPa, Young's modulus of 20 Gpa, and elastic

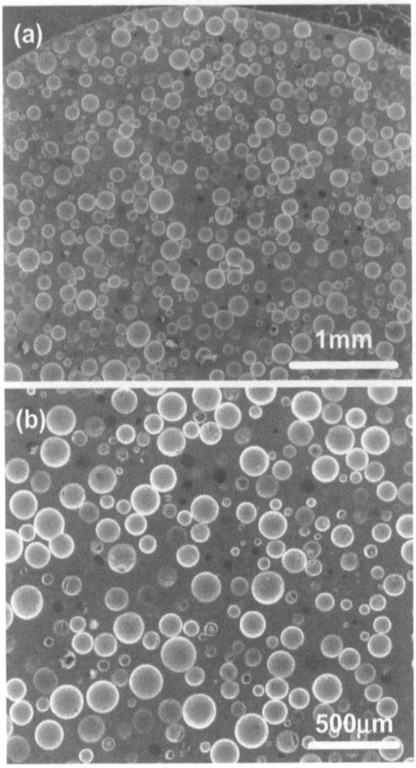

Fig. 15. SEM images of the transverse cross section of the porous glassy $Pd_{42.5}Cu_{30}Ni_{7.5}P_{20}$ alloy rod prepared by water quenching from 833 K in a hydrogen pressure of 1.5 MPa. **a** Low magnification, **b** high magnification

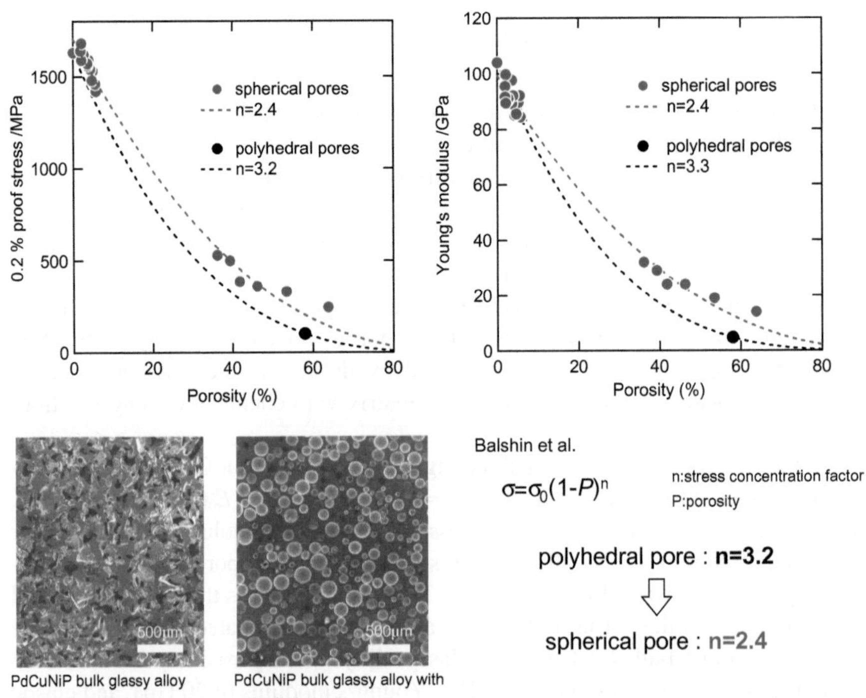

Fig. 16. Relation between 0.2% proof stress or Young's modulus and porosity for porous bulk glassy alloys including spherical or polyhedral pores

Table 4. Mechanical properties, microstructure, and nominal compressive stress–stain curve of a porous Pd-based bulk glassy alloy with a porosity of about 60%

Alloy	Tensile strength, Mpa	Young's modulus, Gpa	Elastic elongation, %
Porous Pd-based glassy alloy	300	20	1.7
Thigh bone	170	19	1.9
Pure Ti	345	102	~0.3
Ti-6Al-4V	910	110	~0.5
SUS316L	515–860	200	~0.3

The data for thigh bone, pure Ti metal, Ti-6Al-4V alloy and SUS316L are also shown for comparison
Microstructure and mechanical property of metallic glass

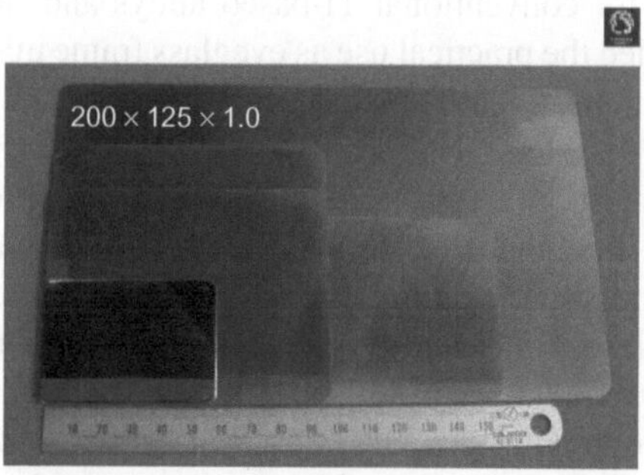

Fig. 17. Outer shape of $Zr_{55}Al_{10}Cu_{30}Ni_5$ glassy alloy sheets with smooth surface, uniform thicknesses and large surface areas produced by the die casting technique

elongation of 1.7%, which are comparable to those (170 MPa, 19 Gpa, and 1.9%, respectively) for thighbone in human beings, as summarized in Table 4.

9. Commercialization of bulk glassy alloys in sheet, rod, ball, and hollow forms

Recently, we have succeeded in developing the production techniques of bulk glassy alloys with various outer shapes of sheet, rod, ball, and hollow forms (XM Wang and A Inoue, 2006, unpublished). Figure 17 shows outer shapes of Zr–Al–Ni–Cu glassy alloy sheets with different dimensions produced by a die casting technique. It is seen that these glassy alloy sheets have uniform thickness, large surface area, and good smooth surface. Further modification of the die casting technique is expected to enable the fabrication of Zr-based glassy alloy sheets with larger dimensions. Very recently, the Zr–Al–Ni–Cu glassy alloy rods with a length of more than 1,000 mm and diameters up to 10 mm have also been produced by utilizing the jointing technique of supercooled liquid melt. The success of the melt jointing technique seems to result in the reduction of the limitation of length for the glassy alloy rods.

10 Summary

We examined the possibility of nanogranular bcc alloys in Ti-based system and bulk glassy alloys in Ti-, Zr-, and Pd-based systems as a new type of biomedical materials. The nanogranular bcc Ti-based alloys in Ti–Zr–Nb and Ti–Zr–Nb–Ta systems exhibit high strength and high ductility in conjunction with low Young's modulus. In addition, the Ti-based nanogranular alloys possessed better corrosion resistance than those for conventional Ti-based alloys and stainless steels. These advantages have enabled the practical use as eyeglass frame materials. The Ti-based bulk glassy alloys without toxic elements were developed in a wide composition range of Ti–Zr–Cu–Pd system. The new Ti-based bulk glassy alloys also exhibit high mechanical strength, large elastic elongation, and high corrosion resistance. Artificial dental roots were made of the Ti–Zr–Cu–Pd glassy alloys, and their performance characteristics were examined in various environmental conditions. The porous bulk glassy alloys in Pd- and Cu-based systems exhibit unique mechanical properties which are comparable to bones of human beings. It is thus concluded that the Ti-based nanogranular bcc alloys as well as the bulk glassy Ti-, Zr-, and Pd-based alloys with fully dense or widely porous state have high potential of practical uses as biomedical materials.

References

1. Tong CJ, Chen MR, Chen AK, et al (2005) Met Mater Trans 36A:1263
2. Hsu US, Hung UD, Yeh JW, et al (2007) Mater Sci Eng A460–461:403–408
3. Cantor B, Chang ITH, Knight P, et al (2004) Mater Sci Eng A375–377:213–218
4. Inoue A, Wang XM Japanese Patent, Application No.2004-217024
5. Inoue A (2003) Advanced technology and application of nano metals. CMC, Tokyo, pp 276–280
6. Wang XM, Inoue A (2003) Achievement Report on NEDO Millennium Project. Development and application of new Ti-Zr-based alloys with high strength and ultra-elasticity for biomaterial
7. Inoue A (1995) Mater Trans JIM 36:866–875
8. Inoue A (2000) Acta Mater 48:279–306
9. Inoue A, Nishiyama N (1996) Mater Trans JIM 37:181–184
10. Nishiyama N, Inoue A (2004) Mater Sci Eng A375–377:359–363
11. Nishiyama N, Inoue A (1999) Mater Trans JIM 40:64–71
12. Zeng Y, Nishiyama N, Wada T, et al (2006) Mater Trans 47:175–178
13. Yokoyama Y, Inoue A (2007) Mater Sci Eng A449–451:621–626
14. Zhang W, Jia F, Zhang Q, et al(2007) Mater Sci Eng A459:330–336

15. Fujita K, Okamoto A, Nishiyama N, et al (2007) J Alloys Comp 434–435:22–27
16. Yang C, Liu RP, Zhan ZJ, et al (2006) Mater Sci Eng A426:298–304
17. Yokoyama Y, Yamasaki T, Liaw PK, et al (2007) Mater Sci Eng A449–451:621–626
18. Kawamura Y, Nakamura T, Kato H, et al (2001) Mater Sci Eng A304–306:674–678
19. Saotome Y, Itoh K, Zhang T, et al (2001) Scripta Mater 44:1541–1545
20. Saotome Y, Inoue A (2000) Proc Int Conf Micro Electro Mechanical Systems 13: p 288
21. Ishida M, Takeda H, Nishiyama N et al (2007) Mater Trans (in press)
22. Zhang T, Inoue A, Masumoto T (1993) Mater Lett 15:379–382
23. Zhu SL, Wang XM, Qin FX, et al (2007) Mater Sci Eng A459:233
24. Qin CL, Oak JJ, Ohtsu N, Asami K, Inoue A (2007) Acta Mater 55:2057–2063
25. Zhu S, Wang XM, Qin F, Inoue A (2007) Mater Trans 48:163–166
26. Wada T, Inoue A, Greer AL, Mater Sci Eng (2007) A449–451:958–961
27. Wada T, Inoue A (2007) Mater Sci Eng A447:254–260
28. Wada T, Inoue A (2003) Mater Trans 44:2228–2231
29. Brothers AH, Scheunemann R, DeFouw JD, Dunand SC (2005) Scripta Mater 52:335–339

Interface Oral Health Science 2007: 21–30

Development of international perspectives in research: applications to oral mucosal biology

S. J. Challacombe*
Department of Oral Medicine and Immunology, King's College London Dental Institute at Guy's Hospital, London, SE1 9RT, UK
*stephen.challacombe@kcl.ac.uk

Abstract. The emergence of new technologies in biology and medicine has given new opportunities for scientific collaboration both in national and international for oral and dental research workers. Technological advances have resulted in further scientific specialisation, and the need for clinicians and basic scientists to work closely to foster innovation in research initiatives. The interests of dental research is best served by a teamwork approach, often working on basic and clinically applied research questions and utilising international contacts and collaborations. The IADR was founded to create an international organisation to further the interests of dental research by combining the talents of clinical and basic research workers and has programmes to encourage such international collaboration. Dental scientists have frequently made observations that have contributed in a major way to medical science. Examples of new technologies in the diagnosis and management of oral diseases include the use of stem cells, proteomics and genomics in oral cancer, potential in growing teeth in vitro and transgenic technologies in plants for the production of oral vaccines. Whilst the oral cavity as a unique body area allows specific oral diseases to be investigated, its anatomical position means that it is not only part of the mucosal immune system and reflective of mucosal immunity but also of dermatological, gastrointestinal and other systemic diseases. International perspectives in research are needed to maximise this potential.

Key words. mucosal biology, international perspectives, genomics, oral diseases

1 Introduction

The emergence of new technologies in biology and the mapping of the human genome in medicine have given new impetus for both national and international collaboration. It is vital that oral and dental researchers take scientific advantage of the new opportunities. Technological advances have also resulted in further scientific specialisation, emphasising the need for clinicians and basic scientists to work closely to foster innovation in research and initiatives in oral and dental health improvement. No longer can researchers expect to be expert in all the techniques needed for modern biological research or expect to be successful by working alone.

The interests and success of dental research is best served by a teamwork approach, often working on basic and clinically applied research questions and utilising international contacts and collaborations. There is a need to expand research in the health care sciences and to develop new models for epidemiological studies to follow the worldwide patterns of oral disease and oral cancer, and to encourage international collaboration [1] .

The IADR was founded with the aim of creating an international organisation to further the interests of dental research by combining the talents of clinical and basic research workers and has programmes to encourage such international collaboration. Dental scientists have frequently made observations that have had a major impact on medical science as a whole. Examples of new technologies in the diagnosis and management of oral diseases include the use of stem cells, proteomics and genomics in oral cancer, and potential in growing teeth in vitro and transgenic technologies in plants for the production of oral vaccines. Whilst the oral cavity as a unique body area allows specific oral diseases to be investigated, its unique anatomical position means that it is not only part of the mucosal immune system and reflective of mucosal immunity but also of dermatological and gastrointestinal diseases. International perspectives in research are needed to maximise this potential, and to achieve the IADR mission "to advance research and increase knowledge for the improvement of oral health worldwide". Perhaps every scientist should remember the maxim of Sir Peter Medawar that "The purpose of science is to make the world a better place to live in" [2].

2 Research

Before one looks at the question of how to enhance international research links, perhaps it is important to appreciate the scope of biomedical research in oral and dental diseases.

This makes us ask the basic question "What is research?" This ranges across the whole spectrum of research and varies from the basic hypothesis-driven investigations to a clinical investigation of disease groups, analysis and classifications, to clinical audit, which is the detached analysis of responses to therapies and the outcomes of treatment. One could argue, however, that the most important aspect of research in medical and dental schools is a philosophy of investigation.

Why do research at all? This remains a pertinent question for the modern clinical academic. Is it: (a) for the benefit of one's profession? (b) for the benefit of one's institution? (c) for the benefit of oneself, to keep oneself occupied? (d) because it is part of the job description? (e) for the benefit of patients or society? John Hunter (the father of odontology) was perhaps the prime example of the driving inquisitiveness to find out all he could about biology. It would be good to be reassured that modern clinical academics had the philosophy, if not the inquisitiveness, of Hunter and the belief that however small a part of research, in general, one's own research is, it will sometime be of benefit to mankind.

3 Hypotheses

Most scientists are now familiar with the eternal circle of observation leading to the generation of preliminary data, leading to a hypothesis, leading to an experimentation to prove or disprove the hypothesis, leading to observations and so on. (Fig. 1) This is actually the circle expounded by Karl Popper. Perhaps hypothesis is the most important mental technique of the investigator and should be used as a tool to uncover new facts rather than an end in itself, according to the philosophy of Karl Popper. One of Popper's central concerns was the growth of knowledge through rational means and that scientific theories can never be proved but only tentatively refuted. His main concern in the philosophy of science was to account for and to promote the growth of knowledge. Popper advocated that one should propound empirically testable theories and then aim to refute them. In this way theories became more robust and thus, given any theory, one should aim to replace it by another theory which is both more general and precise and not refutable.

John Hunter (1728–1793) (Fig. 2) is rightly described as the father of odontology for his enormous contributions to dental science [3, 4]. However, he made an even bigger contribution to medical science (such as discovery of the lymphatic system) and demonstrated no boundaries between them. Over the years, Hunter set about a systematic exploration of the human body. He traced every vessel, probed every cavity and followed every fibre. Once he had established the normal, he moved to disease and compared the normal with the diseased, to work out what went wrong. This remains a cardinal principle in any aspect of clinical research. Hunter probably never thought about the philosophy of what he was doing. His enormous natural scientific curiosity led to an experimental approach in surgery. He would try a traditional method, analyse the outcomes and form a hypothesis aimed at improving that outcome. Hunter's real contribution was not only in formulating these ideas

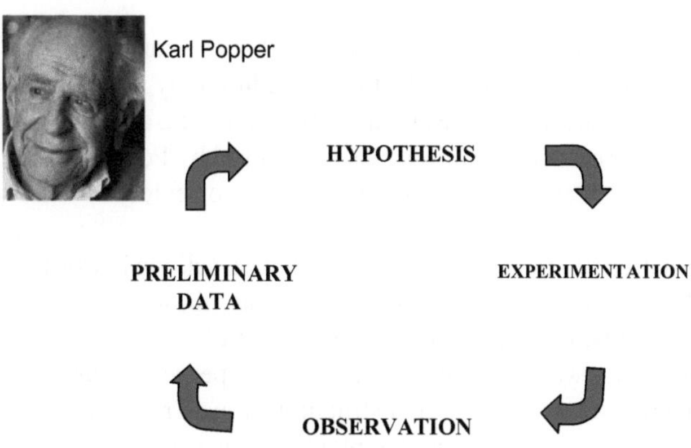

Fig. 1. Hypothesis-based research according to Karl Popper

Fig. 2. John Hunter (1728–1793), Father of Odontology

but also in applying them so rapidly into his clinical practice and through his pupils like Edward Jenner.

Hunter joined the circle usually at the stage of observation, but did indeed generate preliminary data and hypotheses, and many famous experiments, vascular and otherwise, lead in from it. Hunter perhaps was the prime example of the driving inquisitiveness to find out all he could about biology.

4 Experimentation

With regards to experimentation, many forget that the concept is that any experimental group of data is a random selection of an infinitely large hypothetical population. Experiments are all fallible and must be controlled, and most biological facts are only true under certain conditions. As recognised by Popper, there is no perfect experiment and, as recognised by Hunter, absence of evidence is not evidence of absence.

In the eternal circle, the role of unexpected observations is worth consideration. As Charles Nicolle once said, "Chance favours only those who know how to court her." In other words, research can be planned, but discovery cannot, and many of Hunter's discoveries were not only the result of his perpetual curiosity, but also his ability to organise scientific facts in a structured way. Translational research has a different paradigm linking basic and applied aspects (Fig. 3).

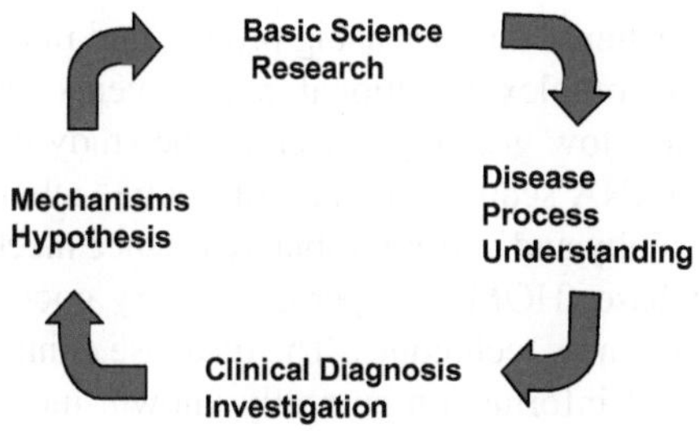

Fig. 3. Translational research: interdependence of basic and clinical research

5 A unique body cavity

The oral cavity is a unique body site and well positioned to exploit the explosion in the new technologies. The oral cavity has its own specialised flora, a specialised epithelia; it has unique diseases of teeth, crevicular fluid and periodontium with a variety of periodontal diseases, and from an immunological viewpoint is an example of a unique meeting of the systemic and mucosal immune systems. Biologically the oral cavity can be a model for various other biological systems and can rightly be described as a window of the body. It has a close relationship with the skin and dermatological diseases, it has a close relationship with the gut and gastrointestinal diseases, it is very closely related with the lungs and via the immune system and is closely related to other body cavities in both health and disease. The oral cavity is therefore a prime example of the interface between oral health and systemic health.

The oral cavity is also in a prime position to exemplify translational research, that relationship between basic science, using immunology, molecular biology, genomics and proteomics etc., and through to their clinical applications. Over the past few years there are many fine examples of the applications of proteomics, micro arrays, transgenics, gene therapy as well as DNA sequencing and real-time PCRs applied to various oral diseases. Perhaps the greatest impetus to biomedical research has been the human genome project which has sequenced the human genome and revealed perhaps 40,000 genes. These proteins can be deduced and it is safe to predict that assessment and diagnosis, prevention, early intervention and management of craniofacial conditions will continue to evolve through the application of new knowledge from the human genome project.

6 Microbiology and immunology

Techniques applied to microbiological diagnosis include the identification of cultured isolates using probes and PCR which have allowed the identification of slow growing bacteria and bacteria with deficient phenotypic identifications. Secondly the direct diagnosis in patients samples using probes, and micro arrays which have revealed bacteria with complex nutritional requirements and particularly non-cultivatable bacteria and slow growing bacteria. The study of new pathogens has been facilitated by 16 sRNA sequencing, and the same techniques can be applied to the study of pathogenicity and anti-microbial resistance mechanisms. The human oral microbiology database (HOMD) is perhaps a very good example of international collaboration using new techniques. The database contains all oral taxa, collating and referencing all information currently known, including gene sequence data, disease associations, phenotype descriptions [5, 6]. This group of international collaborators have now identified more than 800 oral species, although 275 are not yet named and 431 have not yet been cultured, and 91 are minimally characterised groups of strains. This should form an enormously helpful database for future relationships with oral diseases but already it is clear that the distribution of species involved with caries, endodontic and periodontal disease is actually quite different.

The second area of direct involvement is that of *Candida* pathogenesis and its interaction with the host. The key questions of (a) how the *Candida* cause infection, (b) how does the human host protect itself from infection (Fig. 4) can now be answered or investigated by using in vitro models, animal models and human subjects, and applying the new technologies including DNA and micro arrays. Microarrays allow a transcript profiling of the *Candida*–host interaction, and the globular analysis of all 6,000 *Candida* genes can help to identify virulence-associated genes during mucosal infections (Fig. 4) [7]. Recently it has been shown that intranasal immunisation might lead to reduction in colonisation in oral and vaginal sites concurrently (Fig. 5) [8].

An example of translation research is the vaccine against dental caries. Early identification of a virulence factor (Antigen I/II) has been reviewed recently by Russell et al. [9]. This antigen was then used in a monkey model (Fig. 6) and showed that systemic immunisation led to a reduction in caries of 75% (Fig. 6). Ma et al. [10] cloned the monoclonal antibody against streptococcal antigen I/II into tobacco plants, and after extraction this was applied to human volunteers to

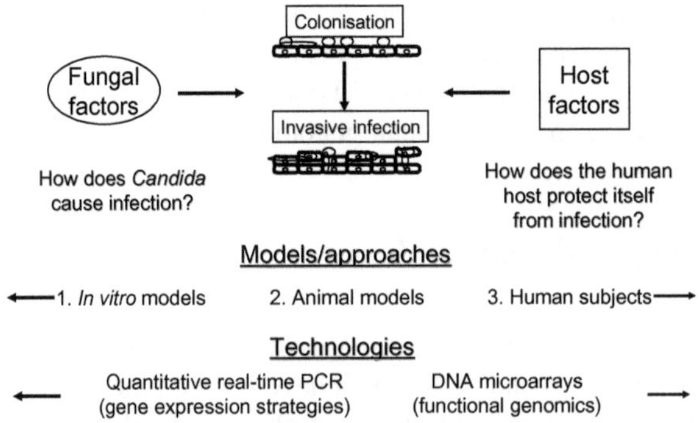

Fig. 4. Concepts of host–pathogen interactions in mucosal *Candida* infection

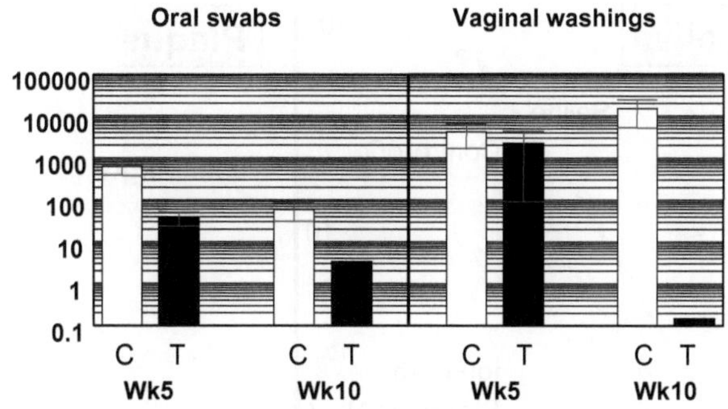

Fig. 5. *Candida albicans* colony forming units in oral swabs and vaginal washings after intranasal immunisation demonstrating a reduction in colonisation at both mucosal sites

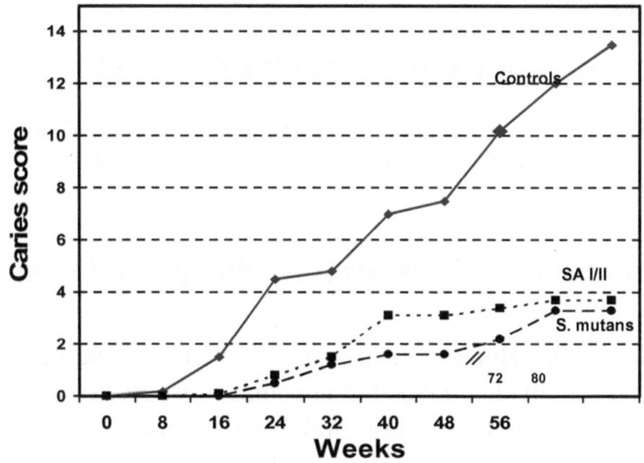

Fig. 6. Prevention of dental caries in rhesus macaques by immunisation with whole cells of *Streptococcus mutans* and purified streptococcal antigen *SA* I/II [18]

allow the first demonstration with application of a secretory IgA antibody inhibiting colonisation of *Streptococcus mutans* in the oral cavity (Fig. 7), and indeed this was the first demonstration of any interference with homeostasis of bacteria in the human body by monoclonal antibodies. Subsequent work has shown that monoclonal IgG antibody to *Porphyromonas gingivalis* can inhibit colonisation in humans [11]. Perhaps a fine example of translational research in the oral cavity has been the application of oral tolerance in the treatment of oral disease. An epitope of heat shock protein 70 was demonstrated as being recognised with Behçet's syndrome [12]; this peptide was later used in an animal model and demonstrated a form of Behçet's syndrome in terms of uveitis [13]. Other experiments showed that the feeding of the peptide prevented this experimental uveitis in the same Lewis rat model [14]. Finally, it was recently demonstrated that intragastric immunisation with the same peptide in patients with Behçet's induced oral tolerance and remission in Behçet's patients [15].

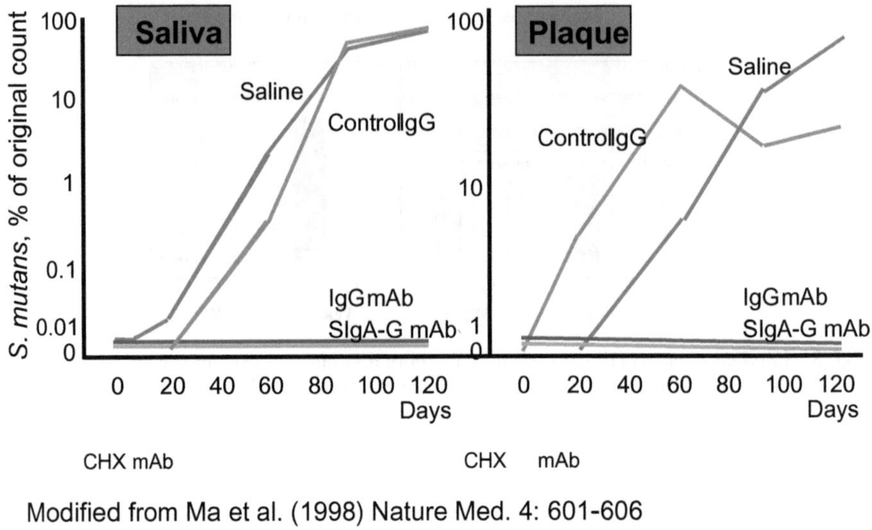

Fig. 7. Recolonisation of *S. mutans* in the oral cavity after treatment with chlorhexidine (*CHX*) and application of monoclonal antibodies against strep antigen I/II [10]. *mAb* monoclonal antibody

7 Contributions of oral research to biology and medicine

From the time of John Hunter dental research workers have not only advanced knowledge in their individual fields but also made discoveries that have impacted on whole areas of science. Examples of milestone observations in mucosal biology include the first recognition of specific adherence of bacteria to epithelia by Williams and Gibbons [16] which has led to a whole field of work identifying the mechanisms which allow the adherence of bacteria to the different anatomical sites in the host area of cellular microbiology, the work of Mestecky et al. [17] in demonstrating that the oral cavity was part of the mucosal immune system in man was pioneering and has subsequently been exploited in mucosal vaccination or in the induction of oral tolerance as above. Lehner et al. [18] showed immunity in primates to the prevention of dental caries, perhaps the first example of the successful vaccination against oral disease. Ma and his colleagues [10] produced the first vaccine in genetically engineered plants and this was against *Streptococcus mutans* and its usage is proof of principle that such vaccines could be active against a variety of microbial diseases and this work continues.

8 Conclusion

Integration of information from the human genome, comparative and microbial genomics, proteomics, bioinformatics, and related technologies will provide the basis for proactive prevention and intervention, and novel and more efficient treatment approaches. Oral health care practitioners will increasingly require knowledge of human genetics and the application of new molecular-based diagnostic and therapeutic technologies. It is important for the dental researchers to be seen as the source of evidence that allows governments to make informed decisions on their strategies for the delivery of oral healthcare and to recognise their role in the provision of robust research and data.

References

1. Challacombe S (2003) Global perspectives for the IADR and dental research. J Dent Res 82(9):671
2. Medawar PB, Medawar JS (1971) Some reflections on science and civilization. Ciba Found Symp 1:9–21
3. Hunter J (1771) The natural history of the human teeth: explaining their structure, use, formation, growth and diseases
4. Hunter J (1778) A practical treatise on the diseases of the teeth: intended as a supplement to the natural history of those parts. Printed for J. Johnson, London
5. Munson MA, Banerjee A, Watson TF, et al (2004) Molecular analysis of the microflora associated with dental caries. J Clin Microbiol 42(7):3023–3029
6. Paster BJ, Olsen I, Aas JA, et al (2000) The breadth of bacterial diversity in the human periodontal pocket and other oral sites. Periodontol 2006;42:80–87
7. Naglik JR, Fostira F, Ruprai J, et al (2006) *Candida albicans* HWP1 gene expression and host antibody responses in colonization and disease. J Med Microbiol. 55(Pt 10): 1323–1327
8. Rahman D, Mistry M, Thavaraj S, et al (2007) Murine model of concurrent oral and vaginal *Candida albicans* colonization to study epithelial host–pathogen interactions. Microbes Infect Microbes Infect 9(5):615–622
9. Russell MW, Childers NK, Michalek SM, et al (2004) A Caries Vaccine? The state of the science of immunization against dental caries. Caries Res 38(3):230–235 (Review)
10. Ma JK, Hikmat BY, Wycoff K, et al (1998) Characterization of a recombinant plant monoclonal secretory antibody and preventive immunotherapy in humans. Nat Med 4(5): 601–606
11. Booth V, Ashley FP, Lehner T (1996) Passive immunization with monoclonal antibodies against *Porphyromonas gingivalis* in patients with periodontitis. Infect Immun 64(2): 422–427
12. Direskeneli H, Hasan A, Shinnick T, et al (1996) Recognition of B-cell epitopes of the 65 kDa HSP in Behcet's disease. Scand J Immunol 43(4):464–468
13. Hu W, Hasan A, Wilson A, et al (1998) Experimental mucosal induction of uveitis with the 60-kDa heat shock protein-derived peptide 336-351. Eur J Immunol 28(8):2444–2455
14. Phipps PA, Stanford MR, Sun JB, et al (2003) Prevention of mucosally induced uveitis with a HSP60-derived peptide linked to cholera toxin B subunit. Eur J Immunol 33(1):224–232
15. Stanford M, Whittall T, Bergmeier LA, et al (2004) Oral tolerization with peptide 336-351 linked to cholera toxin B subunit in preventing relapses of uveitis in Behcet's disease. Clin Exp Immunol 137(1):201–208
16. Williams RC, Gibbons RJ (1972) Inhibition of bacterial adherence by secretory immunoglobulin A: a mechanism of antigen disposal. Science 177(50):697–699
17. Mestecky J, McGhee JR, Arnold RR, et al (1978) Selective induction of an immune response in human external secretions by ingestion of bacterial antigen. J Clin Invest 61(3):731–737
18. Lehner T, Wang Y, Ping L, et al (1999) The effect of route of immunization on mucosal immunity and protection. J Infect Dis 179(Suppl 3):S489—S492 Review

Symposium I:
Host–parasite interface, from oral biofilm to host response in oral mucosa

Interface Oral Health Science 2007: 33–42

Stress and microbial diversity in the oral biofilm

David Beighton*
Department of Microbiology, The Henry Wellcome Laboratories for Microbiology and Salivary Research, King's College London Dental Institute, London Bridge, SE1 9RT, UK
*david.beighton@kcl.ac.uk

Abstract. The oral biofilm harbors in more than 750 species/phylotypes but even greater diversity is apparent when the genotypic diversity of individual species is considered. Analysis of individual species of the genera *Streptococcus*, *Actinomyces*, *Fusobacterium*, and *Porphyromonas* has shown that each species is genotypically diverse and that, except within closely living individuals, no two individuals harbor the same genotype. Data are presented for the relationships between the stress and the diversity of *Streptococcus oralis*, *Actinomyces naelsundii* at interproximal sites in caries-free and caries-active children, below dental restorations and of *Veillonella* spp. on carious and sound surfaces. The idea of "stress" requires consideration; it is relative and usually applied unevenly across the biofilm. Acidic stress increased the diversity of *S. oralis*, but not *A. naelsundii*, at interproximal sites, perhaps as a result of increased habitat diversity. While the genotypic diversity of *Veillonella* spp. was significantly reduced within carious lesions (acidic stress), as was the range of *Veillonella* spp. isolated. The diversity of the microflora surviving below dental restorations, with nutrients of limited composition, was genotypically less diverse than that of the lesion and exhibited similar phenotypic characteristics. The most stressful environments may also be the most stable. The diversity of individual species in the oral biofilm may be increased by the application of a small stress but a larger stress may "flatten" the environment, making it less subject to perturbation, reducing genotypic and phenotypic diversity. It would appear that the mechanism driving increased microbial diversity is continual perturbation of habitat diversity.

Key words. Actinomyces, streptococci, Veillonella, stress, genotyping

1 Introduction

The oral microbiota is very diverse with more than 750 different taxa being identified from cultural and non-cultural studies employing the analysis of 16S rRNA sequence data [1]. Of course it is unlikely that all individuals harbor each of these taxa so that the extent of the range of taxa colonising any individual is not known, although given the array of different intra-oral sites, it must be expected that each individual must harbor in excess of 100 taxa. Taxa here being defined as organisms (cultured or not yet cultured) having a different 16S rRNA sequence but as some species, e.g. *Streptococcus oralis* and *Streptococcus mitis* and *Veillonella* spp. have similar16S rRNA sequences, this might be an underestimation of the species diversity.

The diversity of individual bacterial species may be investigated using a number of different molecular tools, although early measures of diversity used phenotypic traits or bacteriocin susceptibility to characterize isolates. More recently PCR-based genotyping schemes including RFLP [2], REP-PCR [3], and APPCR [4] have been variously applied to many species of oral streptococci, and the more precise and reproducible multilocus sequence typing (MLST) schemes have been applied to a small but growing range of oral bacteria [5]. Such approaches investigate diversity at the individual isolate level but populations maybe investigated using denaturing gradient gel electrophoresis (DGGE) which may enable microbial populations associated with specific health or diseased states to be characterized [6].

The outcome of the application of all these varied phenotypic and molecular measures of biofilm or species diversity is that individuals are unique, and that the microbiota associated with the human oral biofilm is at this simplistic level, not at the level of function, very diverse.

2 Stress in the biofilm: insurance hypothesis

The stresses which have been investigated, usually with respect to bacteria or bacterial populations growing in defined conditions may include acid or base (intrinsic or extrinsic), oxidizing agents, temperature, nutrient availability or nutrient imbalance, co-factor availability, hydration, exposure to toxic agents (intrinsic or extrinsic). Within the oral biofilm many or all of these stresses may operate but individual stresses do not operate continually, and it is the fluctuations in stresses, which perhaps have the most important impact on biofilm diversity.

The application of a constant stress will result in the selection for the fittest phenotype (or genotype), and such an application of stress will result in a simplification of biofilm diversity. The fluctuation of the applied stress or stresses it is hypothesized will result in greatest biofilm diversity. This is known as the "insurance hypothesis" [7]. The insurance hypothesis states that diversity is a good thing; biodiversity insures ecosystems against declines in their functioning because many species provide greater guarantees that some will maintain functioning even if others fail, in response to an externally applied stress. The major insurance effects of biodiversity are (i) a reduction in variance of productivity over time and (ii) an increase in mean of productivity over time.

In an experimental examination of this hypothesis [8] the short-term growth of *P. aeruginosa* phenotypes in biofilms generated extensive genetic diversity in the resident bacteria by a mechanism that required the *recA* gene and most likely involves recombination functions. The authors found that diversity was self-generated within the biofilms they studied, and that the genetic diversity produced in the biofilms produced bacterial subpopulations with specialized functions, which were only apparent upon the application of an external stress. They found that as a consequence of the external applied stress, in this case peroxide, the functional diversity self-generated within the biofilm increased the ability of the biofilm community to withstand the applied stress in accord with the insurance hypothesis.

While the application of an applied fluctuating stress may result in increased phenotypic diversity it may also be the case that the application of the stress may also increase habitat diversity. It is easy to visualize that acid is not produced uniformly across or within the oral biofilm, rather it is produced focally with pH gradients surrounding those bacteria producing acid within the plaque [9]. Along such gradients habitat diversity is increased which may be exploited by different members of the oral biofilm so that the application of an acid stress may lead to increased genotypic or phenotypic diversity. Of course if the acid stress is too great, this cannot be defined as our knowledge is insufficient, no organisms may be able to exploit such changes in habitat diversity or if the applied stress is too great and applied for too long a period similarly no exploitation may occur and phenotypic simplification may follow.

3 Acid stress and oral biofilm diversity

Acid stress may arise in the oral biofilm by a number of distinct means. The most obvious is the consumption of fermentable carbohydrates in food, which results in a rapid and somewhat prolonged reduction in plaque pH. This is illustrated in Fig. 1. which shows the plaque pH change in root surface plaque response to the exposure of the biofilm to a glucose rinse. Plaque pH changes will also be affected by reduced saliva flow arising from prescription of xerostomic drugs or irradiation for head and neck cancer, increased oral clearance times due to the presence of partial or full dentures, orthodontic brackets or cleft palette. While gastric acid reflux and the consumption of acidic drinks will reduce the intra-oral pH and cause

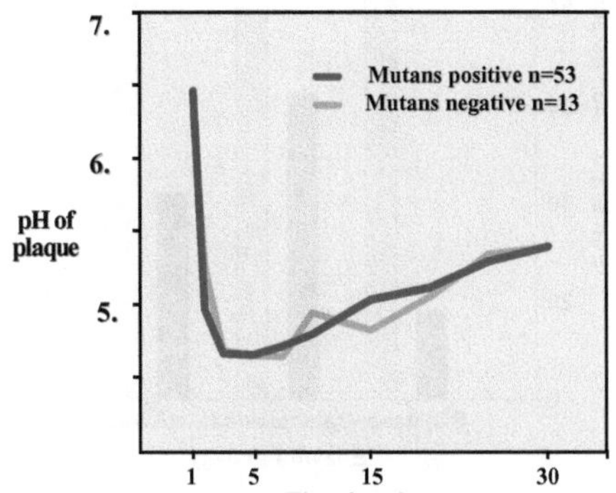

Fig. 1. Changes in mean root surface plaque pH of individuals with or without mutans streptococci following 1 min rinsing with a sucrose mouth rinse (supplied by Professor A. Scheie)

dental erosion, such changes also modify the oral microbiota, and perhaps the most persistent effect mediated by acidic pH is that arising as a consequence of active occlusal or root carious lesions. In carious lesions the reduced pH may be effective for considerably longer periods than in dental plaque associated with non-carious surfaces where the ambient pH will return to near neutrality within 30–60 min following carbohydrate exposure

It is not my purpose to review systematically the effects of acid stress on the oral biofilm as many of these are well known and have been reproducibly established in various in vitro and in vivo studies. Thus acid exposure results in an increase in the numbers and proportions of *S. mutans*, *S. sobrinus*, *Bifidobacterium* spp., *Lactobacillus* spp., and *Candida albicans* while the proportions of *S. sanginius* and *A. naeslundii* are likely to be reduced.

4 Does genotypic diversity mean anything?

We really want to know why an organism survives in a stressed environment. In order to determine this we need to know what specific phenotypic characteristics have been selected for and in most cases we do not know what to test for. As a proxy measure we use genotyping as described previously but the question remains "Does genotyping have any purpose; does it mean anything?". It is well established that for every species investigated each person harbors their own unique genotypes, and that each person may have more than one genotype of any species. These observations have been made for, among other taxa, *S. oralis*, *S. mitis*, *S. mutans*, *F. nucleatum*, *P. gingivalis*, and *S. sanginius*. Occasionally isolates are found in

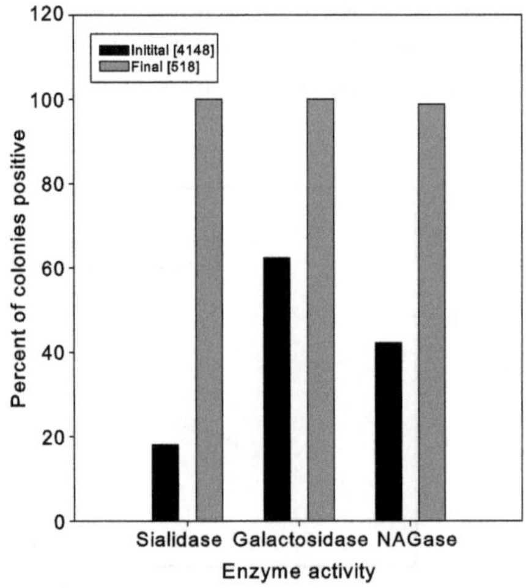

Fig. 2. Comparison of the changes in the proportions of bacteria producing sialidase, β-galactosidase and β-N-acetylglucosaminidase in infected dentine before restoration and in the surviving bacteria recovered 5 months after placement of the restoration [see reference 12]

different individuals who have the same genotype but this is the exception and not the rule.

In a study of aciduricity we [10] were able to show that isolates of *S.oralis* from the same person isolated on different culture media, at pH 5.2 or 7.0, were genotypically distinct (Fig. 2), a finding akin to the earlier observations that non-mutans streptococci from caries-free sites were less aciduric than the same species from carious lesions. Thus the same species from carious lesions may be phenotypically heterogeneous with respect to aciduricity and acidogenicity and also exhibit genotypic heterogeneity, which was correlated with a phenotypic trait. So genotypic diversity may have some meaning but will need to be interpreted with caution.

5 Effects of apparent stresses on biofilm diversity in vivo

In a number of in vivo studies we have applied REP-PCR to the study of the diversity of selected oral biofilm taxa [11]. In the first study we determined the effects of apparent acid stress on the genotypic diversity of *A. naeslundii* and *S. oralis* in the oral biofilm of radiographically caries-free approximal sites in subjects who were either caries-free or had occlusal caries at other sites in the mouth. The genotypic diversity of *A. naeslundii* genospecies 2 (424 isolates) and *S. oralis* (446 isolates) strains isolated from the subjects was investigated. To demonstrate the acid stress experienced by the sites in the caries active subjects we found that they harbored significantly greater proportions of mutans streptococci and lactobacilli and a smaller proportion of *A. naeslundii* organisms than the plaque sampled from the caries-free subjects. These data confirmed that the sites of the two groups of subjects were subjected to different environmental stresses, probably determined by the prevailing or fluctuating acidic pH values. We tested the hypothesis that the microflora of the sites subjected to greater stress (the plaque samples from the caries-active subjects) would exhibit reduced genotypic diversity because the sites would be less favorable. However, this was not demonstrable as the diversity of *A. naeslundii* strains did not change ($\chi^2 = 0.68$; $p = 0.41$) although the proportional representation of *A. naeslundii* was significantly reduced ($P < 0.05$) and the diversity of the *S. oralis* strains increased ($\chi^2 = 11.71$; $p = 0.0006$) and the proportional representation of *S. oralis* did not change.

In a second study we investigated the effects of sealing infected carious dentine below dental restorations on the phenotypic and genotypic diversity of the surviving microbiota [12]. It was hypothesized that the microbiota would be subject to nutrient limitation or nutrient simplification, as it would no longer have access to dietary components or salivary secretion for growth, and that this stress would lead to a reduction in biofilm diversity. Under the restoration the residual bacteria in the infected dentine would have no access to the diet, and the available nutrients would be limited primarily to serum glycoproteins passing from the pulp through the patent dentinal tubules. We found at baseline microbiota were composed primarily of *Lactobacillus* spp., *S. mutans*, *S. parasanguinis*, *A. israelii*, and *A. gerencseriae*.

When the restorations were removed 5 months later none of these taxa was isolated from the dentine. The surviving microbiota consisted of only *A. naeslundii*, *S. oralis*, *S. intermedius*, and *S. mitis*. The microbiota of the final sample exhibited a significantly ($P < 0.001$) increased ability to produce glycosidic enzymes (sialidase, β-N-acetylglucosaminidase, and β-galactosidase), which liberate sugars from serum glycoproteins, and the genotypic diversity, determined using REP-PCR, of *S. oralis* and *A. naeslundii* was significantly ($P = 0.002$ and $P = 0.001$, respectively) reduced in the final samples (Fig. 3).

In a more recent study we investigated the diversity of *Veillonella* spp. in occlusal caries lesions in children and compared this microflora with that isolated from sound occlusal and buccal surface plaque from caries-free children. Here, we found that the genotypic diversity of the Veillonella recovered from the carious lesions was significantly less than that from either of the other sites. Thus a single predominant Veillonella species, forming >85% of the total Veillonella population, was present in 16 of 18 occlusal carious lesions, whereas in the plaque from sound tooth surfaces a single predominant species was apparent in only 7 of 15 sites ($\chi^2 = 9.75$; $p = 0.0018$). When we considered the diversity of genotypes amongst the Veillonella, amongst these isolates the number of genotypes per 100 isolates for the lesions was 4 (median; 2–14 range) compared to 21 (2–100) for the occlusal surfaces and 23 (10–46) for the buccal surfaces ($P < 0.001$ and $P = 0.011$), respectively.

These studies represent a range of different environmental situations in which the simple Darwinian Law of the survival of the fittest must prevail. But how can the outcomes be interpreted? In the first study the stress apparently increased diversity. We propose that under these environmental conditions the diversity and number of niches within the oral biofilm that could be exploited by *S. oralis* increased, resulting in the increased genotypic diversity of this species. Apparently, *A. naeslundii* was

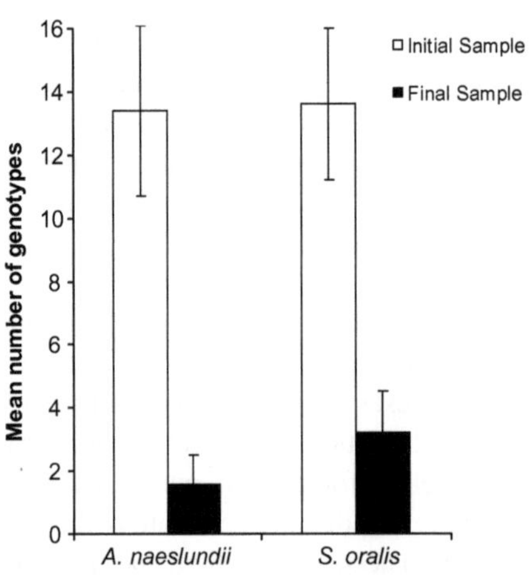

Fig. 3. The mean numbers of genotypes of *S. oralis* and *A. naelsundii* per sample in infected dentine before restoration and in those recovered 5 months after placement of the restoration [see reference 12]

not able to exploit the new niches because the prevailing conditions within the niches may have been deleterious and not supportive of its proliferation. These results suggest that environmental stress may modify a biofilm such that the diversity of the niches is increased, and that these niches may be successfully exploited by some, but not necessarily all, members of the microbial community. The increase in niche diversity in this situation is believed to be driven by changing pH gradients within the oral biofilm. In the third study it is likely that again the diversity of the investigated organisms was determined by the environmental pH; however, here we would propose that the acid stress was applied more uniformly to the infecting biofilm resulting in a reduced diversity of the *Veillonella* spp. The "flatter" is the environment the less diverse will be the resulting biofilm. In the study of interproximal plaque the environment was heterogeneous and not flat, in the lesions the environment was flatter but the environment underneath the restorations we propose was the flattest of all. And the flattest environment will lead to the least diverse microbiota. In this particular situation we propose that the available nutrient, primarily serum proteins, or the relative simplicity and homogeneity of the nutrient supply significantly affected the surviving microbiota. The surviving microbiota was less complex, based on compositional, phenotypic, and genotypic analyses, than that isolated from carious lesions, which were also exposed, to salivary secretions and pH perturbations. It also follows from these observations that the oral biofilm growing on non-carious surfaces will be subjected to a less than flat environment; notably its constant removal and requirement to renew result in a fluctuating environment which may act as a major driver towards increased genotypic and phenotypic diversity.

6 Multilocus sequence typing (MLST)

The PCR-based genotyping schemes presented above are not necessarily reliable in that the results obtained in one laboratory with a set of strains may not be the same from another laboratory. These differences arise from the instrumentation used, methods for culturing strains, and methods of DNA extraction, amongst other variables. To overcome these difficulties a novel typing scheme, multilocus sequence typing has been established [5]. This method relies on absolute characteristics of the strains under investigation and is available for all strains of a given species. These characteristics are the partial sequence data of individual, unlinked house keeping genes; usually at least seven genes are used. MLST schemes are portable typing systems that permit strain comparison between remote laboratories by means of centrally accessible sequence data. These data maybe subject to phylogenic analysis enabling evolutionary relationships between strains to be investigated, and it is most useful for epidemiological studies enabling acquisition and cross-infection to be studied but perhaps most importantly for the identification of pathogenic and epidemic strains within diseases-associated taxa. Much of these data are available on-line and the reader is referred to the following websites; http://www.pubmlst.org/ and http://www.mlst.net/.

This approach to the study of the diversity of oral bacteria has not yet progressed greatly, but the wide use of this methodology in medical microbiology coupled with the reduced cost and increased ease and speed of DNA sequencing will make this approach the method of choice for typing oral taxa in order to study, for example, transmission and acquisition of specific species.

The geographical variation in *P. gingivalis* has been studied using MLST [13]. In this study 40 *P. gingivalis* strains from worldwide sources were analyzed by determining partial sequences (310- to 420-bp) of the eight house-keeping genes, ftsQ, hagB, gdpxJ, pepO, mcmA, recA, pga, and nah. The number of sequences at individual loci ranged from 2 to 19, and a total of 33 allelic profiles, or sequence types (STs), were identified amongst the 40 strains so that identical or similar STs were found in isolates from different regions. These data indicate that the species showed signs of a clonal population structure but that the level of recombination was not as high as previously suggested. As had been found with PCR-based genotyping data for the species isolates from individual patients were genetically heterogeneous. Since this study there are now 99 STs in the MLST database (http://www.pubmlst.org/pgingivalis/).

7 Denaturing gel electrophoresis (DGGE)

An overall assessment of the diversity of oral biofilm samples can be gained using DGGE [6], a technique in which partial sequences of individual genes (e.g. rRNA, rpoB, and amoA) amplified by PCR with a GC-clamp attached. The amplicons are separated according to G+C mol% in a denaturing gel, and the resulting patterns represent the predominant taxa harboring the gene of interest; using the 16S rRNA gene the predominant microbiota are examined, and a comparison of the patterns may be used to estimate the diversity of the sampled population. Individual bands maybe excised from the gel and sequenced to identify the origin of the amplicons. In some instances individual bands may consist of more than one amplicons.

Using this methodology, treatment of periodontitis has been shown to cause shifts in composition and diversity of the microbial population, and that treatment resulted in a decrease in the diversity of the population [14]. While the microbial diversity and complexity of the microbial biota in dental plaque are significantly less in children with severe childhood caries than in caries-free children [15].

8 Conclusions

The extent of the diversity of the oral biofilm continues to be appreciated as a consequence of the various 16S rRNA sequencing projects, which have investigated the flora of healthy and diseased intra-oral sites in humans. To date some 700 taxa have been identified on the basis of a comparison of the 16S rRNA sequence data, and it is most likely that this number will continue to increase and the absolute number may never be known as all individuals examined to date have been found to harbor at least one novel taxa. Of course not everyone harbors all of these taxa but it is not known how many individual taxa an individual may harbor.

When the genotypic diversity of individual species is added into the mix the total diversity of the human oral microbiota becomes vast. However, it is good to remember that genotypic diversity may not mean functional diversity: different genotypes in different individuals may fulfil the same function in the biofilm. But as we have seen here different genotypes of the same species may have different phenotypes. We are just starting to investigate the relationships between genotypic diversity and function.

The insurance hypothesis suggests that the oral biofilm may exhibit a high level of "self-generated" diversity at the species level, as well as at the genotypic level. Diversity is a good thing. Examination of the effects of stress were not always predictable because a "small stress" may increase niche diversity and consequently increase genotypic diversity, whereas a "large stress" reduces niche diversity and may decrease genotypic and phenotypic diversity (simplification). The idea of a large stress "flattening" the environment reducing niche diversity and consequently genotypic and phenotypic diversity clearly illustrates the phenomenon. In the end the "fittest" taxa proliferate.

References

1. Aas JA, Paster BJ, Stokes LN, et al (2005) Defining the normal bacterial flora of the oral cavity. J Clin Microbiol 2005 43:5721–5732
2. Caufield PW, Walker TM (1989) Genetic diversity within *Streptococcus mutans* evident from chromosomal DNA restriction fragment polymorphisms. J Clin Microbiol 27:274–278
3. Versalovic J, Koeuth T, Lupski R (1991) Distribution of repetitive DNA sequences in eubacteria and application to fingerpriting of bacterial genomes. Nucleic Acids Res 19:6823–6831
4. Saarela M, Hannula J, Mättö J, et al (1999) Typing of mutans streptococci by arbitrarily primed polymerase chain reaction. Arch Oral Biol 41:821–826
5. Maiden MC, Bygraves JA, Feil E, et al (1998) Multilocus sequence typing: a portable approach to the identification of clones within populations of pathogenic microorganisms. Proc Natl Acad Sci USA 95:3140–3145
6. Muyzer G, de Waal EC, Uitterlinden AG (1993) Profiling of complex microbial populations by denaturing gradient gel electrophoresis analysis of polymerase chain reaction-amplified genes coding for 16S rRNA. Appl Environ Microbiol 59:695–700
7. Yachi S, Loreau M (1999) Biodiversity and ecosystem productivity in a fluctuating environment: the insurance hypothesis. Proc Natl Acad Sci USA 96:1463–1468
8. Boles BR, Thoendel M, Singh PK (2004) Self-generated diversity produces "insurance effects" in biofilm communities. Proc Natl Acad Sci USA 101:16630–16635
9. Vroom JM, De Grauw KJ, Gerritsen HC, et al (1999) Depth penetration and detection of pH gradients in biofilms by two-photon excitation microscopy. Appl Environ Microbiol 65:3502–3511

10. Alam S, Brailsford SR, Adams S, et al (2000) Genotypic heterogeneity of *Streptococcus oralis* and distinct aciduric subpopulations in human dental plaque. Appl Environ Microbiol 66:3330–3336
11. Paddick JS, Brailsford SR, Kidd EA, et al (2003) Effect of the environment on genotypic diversity of *Actinomyces naeslundii* and *Streptococcus oralis* in the oral biofilm. Appl Environ Microbiol 69:6475–6480
12. Paddick JS, Brailsford SR, Kidd EA, et al (2005) Phenotypic and genotypic selection of microbiota surviving under dental restorations. Appl Environ Microbiol 71:2467–2472
13. Enersen M, Olsen I, van Winkelhoff AJ, et al (2006) Multilocus sequence typing of *Porphyromonas gingivalis* strains from different geographic origins. J Clin Microbiol 44:35–41
14. Zijnge V, Harmsen HJ, Kleinfelder JW, et al (2003) Denaturing gradient gel electrophoresis analysis to study bacterial community structure in pockets of periodontitis patients. Oral Microbiol Immunol 18:59–65
15. Li Y, Ge Y, Saxena D, Caufield PW (2007) Genetic profiling of the oral microbiota associated with severe early-childhood caries. J Clin Microbiol 45:81–87

Interface Oral Health Science 2007: 43–52

Characterisation of the human oral microbiome and metagenome

William G. Wade*
King's College London Dental Institute, Infection Research Group, London, SE1 9RT, UK
*william.wade@kcl.ac.uk

Abstract. Members of the oral microbiota are responsible for the commonest bacterial diseases of man: dental caries and the periodontal diseases. Although some specific organisms have been implicated in the pathogenesis of these conditions, it is now recognised that they are not classical infectious diseases but, rather, complex diseases resulting from a breakdown in the normal homeostasis between the human host and its commensal microbiota. The first goal in understanding the mechanisms involved must be the comprehensive characterisation of the oral bacterial community. Since around half of oral species cannot be cultured on artificial media, molecular techniques have been developed based on PCR, cloning and sequencing of housekeeping genes such as that encoding 16S rRNA. These methods have revealed numerous novel bacterial lineages, including deep branches of the phyla *Bacteroidetes* and *Firmicutes* and novel phyla such as *Synergistes*. Taxa belonging to these newly described lineages are associated with oral diseases. In addition, novel genetic typing methods, such as multi locus sequence typing, are showing that intra-specific diversity is far greater than previously thought. Given this level of complexity, investigation of the oral bacterial metagenome is under way, which treats the oral bacterial community and its constituent genes as a whole. Functional screening of metagenomic libraries is a powerful tool for the identification of novel genes of interest in oral bacteria. These new molecular tools are revolutionising concepts of oral infectious disease pathogenesis and will enable new treatments to be developed based on modification of the composition and function of the oral microbiota.

Key words. metagenome, microbiome, periodontitis, dental caries

1 Bacterial specificity in oral bacterial diseases

The human oral cavity is host to a diverse microbiota of more than 800 species which is responsible for the two commonest bacterial diseases of man—dental caries and periodontal disease. Whether any specific bacteria are responsible for these diseases has been under investigation since the "golden age" of medical microbiology in the last quarter of the nineteenth century when bacteria could be reliably cultured for the first time. Associations were made between individual diseases and specific bacteria which could be shown to satisfy the postulates proposed by Koch. Initially, no single, or small group, of bacterial species were found to be associated with oral infections, and the diseases were thought to result from the non-specific action of the oral bacterial community as a whole. However, detailed studies performed in the mid-twentieth century were summarised by Loesche in the form of the specific plaque hypothesis [1]. This was based on the observation that the bacterial communities associated with oral infections differed from those seen in health; for example, periodontitis was associated with elevated levels of Gram-negative anaerobic rods. Loesche proposed that microbiological diagnosis should be performed and appropriate treatment, mechanical, chemical or both, instigated to restore the dental plaque flora to a healthy state. Later evidence for bacterial specificity in periodontitis was provided in a longitudinal study that demonstrated that a small group of organisms: *Porphyromonas gingivalis*, *Fusobacterium nucleatum*, *Tannerella forsythensis* and *Campylobacter rectus*, was found at sites prior to loss of alveolar bone [2].

Attention was then focused on these species, *P. gingivalis* in particular, to identify virulence determinants which would give them the potential to cause disease [3]. It was found to possess a wide range of such factors including the ability to evade the host defence by forming a capsule, inhibiting neutrophil chemotaxis, production of superoxide dismutase to prevent killing by the oxygen radicals produced by neutrophils, and production of immunoglobulin proteases to combat humoral immunity. Factors capable of damaging the host include the ability to invade cells and damage them by producing a range of proteases including an Arg-1 protease which has been extensively characterised. The lipopolysaccharide of *P. gingivalis* was also shown to stimulate bone resorption.

The numerical association between *P. gingivalis* and periodontitis and the organism's armoury of virulence factors has sometimes led to the assumption that *P. gingivalis* is a specific pathogen that causes the disease. However, it is very difficult to distinguish between an organism that colonises a site and then actively causes disease there, from one whose preferred habitat is a diseased site and colonises sites already undergoing breakdown. *P. gingivalis* requires haemin for growth and thus is likely to prefer a habitat rich in haemin such as inflamed and bleeding gingivae. It is also a possibility, though, that organisms such as *P. gingivalis* might colonise already inflamed sites and then exacerbate the damage to the host with their proteases and toxins. It is important then to distinguish between periodontal pocket formation, the initial breakdown of attachment to form the pocket, and the development of the chronic lesion, heavily colonised with anaerobic bacteria and neutrophils and lymphocytes. Neutrophils, in particular, produce proteases which contribute to the destruction of the local tissue [4].

The periodontal microbiota is highly complex with more than 350 different species which are able to colonise this habitat. In many ways, periodontitis does not resemble the classical infectious diseases for with Koch constructed his postu-

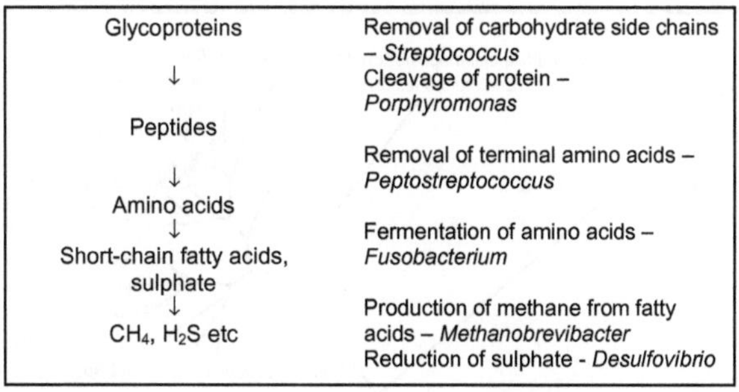

Fig. 1. The periodontal pocket as a tissue digestor—sequential degradation of glycoproteins by oral bacteria

lates. As with other complex diseases affecting mucosal surfaces with an associated normal microbiota, no single pathogen causes the disease. Indeed, all of the organisms implicated in periodontitis to date can be found at healthy sites. What is perhaps more surprising given the ubiquity of the periodontal pathogens is that so few individuals suffer from the disease. The disease might best be described as one where the normal homeostasis between the human host and its associated normal microbiota has broken down, resulting in an inappropriate inflammatory response from the host given the level of bacterial challenge. The two phases of the disease can then be separated into initiation, where an inappropriate host response to the normal microbiota leads to loss of attachment and the formation of a periodontal pocket and progression, where the new anatomy formed is colonised by a consortium adapted to that habitat which work together to cooperatively degrade the tissue of the periodontium. The periodontal pocket can thus be regarded as an ecosystem in which glycoproteins are progressively degraded to simple molecules such as CH_4, H_2S, H_2O, and so on. Some key possible steps in this process and examples of organisms likely to be responsible are shown in Fig. 1.

2 Unculturable bacteria

The primary obstacle to our understanding of the role of bacteria in oral disease is our inability to culture around half of the bacteria present on artificial media in the laboratory. In recent years, however, methods have become available for the study of unculturable bacteria based on the amplification, cloning and sequencing of housekeeping genes such as that encoding the small subunit (16S) ribosomal RNA molecule. This gene has been highly conserved throughout evolution and thus has regions invariant among most bacteria which are useful for the design of PCR

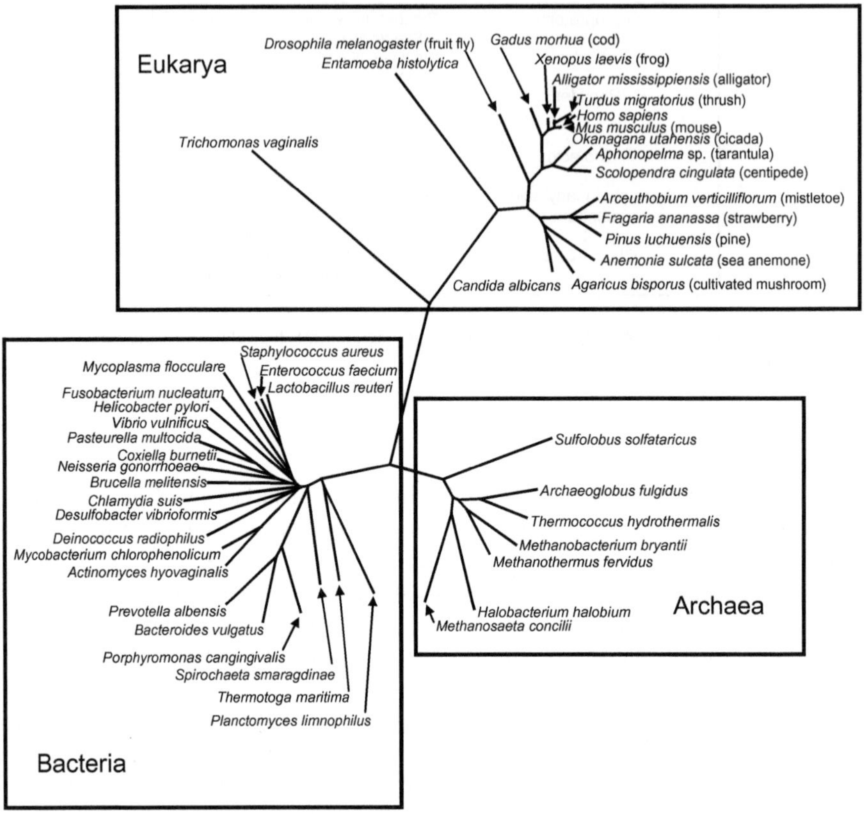

Fig. 2. Phylogenetic tree showing the three domains of cellular life on Earth based on comparisons of small sub-unit ribosomal RNA

primers of broad specificity. At the same time other regions of the gene are far more variable and can be used to distinguish between taxa at species level, in most cases. Woese [5] compared small subunit rRNA molecules from a wide range of prokaryotes and eukaryotes and used the data to construct an evolutionary tree; the so-called tree of life. An example of such a tree is shown in Fig. 2. As expected, prokaryotes and eukaryotes constituted separate branches of the tree. What was not expected was that two Kingdoms were present among the prokaryotes, now known as the domains *Archaea* and *Bacteria*. *Archaea* were originally thought to be those bacteria found in hostile conditions such as volcanoes and the bottom of the oceans but, in fact, they are widely distributed and have representatives, members of the genus *Methanobrevibacter*, in the human mouth.

The availability of the phylogenetic tree means that only the DNA of an organism is needed to identify it. The ribosomal RNA gene can be amplified and sequenced and then compared to the DNA databases to obtain an identification.

The general procedure for the characterisation of complex bacterial communities is as follows. DNA is extracted directly from the sample of interest using standard methods. The 16S rRNA genes are then amplified by PCR with universal primers, and the mixed genes are then joined to a plasmid vector by ligation and the vector plus insert is used to transform competent *Escherichia coli* cells to create a library whereby each *E. coli* culture has a 16S rRNA gene from one of the organisms in the original sample. The cloned genes can then be sequenced and identified to determine the composition of the sample. These techniques have now been used for a wide range of different oral bacterial communities and disease states including dental abscesses [6], endodontic infections [7], dentinal carious lesions [8], and periodontitis [9]. Each study confirmed that around half of the microbiota present was uncultivable, and that numerous hitherto undescribed organisms were present. For example, de Lillo et al. [9] showed that among 177 bacterial isolates and 417 cloned genes isolated from three subgingival plaque samples, 137 taxa were identified of which 86 were found by the molecular technique alone. Many of these novel taxa are found in lineages with no cultivable representatives. Lineages that are especially prominent among the oral microbiota include branches related to the genera *Bacteroides* and *Porphyromonas* in the phylum *Bacteroidetes* and a branch of the family *Lachnospiraceae* within the *Firmicutes*.

There are even entire phyla without cultivable representatives. These include candidate Division TM7 which is widespread in the environment and has been detected in peat bogs, soil, reactor sludges, mouse faeces as well as the human mouth. Hugenholtz et al. [10] used specific fluorescent-labelled oligonucleotide probes to perform fluorescent in-situ hybridisation (FISH) to determine the morphology of TM7 cells and found them to be sheathed filaments with a Gram-positive cell wall; only the third phylum of the 37 of the domain Bacteria to include Gram-positive organisms.

The use of molecular methods to characterise oral bacterial communities has not only allowed the detection and identification of organisms refractory to cultivation but has also allowed far more precise identifications of individual organisms to be obtained than previously possible by phenotypic methods alone. A meta-analysis of the community structure of the microbiota associated with dental caries, endodontic infections and periodontitis showed that the community associated with each condition was remarkably specific [11]. The majority of species found in each disease were only found in that disease and of 268 species identified in total, only three were found in all three diseases.

As discussed above, a large number of novel taxa have been discovered through the use of culture-independent molecular methods. The next task is to determine their role in oral health and disease, and the first stage is to seek associations between specific taxa and disease. Kumar et al. [12] used PCR to determine prevalence ratios between 39 oral taxa in subjects with periodontitis and healthy controls. At a level equal or greater than the "red cluster" of organisms described by Socransky et al. [13]: *Porphyromonas gingivalis*, *Tannerella forsythensis*, and *Treponema denticola*, and traditionally associated with periodontitis, they found six taxa. These were *Eubacterium saphenum*, *Porphyromonas endodontalis*, *Bacteroides* phylotype AU126, *Synergistes* phylotype BH017, *Treponema lecithinolyticum*, and *Filifactor alocis*. The number and range of organisms associated with oral disease is therefore expanding and will allow greater insight into the disease process and provide additional targets for prevention and treatment of disease.

3 Sub-specific diversity

All of the studies described thus far have taken the species to be the "unit of virulence" with a working assumption that all strains of a species are of equal virulence. However, it has long been known that strains vary in their virulence. Virulence genes are commonly found within chromosomal pathogenicity islands or on plasmids which can be transferred between strains. The advent of bacterial genome sequencing has made additional data available on the intra-specific genetic differences between strains. These have shown that the variation is far greater than previously thought. For example, in one study three strains of *E. coli* were compared; the harmless laboratory strain K12, an enterohaemorrhagic strain of the O157 serotype associated with beef products, and a uropathogenic strain. It was found that they had only 39% of their genes in common [14]. Tettelin et al. [15] compared the genomes of six strains of *Streptococcus agalactiae*, a Group B *Streptococcus* which is a major cause of morbidity and/or mortality in neonates and of invasive infection in the elderly. Around 80% of genes were common to all strains but the remaining 20% of each genome was highly variable. An estimate was made of the number of strains that would have to be sequenced to determine all of the genetic variability in *S. agalactiae* but was found to be infinite! This and other studies has led to the description of the concepts of the pan-genome, all of the genes that can be found among members of a species; the core genome, those genes found in all strains normally encoding housekeeping and other essential functions, and the peripheral or dispensable genome encoding strain-specific characteristics including virulence and antimicrobial resistance.

4 The metagenome

It follows then, that if strains of a species can vary so markedly in their genetic composition, that the value of investigations of pathogenesis and microbiological diagnosis dependent on species identifications are limited. An alternative approach is to consider the bacterial community found at a given site in its entirety and to perform genetic analyses on all the genomes together, the so-called metagenome [16]. Metagenomic analyses are of two main types. In both types, DNA is extracted from the sample and fragments incorporated into vectors by ligation. In studies to generate sequence data from the metagenome, small fragments of DNA are cloned into plasmid vectors which can be easily sequenced. Alternatively, fosmids or bacterial artificial chromosomes can be used to clone large fragments of DNA that can be expressed and screened for functions of interest. Since most naturally occurring antibiotics are produced by soil bacteria, the metagenome of the soil has been cloned and screened for the production of new antimicrobials. Turbomycin A and B were found in this way [17]. Similar studies are underway screening metagenomic libraries for functions of relevance to oral infections.

A major goal is to obtain complete genome sequence data from as yet uncultivable organisms. One approach is to randomly sequence DNA isolated and cloned from a bacterial community and then attempt to reconstruct the genomes of the constituent organisms in the community. This has been attempted for a relatively simple community found in a subterranean acid mine drainage site, and although successful, near complete genome sequences could only be reconstructed for two organisms [18]. Alternatively, individual uncultivable organisms can be targeted. 16S rDNA sequence data from such organisms is available from the molecular ecology studies described above, and this can be used to design oligonucleotide probes that are specific for the organism of interest. These can be labelled with fluorescent compounds and used in conjunction with flow cytometry/cell sorting instruments to physically isolate the organisms. Typically, only a small number of cells are recovered which would yield insufficient DNA for genome sequencing. However, the recent introduction of whole-genome amplification techniques has made it possible to obtain sufficient DNA for genome sequencing from a single cell [19], and the successful partial sequencing of a representative of the uncultivable TM7 Division isolated from a complex community has recently been reported [20].

5 Conclusions

It is now clear that oral infections are complex bacterial diseases of multifactorial aetiology arising from a breakdown in the normal homeostasis that exists between the host and its commensal microbiota, arising from susceptibility in the host interacting with environmental factors and the bacteria themselves. The oral microbiota is a highly diverse bacterial community with high proportion of unculturable organisms. The new methods described in this chapter are allowing the dissection of this community and the determination of its role in health and disease. The high level of genomic diversity among strains of the same species means that species-level identification must be regarded as of limited relevance to diagnosis. However, the new discipline of metagenomics, that treats all of the genomes of the organisms in a community as a single entity allows new insight into the metabolic and virulence potential of the oral microflora which, in turn, will allow new methods for the prevention and treatment of oral disease to be developed.

References

1. Loesche WJ (1976) Chemotherapy of dental plaque infections. Oral Sci Rev 9:65–107
2. Tanner AC, Haffer C, Bratthall GT, et al (1979) A study of the bacteria associated with advancing periodontitis in man. J Clin Periodontol 6:278–307
3. Cutler CW, Kalmar JR, Genco CA (1995) Pathogenic strategies of the oral anaerobe, *Porphyromonas gingivalis*. Trends Microbiol 3:45–51
4. Buchmann R, Hasilik A, Van Dyke TE, et al (2002) Amplified crevicular leukocyte activity in aggressive periodontal disease. J Dent Res 81:716–721
5. Woese CR (1987) Bacterial evolution. Microbiol Rev 51:221–271
6. Dymock D, Weightman AJ, Scully C, et al (1996) Molecular analysis of microflora associated with dentoalveolar abscesses. J Clin Microbiol 34:537–542
7. Munson MA, Pitt-Ford T, Chong B, et al (2002) Molecular and cultural analysis of the microflora associated with endodontic infections. J Dent Res 81:761–766
8. Munson MA, Banerjee A, Watson TF, et al (2004) Molecular analysis of the microflora associated with dental caries. J Clin Microbiol 42:3023–3029
9. de Lillo A, Ashley FP, Palmer RM, et al (2006) Novel subgingival bacterial phylotypes detected using multiple universal polymerase chain reaction primer sets. Oral Microbiol Immunol 21:61–68
10. Hugenholtz P, Tyson GW, Webb RI, et al (2001) Investigation of candidate division TM7, a recently recognized major lineage of the domain *Bacteria* with no known pure-culture representatives. Appl Environ Microbiol 67:411–419
11. Wade WG, Munson MA, de Lillo A, et al (2005) Specificity of the oral microflora in dentinal caries, endodontic infections and periodontitis. In: Interface oral health science, International Congress Series. Elsevier, Amsterdam, pp 150–157
12. Kumar PS, Griffen AL, Barton JA, et al (2003) New bacterial species associated with chronic periodontitis. J Dent Res 82:338–344
13. Socransky SS, Haffajee AD, Cugini MA, et al (1998) Microbial complexes in subgingival plaque. J Clin Periodontol 25:134–144
14. Welch RA, Burland V, Plunkett G, et al (2002) Extensive mosaic structure revealed by the complete genome sequence of uropathogenic *Escherichia coli*. Proc Natl Acad Sci USA 99:17020–17024
15. Tettelin H, Masignani V, Cieslewicz MJ, et al (2005) Genome analysis of multiple pathogenic isolates of *Streptococcus agalactiae*: implications for the microbial "pan-genome". Proc Natl Acad Sci USA 102:13950–13955
16. Rondon MR, August PR, Bettermann AD, et al (2000) Cloning the soil metagenome: a strategy for accessing the genetic and functional diversity of uncultured microorganisms. Appl Environ Microbiol 66:2541–2547
17. Gillespie DE, Brady SF, Bettermann AD, et al (2002) Isolation of antibiotics turbomycin A and B from a metagenomic library of soil microbial DNA. Appl Environ Microbiol 68:4301–4306
18. Tyson GW, Chapman J, Hugenholtz P, et al (2004) Community structure and metabolism through reconstruction of microbial genomes from the environment. Nature 428:37–43

19. Kvist T, Ahring BK, Lasken RS, et al (2007) Specific single-cell isolation and genomic amplification of uncultured microorganisms. Appl Microbiol Biotechnol 74:926–935
20. Podar M, Abulencia CB, Walcher M, et al (2007) Targeted access to the genomes of low-abundance organisms in complex microbial communities. Appl Environ Microbiol 73:3205–3214

Interface Oral Health Science 2007: 53–62

Novel functions of adhesins encoded by gingipain genes of *Porphyromonas gingivalis*

Koji Nakayama*
Division of Microbiology and Oral infection, Department of Molecular Microbiology and Immunology, Nagasaki University Graduate School of Biomedical Sciences, Nagasaki 852–8588, Japan
*knak@nagasaki-u.ac.jp

Abstract. The oral anaerobic bacterium *Porphyromonas gingivalis* has been implicated as a major pathogen for chronic periodontitis. The microorganism produces strong proteinases, Arg-gingipain (Rgp) and Lys-gingipain (Kgp), on the cell surface and in extracellular milieu. Gingipain genes encode polyproteins consisting of four parts: signal peptide, propeptide, proteinase, and a C-terminal adhesin domain region. The C-terminal adhesin domain region encoded by Rgp-encoding gene A (*rgpA*) comprises four domains (Hgp44, Hgp15, Hgp17, and Hgp27). A number of studies have indicated that a major hemagglutinin is derived from the gingipain genes; however, direct evidence has not yet been obtained. We showed that a fully processed recombinant Hgp44 had hemagglutinating activity, and that Hgp44 hemagglutinin bound to glycophorin on the erythrocyte membrane. *P. gingivalis* cells have the ability to aggregate platelets. We showed that *P. gingivalis* cell-mediated platelet aggregation required Hgp44 hemagglutinin on the bacterial cell surface, *P. gingivalis*-reactive antibody in plasma, and FcγRIIa, GPIIa/IIIb and GPIb-IX–V receptors on platelets. We also showed that a major protein of the culture supernatant of *P. gingivalis*, Hgp15, suppressed in vitro receptor activator of nuclear factor-κB (NF-κB) ligand (RANKL)-mediated osteoclastogenesis. Hgp15 inhibited RANKL-mediated induction of c-Fos and NFATc1. These results suggest that the C-terminal adhesin domains have multiple functions associated with virulence of *P. gingivalis*.

Key words. gingipain, hemagglutination, osteoclastogenesis, platelet aggregation, *Porphyromonas gingivalis*

1 Introduction

Periodontitis is chronic inflammatory conditions of supporting tissues in teeth. Evidence has revealed that the initiation and progression of periodontitis are closely associated with the quantitative and qualitative changes of microorganisms in the subgingival biofilm. Among them, the Gram-negative anaerobic bacterium *Porphyromonas gingivalis* has been implicated as one of the major etiological agents of chronic adult periodontitis. *P. gingivalis* colonizes in periodontal pockets and spreads into deeper tissues, including connective and bone tissues [1–3]. The bacterium possesses a number of virulence factors, such as fimbriae, lipopolysaccharide (LPS), and cysteine proteinases named gingipain [1, 2]. In general, LPS displays multiple biological and immunological activations via Toll-like receptors (TLRs) and potently stimulates bone resorption [4, 5]; however, *P. gingivalis* LPS has a weak ability to activate inflammation [6, 7]. Fimbriae mediate bacterial adhesion to human epithelial cells and have the ability to induce proinflammatory responses through TLRs [8, 9]. Arginine-specific gingipains (Arg-gingipain-A, RgpA and Arg-gingipain-B, RgpB) and lysine-specific gingipain (Lys-gingipain, Kgp) degrade a number of proteins, including extracellular matrix proteins, cytokines, complements, antibodies, and proteinase inhibitors [10–12]. Rgp also functions in maturation of fimbrilin, a unit protein of fimbriae [13]. *rgpA* and *kgp* have similar gene structures, consisting of an N-terminal propeptide region, a proteolytic domain, and C-terminal adhesin domains [Hgp15 (HbR), Hgp17, Hgp27 and Hgp44]. The adhesin domains are also encoded by the hemagglutinin-encoding gene *hagA*; these regions are similar to C-terminal regions (Hgp44 and Hgp15) of *rgpA* and *kgp* [14]. This review article describes the novel functions of the adhesin domains that are associated with virulence of *P. gingivalis.*

2 Role of Hgp44 domain protein in *P. gingivalis*-induced hemagglutination

Previous workers have attempted to identify the hemagglutinin molecule. First Okuda and Takazoe [15] reported that bacterial surface components of the bacterium have a hemagglutinating activity. It was reported that *P. gingivalis* 381 FimA fimbriae and their oligopeptide segments have activity to agglutinate erythrocytes [16], whereas other researchers found that purified FimA fimbriae from *P. gingivalis* 381 exhibited no hemagglutinating activity [17] and also removal of fimbriae from *P. gingivalis* W12 cells had no effect on the hemagglutinating activity [18]. Hemagglutinins with apparent molecular masses of about 24,000, 37,000, and 44,000 were partially purified by Inoshita et al. [19] from a culture supernatant of *P. gingivalis* 381. Okuda et al. [20] also partially purified hemagglutinins and obtained similar results. They reported that the hemagglutinating activity is not inhibited by sugar but is inhibited by arginine and arginine-containing peptides. Others have also partially purified hemagglutinins [21, 22]. Monoclonal antibodies that inhibited hemagglutination of *P. gingivalis* were found to recognize a peptide within the Hgp44 domain encoded by *rgpA*, *kgp*, and *hagA* [23, 24]. Our previous studies [25, 26] suggested that all three genes are responsible for hemagglutination. Kgp proteinase–adhesin complexes have hemagglutinating activity [27], while a single-chain 50-kDa form of RgpA has no such activity [28], suggesting that the proteinase domain alone is not sufficient for hemagglutination.

We initially constructed a GST–$Hgp44_{720–1138}$ fusion protein and purified $Hgp44_{720–1138}$ after proteolytic digestion (Fig. 1), and we examined the ability of the purified protein to agglutinate human erythrocytes; however, unexpectedly, the

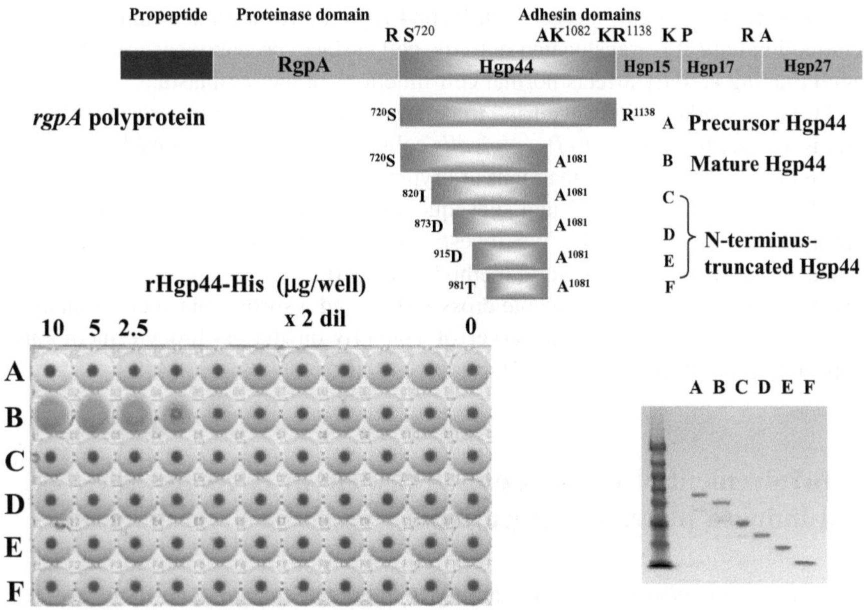

Fig. 1. Ability of mature Hgp44 protein (Hgp44B) to agglutinate human erythrocytes

protein was unable to agglutinate them. Veith et al. [29] reported that the C-terminal end of Hgp44 domain protein on the cell surface of *P. gingivalis* was predicted at Ala$_{1081}$. We therefore constructed and purified a C-terminal-deleted protein (Hgp44B, Hgp44$_{720-1081}$), N-terminal-deleted proteins with C-terminal deletion (Hgp44C, Hgp44$_{820-1081}$; Hgp44D, Hgp44$_{873-1081}$; Hgp44E, Hgp44$_{915-1081}$; Hgp44F, Hgp44$_{982-1081}$), and a whole-length Hgp44 (Hgp44A, Hgp44$_{720-1138}$) using a His-tagged recombinant protein construction and purification system, and we determined their ability to agglutinate erythrocytes (Fig. 1). Hgp44B had the ability to agglutinate erythrocytes, whereas Hgp44A, Hgp44C, Hgp44D, Hgp44E, and Hgp44F could not agglutinate erythrocytes.

Kelly et al. [23] mapped the hemagglutinating epitope of Hgp44 region on residues 1073–1112 because a synthetic peptide corresponding to residues 1083–1102 inhibited *P. gingivalis* cell-mediated hemagglutination efficiently, and two peptides corresponding to residues 1073–1092 and 1093–1112 inhibited it with less efficiency. We found that Hgp44B without the hemagglutinating epitope induced hemagglutination, whereas Hgp44A with the hemagglutinating epitope did not induce hemagglutination. Taken together, the results suggest that the inter-domain amino acid region may not function as a hemagglutinating epitope but as a suppressor of hemagglutination. We also found that the peptide corresponding to residues 1083–1102 inhibited Hgp44B-mediated hemagglutination with human erythrocytes and that the peptide bound Hgp44B in a dose-dependent manner. These results indicate that in the unprocessed *rgpA* polyprotein the inter-domain regional residues might interact with Hgp44 regional residues and inhibit hemagglutination until it is translocated onto the cell surface, probably because the hemagglutinating activity affects normal cell functions in the periplasm.

Previous workers [30] have tried to determine the erythrocyte surface molecule that is a target for binding to *P. gingivalis* cells. Band 3 has been considered as one of the targets, because the treatment of erythrocytes with chymotrypsin resulted in the disappearance of band 3 and subsequently the inhibition of hemagglutinating ability of erythrocytes [31]. We investigated an Hgp44B-targeted molecule on human erythrocytes using several methods with proteinase treatment, a solid phase overlay, a membrane-impermeable cross-linker, and a surface plasmon resonance. The results suggested that the target of Hgp44B on the erythrocyte membrane appeared to be glycophorin A [32].

3 Involvement of Hgp44 domain protein in *P. gingivalis* cell-induced platelet aggregation

Porphyromonas gingivalis has the ability to aggregate human platelets in platelet-rich plasma (PRP) [33]. A proteinase preparation from *P. gingivalis* (protease I), which seemed to be Rgp and Kgp, was able to induce platelet aggregation with purified platelets [34]. Lourbakos et al. [35] found that Rgp has the ability to induce platelet aggregation with purified platelets through activation of the protease-activated receptors, (PAR)-1 and -4, expressed on the surface of platelets. The extracellular N-terminal regions of PARs contain thrombin cleavage sites and cleavage of the N-terminal regions by thrombin causes activation of PAR receptors. Rgp-induced PAR activation appears to be similar to that of thrombin. We examined whether *P. gingivalis* cell-induced platelet aggregation was also attributable to Rgp and/or Kgp [36].

To examine the possibility, a proteinase inhibitor that had the potential to inhibit Rgp or Kgp was added to the mixture of *P. gingivalis* cell suspension and PRP. Addition of Rgp-specific and Kgp-specific inhibitors showed no significant effect on *P. gingivalis*-induced platelet aggregation, indicating that platelet-aggregating factors other than gingipains might be present in *P. gingivalis.* Subcellular fractionation revealed that the aggregating factors were localized on the outer membrane fraction. We have constructed various mutant strains concerning gingipains. When using PRP, the Rgp-null mutant KDP133 (*rgpA rgpB*) showed platelet aggregation with no significant difference from the wild type parent strain, that was consistent with the fact that the Rgp-specific proteinase inhibitor failed to inhibit *P. gingivalis*-induced platelet aggregation in PRP. The Kgp-null mutant KDP129 (*kgp*) also showed platelet aggregation in PRP, whereas the Rgp/Kgp-null mutant KDP136 (*rgpA rgpB kgp*) showed no platelet aggregation. The *rgpA* and *kgp* genes encode the adhesin domain proteins at their 3′-terminal regions in addition to encoding Rgp and Kgp, respectively. The adhesin-null mutant KDP137 (*rgpA kgp hagA*) that had *rgpB*-derived Rgp activity showed no platelet aggregation. These results suggested that the adhesin proteins might be involved in *P. gingivalis*-induced platelet aggregation in PRP. Treatment of KDP136 cells with trypsin induced processing of HagA proprotein on the bacterial cell surface and platelet aggregation with PRP. Cells of Rgp/Kgp/adhesin-null mutant (*rgpA rgpB kgp hagA*; KDP153) were pretreated with the Hgp44B adhesin, and the resulting mixture was subjected to platelet aggregation in PRP. KDP153 cells or the Hgp44B adhesin alone failed to induce platelet aggregation in PRP, whereas Hgp44B-pretreated KDP153 induced it, indicating that Hgp44B is involved in *P. gingivalis* cell-induced platelet aggregation.

In *Streptococcus sanguis*-induced platelet aggregation, platelets are stimulated through GPIbα receptor on the platelet surface [37]. After that, GPIIb/IIIa is activated by inside-out signaling. Addition of an antagonist of GPIIb/IIIa to the mixtures of *P. gingivalis* cells and PRP completely inhibited platelet aggregation, indicating that *P. gingivalis* cell-induced platelet aggregation requires platelet activation. Various monoclonal antibodies directed against different N-terminal regions of GPIbα receptor were examined for their ability to inhibit the aggregation. Among them the monoclonal antibody SZ-2 significantly inhibited *P. gingivalis*-induced platelet aggregation. To confirm the interaction of GPIbα with *P. gingivalis* cell-induced platelet aggregation, PRP were pretreated with mocarhagin, a cobra venom metalloproteinase that specifically cleaved an N-terminal region of GPIbα and abolished signals via GPIb-IX–V complex. Platelet aggregation in PRP with *P. gingivalis* cells was partially inhibited by treatment with mocarhagin, indicating that platelet activation via GPIbα contributed to *P. gingivalis*-induced platelet aggregation in PRP. The previous reports concerning *Helicobacter pylori* and *Streptococcus sanguis* suggested contribution of immunoglobulin to induce platelet aggregation [37, 38]. FcγRIIa is the sole immunoglobulin receptor on human platelets, and can bind polymeric IgG. GPIb-IX–V complex and FcγRIIa may associate in physical proximity, resulting in functional interplay on human platelet activation [39, 40]. Trypsin-treated KDP136 cells could induce aggregation with purified platelets in the presence of IgG but not IgG-depleted platelet-poor plasma, suggesting that IgG was essential for *P. gingivalis* cells to induce platelet aggregation. Anti-FcγRIIa monoclonal antibody completely inhibited *P. gingivalis* cell-induced platelet aggregation in PRP. These results strongly indicated that plasma IgG and FcγRIIa receptor were indispensable for *P. gingivalis*-induced platelet aggregation in plasma. Human IgG lacking *P. gingivalis* cell-reactive antibody failed to induce platelet aggregation with purified platelets and trypsin-treated KDP136 cells. Interestingly, human IgG pretreated with cells of KDP137 or KDP136 also lost the aggregation-inducing ability, indicating that *P. gingivalis* cell-reactive IgG was required for *P. gingivalis*-induced platelet aggregation, but it was not necessary for the IgG to be reactive to adhesins such as Hgp44.

Secretion of fibrinogen, fibronectin, and von Willebrand Factor (vWF) from activated platelets may account for no requirement of vWF, fibrinogen or

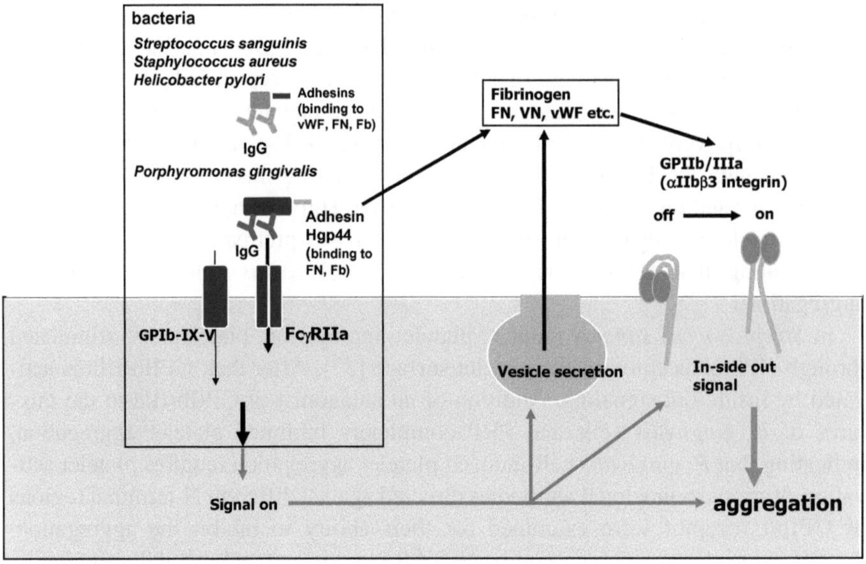

Fig. 2. Mechanism of bacterial cell-induced platelet aggregation

fibronectin as a plasma factor for *P. gingivalis* cell-induced platelet aggregation using purified platelets. The Hgp44B adhesin has the ability to bind fibrinogen and fibronectin. Interaction of these extracellular matrix (ECM) proteins with adhesins such as Hgp44B and cluster formation of *P. gingivalis*-reactive IgG may take place on the same *P. gingivalis* cell surface when the bacterial cells are mixed with PRP. Clustered IgG on the bacterial cell surface may bind to FcγRIIa activate platelets, and ECM proteins anchored to the same bacterial cell by adhesins may bind to activated GPIIb/IIIa and further activate the platelets, resulting in development of platelet aggregation [41] (Fig. 2).

4 Inhibitory effect of Hgp15 domain protein on RANKL/M-CSF-induced osteoclast formation

Alveolar bone resorption is a major clinical characteristic of periodontitis including chronic adult periodontitis [42]. Osteoclasts in charge of bone resorption are derived from hematopoietic stem cells. Two molecules, macrophage colony stimulating factor (M-CSF) and receptor activator of nuclear factor-κB (NF-κB) ligand, are essential for differentiation to osteoclasts [43]. M-CSF, which is indispensable for macrophage maturation, binds to its receptor in early osteoclast precursors, thereby providing signals required for their survival, proliferation, and differentiation to osteoclasts. RANKL, belonging to the tumor necrosis factor-α (TNF-α) family, binds to their receptor, receptor activator of NF-κB (RANK) and activates several intracellular signaling pathways, leading to osteoclastic differentiation and activation. Infection with viable *P. gingivalis* wild type strain induced RANKL expression in osteoblasts, whereas infection with the Rgp/Kgp-null (*rgpA rgpB kgp*) mutant did not [44]. This effect of *P. gingivalis* appears to induce osteoclastogenesis. Then, to examine whether *P. gingivalis* directly affect osteoclast formation, we investigated the effect of the culture supernatant of *P. gingivalis* on RANKL/M-CSF-induced osteoclast formation from bone marrow macrophages [45]. When the LPS-free culture supernatant of *P. gingivalis* was added to the cell culture in the presence of RANKL and M-CSF, TRAP-positive multinuclear cells were markedly decreased, indicating an inhibitory effect of the culture supernatant on osteoclast formation. Since the culture supernatant contained a few proteins and a major protein was identified as the Hgp15 domain protein, we examined the effect of Hgp15 on RANKL/M-CSF-induced osteoclast formation from bone marrow macrophages. Hgp15 markedly inhibited formation of TRAP-positive multinuclear cells in a dose-dependent fashion. Since viability of bone marrow macrophages did not change during the treatment, Hgp15-mediated inhibition of RANKL-induced osteoclast formation was not due to cell death. Hgp15 suppressed RANKL-induced Akt phosphorylation and inhibited expression of c-Fos and NFATc1.

5 Conclusion

The *rgpA, kgp* and *hagA* genes initially produce polyproteins that consist of gingipains and adhesins. Most of matrix metalloproteinases (MMPs) comprise domain structures. A substrate-binding exosite domain, the hemopexin domain, is located at the COOH-terminal. In MMP-2, the hemopexin domain binds to monocyte chemoattractant protein-3 (MCP-3) to elicit degradation of MCP-3 by the proteinase domain of MMP-2 [46]. Binding property of the adhesin domains encoded by *rgpA* and *kgp* is, therefore, very important to know target proteins for Rgp and Kgp proteinases. Now, the Hgp44 domain protein is found to have the ability to agglutinate human erythrocytes and the target molecule of Hgp44 is glycophorin A. Since heme markedly accelerates growth of *P. gingivalis*, Hgp44-mediated hemagglutination can help the bacterium to survive in the gingival crevice. Hgp44 is also involved in platelet aggregation with human plasma. The Hgp15 domain protein has the ability to bind hemoglobin and lactoferrin [47]. Hgp15 can inhibit osteoclastogenesis from bone marrow macrophages. Since rapid destruction of the alveolar bone causes tooth loss, resulting in loss of the gingival crevice that is an anatomical niche for *P. gingivalis*, the suppressive effect of Hgp15 on osteoclast formation may benefit the bacterium in survival in the niche.

References

1. Holt SC, Kesavalu L, Walker S, et al (1999) Periodontol 2000 20:168–238
2. Lamont RJ, Jenkinson HF (1998) Microbiol Mol Biol Rev 62:1244–1263
3. Saglie FR, Marfany A, Camargo P (1988) J Periodontol 59:259–265
4. Dumitrescu AL, Abd-El-Aleem S, Morales-Aza B, et al (2004) J Clin Periodontol 31:596–603
5. Hausmann E, Raisz LG, Miller WA (1970) Science 168:862–864
6. Reife RA, Shapiro RA, Bamber BA, et al (1995) Infect Immun 63:4686–4694
7. Yoshimura A, Kaneko T, Kato Y, et al (2002) Infect Immun 70:218–225
8. Asai Y, Ohyama Y, Gen K, et al (2001) Infect Immun 69:7387–7395
9. Harokopakis E, Hajishengallis G (2005) Eur J Immunol 35:1201–1210
10. Curtis MA, Kuramitsu HK, Lantz M, et al (1999) J Periodontal Res 3:4464–4472
11. Kadowaki T, Nakayama K, Okamoto K, et al (2000) J Biochem 128: 153–159
12. Potempa J, Banbula A, Travis J (2000) Periodontol 2000:24153–24192
13. Shoji M, Naito M, Yukitake H, et al (2004) Mol Microbiol 52:1513–1525
14. Han N, Whitlock J, Progulske-Fox A (1996) Infect Immun 64:4000–4007
15. Okuda K, Takazoe KI (1974) Arch Oral Biol 19:415–416
16. Ogawa T, Hamada S (1994) Infect Immun 62:3305–3310
17. Yoshimura F, Takahashi K, Nodasaka Y, et al (1984) J Bacteriol 160:949–957
18. Boyd J, McBride BC (1984) Infect Immun 45:403–409
19. Inoshita E, Amano A, Hanioka T, et al (1986) Infect Immun 52:421–427
20. Okuda K, Ono M, Kato T (1989) Infect Immun 57:1635–1637
21. Nishikata M, Yoshimura F (1991) Biochem Biophys Res Commun 178:336–342
22. Hayashi H, Nagata A, Hinode D, et al (1992) Oral Microbiol Immunol 7:204–211
23. Kelly CG, Booth V, Kendal H, et al (1997) Clin Exp Immunol 110: 285–291
24. Shibata Y, Hayakawa M, Taniguchi H, et al (1999) J Biol Chem 274:5012–5020
25. Nakayama K, Kadowaki T, Okamoto K, et al (1995) J Biol Chem 270:23619–23626
26. Shi Y, Ratnayake DB, Okamoto K, et al (1999) J Biol Chem 274:17955–17960
27. Pike RN, Potempa J, McGraw W, et al (1996) J Bacteriol 178:2876–2882
28. Pike R, McGraw W, Potempa J, et al (1994) J Biol Chem 269:406–411
29. Veith PD, Talbo GH, Slakeski N, et al (2002) Biochem J 363:105–115
30. Mouton C, Chandad F (1993) Hemagglutination and hemagglutinins. In: Shah HN, Mayrand D, Genco RJ (eds) Biology of the species *Porphyromonas gingivalis*, CRC Press, Boca Raton, pp 199–217
31. Hayashi H, Nagata A, Hinode D, et al (1992) Oral Microbiol Immunol 7:204–211
32. Sakai E, Naito M, Sato K, et al (2007) J Bacteriol 189:3977–3986
33. Herzberg MC, MacFarlane GD, Liu P, et al (1994) The platelet as an inflammatory cell in periodontal diseases: interaction with *Porphyromonas gingivalis*. In: Genco R, Hamada S, Lehner T, et al (eds) Molecular pathogenesis of periodontal disease. American Society for Microbiology, Washington, DC, pp 247–255
34. Curtis MA, Macey M, Slaney JM, et al (1993) FEMS Microbiol Lett 110:167–173
35. Lourbakos A, Yuan YP, Jenkins AL, et al (2001) Blood 97:3790–3797
36. Naito M, Sakai E, Shi Y, et al (2006) Mol Microbiol 59:152–167
37. Kerrigan SW, Douglas I, Wray A, et al (2002) Blood 100:509–516
38. Byrne MF, Kerrigan SW, Corcoran PA, et al (2003) Gastroenterology 124:1846–1854
39. Sullam PM, Hyun WC, Szollosi J, et al (1998) J Biol Chem 273:5331–5336
40. Wu Y, Suzuki-Inoue K, Satoh K, et al (2001) Blood 97:3836–3845
41. Fitzgerald JR, Foster TJ, Cox D (2006) Nat Rev Microbiol 4:445–457
42. Schwartz Z, Goultschin J, Dean DD, et al (1997) Periodontol 2000 14:158–172

43. Yasuda H, Shima N, Nakagawa N, et al (1998) Proc Natl Acad Sci USA 95:3597–3602
44. Okahashi N, Inaba H, Nakagawa I, et al (2004) Infect Immun 72:1706–1714
45. Fujimura Y, Hotokezaka H, Ohara N, et al (2006) Infect Immun 74:2544–2551
46. McQuibban GA, Gong J, Tam EM, et al (2000) Science 289:1202–1206
47. Shi Y, Kong W, Nakayama K (2000) J Biol Chem 275:30002–30008

Interface Oral Health Science 2007: 63–72

Implication of immune interactions in bacterial virulence: is *Porphyromonas gingivalis* an "Invader" or "Stealth Element" in periodontal lesions?

Hidetoshi Shimauchi[1]* **and Tomohiko Ogawa**[2]*
[1]*Division of Periodontology and Endodontology, Tohoku University Graduate School of Dentistry, Sendai 980-8575;* [2]*Division of Oral Microbiology, Asahi University School of Dentistry, Mizuho 501-0296; Japan*
*simauti@mail.tains.tohoku.ac.jp, tomo527@dent.asahi-u.ac.jp

Abstract. *Porphyromonas gingivalis* (Pg), a Gram-negative anaerobic black-pigmented rod bacterium, has been recognized as the most potent etiologic bacterium in human chronic periodontitis. It possesses a variety of putative virulence factors providing both tissue destruction and host evasion including lipopolysaccharides (LPS), fimbriae, various proteinases, etc. These factors actively participate in periodontal tissue destruction. However, recent evidence suggests that Pg has also evolved mechanisms to inhibit or confuse host immune systems. Thus, Pg is suggested to behave not only like an "active invader", but also like a "stealth element" in periodontal lesions. In the present study, repeated exposure of Pg components induced tolerance resulting in selective inhibition of cytokine production of both monocytes and gingival fibroblasts in a different fashion from that described for LPS in *Escherichia coli*. It was also revealed that Pg LPS induced a unique dendritic cell subset with a $CD14^{+}CD16^{+}$ phenotype that exhibited weak maturation. In animal studies, administration of live Pg or its LPS exerted a regulatory effect on systemic markers such as triglycerides or adiponectin. Taken together, these findings suggest that Pg may be able to adapt to the local immune defense, contributing to the connection between systemic and periodontal disease.

Key words. adiponectin, dendritic cells, *Porphyromonas gingivalis*, triglycerides, virulence factors

1 Introduction

Periodontitis is a chronic inflammatory disease caused by oral bacteria that leads to the destruction of tooth-supporting tissues and is a major cause for tooth loss in human adults. A large number of different oral microfilm bacteria have been recognized as members of the periodontal environment, and some bacterial species in the subgingival niche are now proposed to be periodontopathic. Individual periodontal pathogens elicit the release of a large number of biological molecules that can act on host tissues resulting in the destruction of soft tissue and bone. These molecules are called "virulence factors" and include bacterial cell components such as lipopolysaccharides (LPS) and fimbriae, as well as proteases and metabolic end products. However, in the current paradigm, virulence factors also function in the establishment of symbiotic or parasitic relationships between bacteria and the host, resulting in the invasion and maintenance of species associated with or within the confines of the host.

Porphyromonas gingivalis (Pg), a Gram-negative anaerobic black-pigmented rod bacterium, has been recognized as the most potent etiologic agent of human chronic periodontitis (formerly adult periodontitis). The number of Pg cells is decreased during the return of oral health after disease, suggesting the importance of this bacterium in controlling disease activity [1]. Classical virulence factors of Pg have been shown to provoke inflammatory immune responses as well as direct tissue destruction. However, Pg LPS and its active center lipid A both showed weak biological activity due to their unique chemical structures, and they acted as antagonists for enterobacterial LPS/ lipid A [2]. Further, recent reports have suggested that Pg and several other periodontal pathogens utilize components of so-called "stealth technology" to gain access to immune privileged sites [3]. This technology includes invasion into gingival epithelial cells [4] and down-regulation of immune responses by gingipains, trypsin-like cysteine proteinases [5].

The immune evasion of Pg not only facilitates sustained colonization in the diseased lesion, but also contributes intravenous access, resulting in Pg-mediated bacteremia. Pg has actually been detected by immunostaining and polymerase chain reaction (PCR) in some human carotid and coronary atheromas [6, 7]. The new links of periodontitis to another systemic conditions/diseases have been reported; these are obesity and hyperlipidemia [8]. These disorders are known to increase the risk for diabetes and cardiovascular disease and are collectively called metabolic syndrome [9]. However, the precise mechanisms for these new associations have not yet been elucidated, and also the role of Pg infection in these links is not understood.

In this paper, we first focus on the effects of chronic exposure of Pg cell wall components on host cells. As periodontitis is a chronic inflammatory disease associated with a long-term infection of pathogens, periodontal tissue and infiltrated immune cells are chronically exposed to Pg virulence factors. Dendritic cells (DCs) sequentially induce innate and immune responses against invading pathogens. Detection of pathogens by DC is mediated by surface Toll-like receptors (TLR), and the usage of TLRs affects the phenotype of DC resulting in a Th1/Th2 polarization [10]. We therefore explored the pattern of DC maturation induced by Pg antigens and their relationship to TLR usage. To understand the role of Pg infection as a risk factor for systemic diseases, we administered live Pg or its LPS in a rat model and investigated the effects on systemic markers of metabolic syndrome.

2 Chronic exposure to Pg components down-regulates cytokine production in human immune and periodontal tissue cells

It has been well established that Pg components, such as LPS and fimbriae, induce a panel of inflammatory cytokine production in host cells that includes interleukin (IL)-1, IL-6, IL-8. However, the most prominent nature of periodontitis is chronic inflammation induced by the persistent presence of periodontal pathogens including Pg, suggesting that repeated exposure to virulence factors may occur in the lesion. In this respect, repeated administration of LPS in vivo induces resistance to its pathophysiological effects known as LPS tolerance or desensitization [11]. We hypothesized that chronic exposure to LPS/fimbriae from Pg may mobilize host cells for cytokine production and create an imbalance of inflammatory responses in periodontal lesions.

2.1 Down-regulation of selected cytokine production by Pg LPS-tolerant human monocytes

We first assessed the production of the inflammatory cytokines IL-6 and IL-8 by monocytes pretreated with a low concentration of Pg LPS [12]. Human peripheral blood monocytes were preincubated for 12 h with graded concentrations (0–100 ng/ml) of Pg LPS. After washing, cells were then restimulated with 1 μg/ml of the same LPS and assayed for IL-6 and IL-8 production.

As shown in Fig. 1a, a relatively low concentration (~0.1 ng/ml) of LPS caused a significant reduction in IL-6 production (up to 70%), but did not affect IL-8

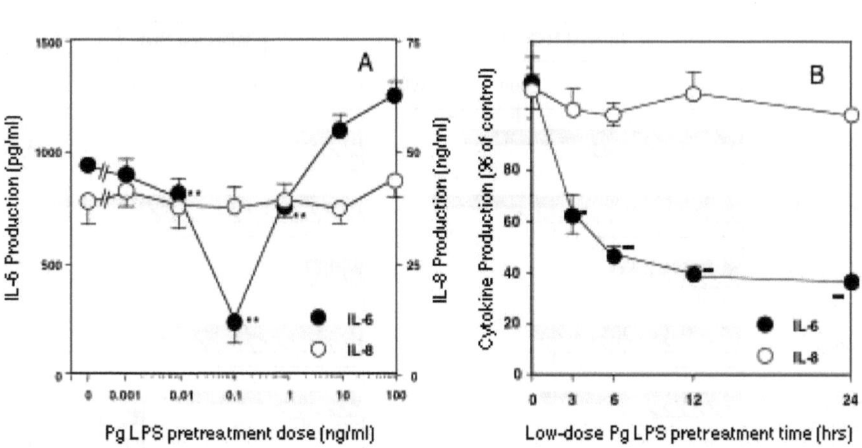

Fig. 1. Modulation of IL-6 and IL-8 response in Pg lipopolysaccharide (*LPS*)-pretreated human monocytes (**A**); time course of effects of Pg LPS pretreatment on IL-6 and IL-8 production after restimulation with the same LPS (**B**). *$P < 0.05$, ** $P < 0.01$; significantly different from the control culture

production when compared to nontreated control, whereas a higher concentration (100 ng/ml) of LPS reversed this decrease. The time requirements for Pg LPS-mediated IL-6 down-regulation were further investigated (Fig. 1b). Significant suppression of IL-6 was observed in monocytes pretreated with 0.1 ng/ml of Pg LPS for 3 and 24 h after restimulation, suggesting time-dependence of pre-exposure. We further investigated the mechanism for this selective down-regulation. Our data showed enhanced IL-10 production in monocytes receiving the same treatment, and this cytokine was essential for autocrine inhibition of IL-6 production (data not shown).

2.2 Down-regulated cytokine production in HGF isolated from chronically inflamed tissues

To test whether chronic exposure of human gingival fibroblasts (HGFs) to Pg components could alter cellular responses, HGFs were obtained from periodontitis patients and healthy volunteers, and IL-8 production levels were compared [13]. As shown in Fig. 2, inflamed HGFs from diseased sites showed reduced IL-8 production when compared to normal HGFs after stimulation with Pg LPS and its lipid A, but not LPS or lipid A from other species including *Escherichia coli* (506) and

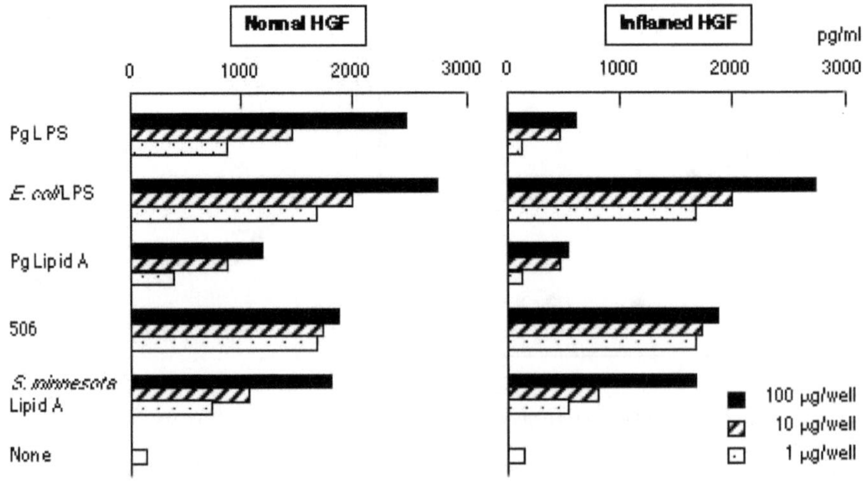

Fig. 2. IL-8 production in *normal* and *inflamed* human gingival fibroblasts (*HGFs*)

Salmonella-type synthesized lipid A. A similar phenomenon was observed when both HGFs were stimulated with Pg fimbriae. The reduced response of IL-8 production to Pg components was reproduced in vitro upon pretreatment of normal HGFs (data not shown).

Taken together with our previous findings [12, 13], prolonged or continuous exposure to Pg components may down-regulate the function of host cells including infiltrated and component cells of periodontally diseased gingiva. The possible clinical significance of "tolerance" to Pg components in the development of periodontal diseases is unknown; however it is attractive to hypothesize that chronic exposure to bacterial components and periodontal pathogens may mobilize host cell functions and create an imbalance of inflammatory responses, finally resulting in the immune evasion of the pathogens.

3 Induction of a unique DC subset by Pg components

Dendritic cells are widely distributed in both lymphoid and non-lymphoid organs including the periodontal tissues, and they work as efficient antigen presenting cells (APC) that can activate naïve T cells. When the maturation is triggered, DCs dramatically up-regulate expression of surface major histocompatibility complex (MHC) class I, class II, and co-stimulatory molecules, and they also secrete various cytokines and chemokines [14]. The precise mechanism and role of Pg components-mediated activation of DCs have not been fully elucidated. We stimulated in vitro-generated human immature DCs with Pg LPS and fimbriae, and compared the induced phenotype with those subsets after stimulation with *E. coli* LPS (TLR4 ligand) and peptidoglycan (PGN; TLR2 ligand) [15, 16]. As shown in Fig. 3, Pg LPS and fimbriae, as well as PGN, were found to induce a unique DC subset that expressed CD14 and CD16. As it was previously reported that Pg LPS and fimbriae were TLR2 ligands [17, 18], we further tested whether an anti-TLR2 antibody could abrogate the induction of the $CD14^{+}CD16^{+}$ subset. Pretreatment with an anti-TLR2 antibody successfully inhibited the up-regulation of CD14 and CD16 expression after stimulation with Pg LPS, fimbriae, and PGN, suggesting that TLR2 signaling is essential for the induction of this subset [16]. Pg LPS and fimbriae-stimulated DCs exhibited weak allogenic T cell stimulation when compared to *E. coli* LPS-triggered cells, possibly due to weaker expression of cytokines and co-stimulatory molecules.

Our results clearly demonstrate impaired DC maturation mediated by Pg components. However, the pathophysiological role of these unique phenomena in the development of periodontitis still remains unclear. Monocytes with a similar phenotype were increased in the peripheral blood of patients with autoimmune disease, sepsis, asthma, AIDS, and also in patients undergoing hemodialysis [19]. Therefore, $CD14^{+}CD16^{+}$ cells are probably associated with chronic infection/inflammatory disease states including periodontal disease.

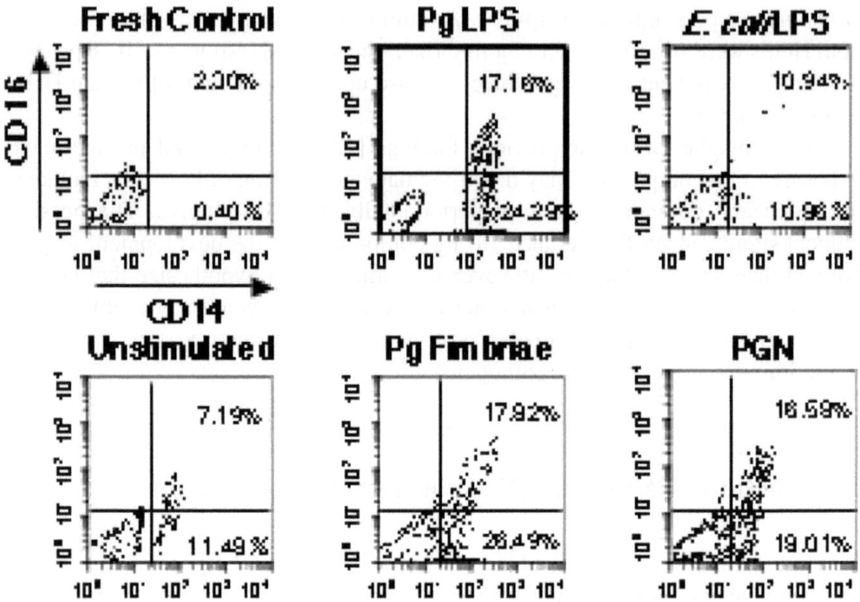

Fig. 3. Induction of a CD14⁺CD16⁺ dendritic cell (DC) subset after stimulation with Pg components

4 Possible mechanisms of the periodontal-systemic link: the role of Pg infection and its virulence factors

Mounting evidence has accumulated over the most recent decade that supports a role for periodontitis and Pg infection as potential risk factors for systemic diseases. Recently, the focus of this link has been shifting from a purely epidemiological association towards a biological mechanism for a causal relationship. Four putative working models have emerged to link mechanisms governing periodontitis-accelerated systemic diseases: (1) direct invasion of periodontal pathogens, (2) immunological surroundings, (3) pathogen trafficking, and (4) autoimmunity [20]. Biopsy studies have demonstrated the presence of periodontal pathogens including Pg in atheromas, supporting several of the working models [6, 7].

To accurately consider the risk for cardiovascular diseases, it should be emphasized that obesity, hyperlipidemia, and diabetes have also been nominated as classical risk factors for atherosclerosis, forming a pathophysiological link that finally leads to cardiovascular death. In this respect, the concept of metabolic syndrome (MetS) has been constantly evolving and has shown a recent and dramatic surge in interest [21]. MetS can be recognized clinically by a cluster of simple clinical measures including waist circumference, blood pressure, triglycerides, high-density lipoproteins, and glucose. This clustering appears to depend on two major factors:

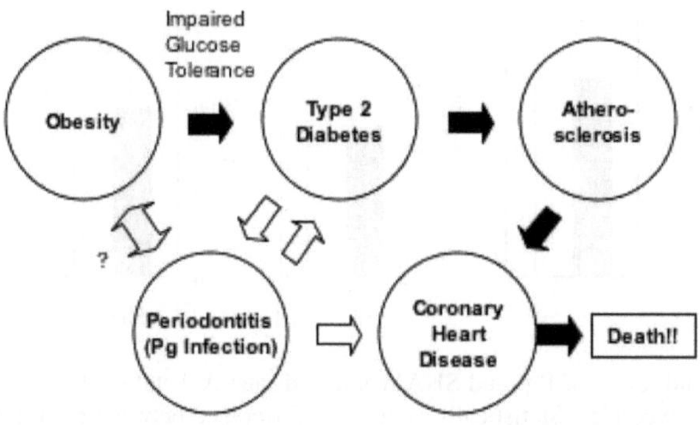

Fig. 4. Association of periodontitis with the "metabolic domino effect". *Solid arrows* indicate the consequences of the metabolic domino effect. *Open arrows* indicate the association of periodontitis with each systemic condition/disease

excess body fat and metabolic susceptibility (e.g. insulin signaling defects, adipose tissue disorders, physical inactivity, aging, drugs). Obesity is known to be the predominant driving force behind this syndrome [22]. Therefore, the sequential scheme of "Metabolic Domino Effect" shown in Fig. 4 has been proposed, starting at obesity and finally resulting in death due to coronary heart disease. Periodontitis and Pg infection have been shown to be the risk factors for diabetes (type 1 and 2) and coronary heart diseases [21]. Previous reports also revealed hypertriglyceridemia in periodontitis patients and an increased risk of periodontitis in obese people with a high body mass index (BMI) [8]. These findings led us to hypothesize that periodontal diseases and Pg infection may increase susceptibility to MetS.

4.1 The possible role of Pg infection in hyperlipidemia

Periodontal disease can lead to low-level bacteremia, and also low doses of endotoxin (LPS) were shown to induce hyperlipidemia and myeloid cell hyperactivity in an animal septic shock model. To confirm the mechanism of the link between periodontitis and hyperlipidemia, especially the association of Pg infection with the link, we examined the serum lipid levels of rats with experimental periodontitis in the presence or absence of Pg infection.

Twenty-four male Wister rats received elastic rings in the cervix of the mandibular M1 to induce experimental periodontitis, and they were then divided into two groups: 12 infected with Pg W83 and 12 sham-infected. Animals were killed after 5 weeks, and peripheral blood samples were collected. Serum levels of Pg-specific antibody, triglycerides (TG), cholesterol (CHO), low-density lipoprotein (LDL), and glucose (GLU) were measured. Although alveolar bone loss was not statistically different between groups, serum TG (Fig. 5c) and anti-Pg antibody levels were significantly higher in the infected group (177.9 ± 36.5 mg/ml vs. 125.9 ±

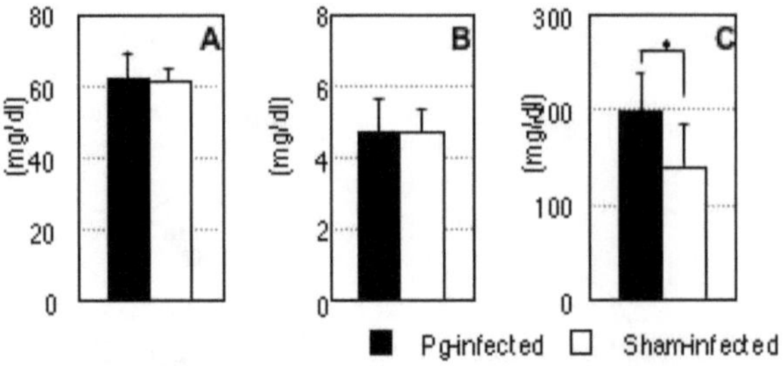

Fig. 5. Serum lipid levels of Pg- and SHAM-infected rats; **A** Total cholesterol; **B** Low-density lipoprotein; **C** Triglycerides. Statistically significant difference between both groups of rats (*$P < 0.005$)

42.5 mg/ml and 6.5 ± 5.5 mg/ml vs. 1.2 ± 1.5 mg/ml, respectively). CHO, LDL (Fig. 5a, b) and GLU (not shown) levels were not different. These results suggest that Pg infection directly induces elevated serum TG levels in periodontitis patients.

4.2 Effect of Pg LPS on serum adipocytokine levels

In obese persons, excess adipose tissue releases a variety of factors that likely contribute to metabolic risk. These factors include adiponectin, leptin, plasminogen activator inhibitor-1 (PAI-1), resistin, angiotensins, etc. These factors are collectively called adipocytokines, and their output is usually increased with obesity [23]. Negative correlations have been reported between serum adiponectin levels and obesity-related characteristics including waist circumference, visceral fat area, serum TG, fasting plasma insulin, and high blood pressure, suggesting that adiponectin level may be a useful biomarker for MetS [24].

We hypothesized here that Pg and its virulence factors may affect serum adiponectin levels. To test this hypothesis, we administered Pg LPS and *E. coli* LPS to rats by *i.p.* injection and compared their effects on serum adiponectin levels. *E. coli* LPS induced higher TG and TNF-α production, although Pg LPS injection significantly decreased adiponectin levels (Fig. 6). These results suggest that after invasion, Pg and its virulence factors may possibly work as metabolic risk factors via the down-regulation of serum adiponectin.

5 Conclusions

Our results demonstrated that Pg components have an ability to down-regulate functions of host cells that might be involved in the establishment of chronic inflammation and tissue invasion. Once they invade the periodontal tissue and systemic

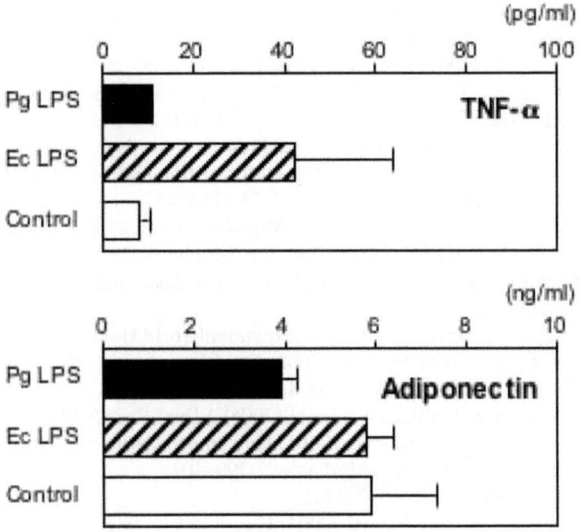

Fig. 6. Effect of i.p. injected Pg LPS on serum adipocytokine levels

circulation, Pg and its components may contribute to systemic events at the molecular level, leading to the subsequent development of metabolic syndrome. Therefore, Pg virulence factors have unique characteristics that allow them to escape from "immune sentinels" in periodontal lesions, and they may therefore play a critical role in the periodontal/systemic link. Here, we define the terms "Invader" and "Stealth Element" as follows: "Invader" is a microbe that enters and actively achieves injurious or destructive effects using virulence factors, and "Stealth Element" is a microbe that is difficult to detect by immune surveillance systems and can therefore intrude deeply into body tissue and body. We can conclude that Pg may act as a "Stealth Element" in periodontal lesions but also like an active "Invader" after systemic intrusion into the body.

Acknowledgment. We are grateful to Dr. Eiji Nemoto, Dr. Sousuke Kanaya, Dr. Yumi Itagaki, and Dr. Yuka Kataoka-Kikuchi (Tohoku University Graduate School of Dentistry) for their excellent work in the studies. This work was in part supported by a Grant-in Aid from Japan Society for the Promotion of Science (No. 16390611 to H. S.).

References

1. Holt SC, Kensavalu L, Walker S, et al. (1999) Periodontol 2000 20:168–238
2. Ogawa T, Asai Y, Makimura Y, et al. (2007) Front Biosci 12:3795–3812
3. Darveau RP, Belton CM, Reife RA et al. (1998) Infect Immun 66:1660–1665
4. Sagile FR, Smith CT, Newman MG, et al. (1986) J Periodontol 57:492–500
5. Sugawara S, Nemoto E, Tada H, et al. (2000) J Immunol 165:411–418

6. Haraszthy VI, Zambon JJ, Trevisan M, et al (2000) J Periodontol 71:1554–1560
7. Carvrini F, Sambri V, Moter A, et al. (2005) J Med Microbiol 54:93–96
8. Saito T, Shimazaki Y. (2007) Periodontol 2000 43:254–266
9. Grumdy SM, Cleeman JI, Daniels SR, et al (2005) Circulation 112:2735–2752
10. Mazzoni A, Segal DM (2004) J Leukoc Biol 75:721–730
11. Fan H, Cook JA (2004) J Endotoxin Res 10:71–84
12. Shimauchi H, Ogawa T, Okuda K, et al (1999) Infect Immun 67:2153–2159
13. Ogawa T, Ozaki A, Shimauchi H, et al (1997) FEMS Immunol Med Microbiol 18:17–30
14. Steinmann RM, Hemmi H (2006) Curr Top Microbiol Immunol 311:17–58
15. Kanaya S, Nemoto E, Ogawa T, et al (2004) Eur J Immunol 34:1451–1460
16. Kanaya S, Nemoto E, Ogawa T, et al (2005) In: Interface Oral Health Science. Elsevier, Amsterdam, pp169–174
17. Hashimoto M, Asai Y, Ogawa T. (2004) Int Immunol 16:1431–1437
18. Ogawa T, Asai Y, Hashimoto M, et al. (2002) Eur J Immunol 32:2543–2540
19. Ziegler-Heitbrock L (2007) J Leukoc Biol 81:584–592
20. Gibson III FC, Yumoto H, Takaishi Y, et al (2006) J Dent Res 85:106–121
21. Laclaustra M, Collera D, Ordovas JM (2007) Nutr Metab Cardiovasc Dis 17:125–139
22. Grundy SM (2007) J Clin Endocrinol Metab 92:399–404
23. Scherer PE (2006) Diabetes 55:1537–1545
24. Miwa R, Nakamura T, Kihara S, et al (2004) Circ J 68:975–981

Symposium II:
Biomaterials: Novel dental biomaterials

Interface Oral Health Science 2007: 75–84

Multifunctional low-rigidity β-type Ti–Nb–Ta–Zr system alloys as biomaterials

Mitsuo Niinomi*
Department of Biomaterials Science, Institute for Materials Research, Tohoku University, Sendai 980–8574, Japan
*niinomi@imr.tohoku.ac.jp

Abstract. Low-rigidity β-type titanium alloys composed of nontoxic and allergy-free elements are receiving considerable attention as biomaterials. Ti-29Nb-13Ta-4.6Zr (TNTZ) composed of nontoxic and allergy-free elements exhibits low Young's modulus of approximately 60 GPa; it is effective in inhibiting bone resorption and enhancing the remodeling of bones, which may be due to the excellent stress transmission between the bone and the implant. TNTZ exhibits multifunctional characteristics such as super elasticity and shape memory effect. It can be applicable for not only dental implants but also for dental prostheses.

Key words. Ti–Nb–Ta–Zr system alloy, low Young's modulus, super elasticity, shape memory effect, fatigue strength

1 Introduction

β-type Ti–Nb—Ta—Zr system alloys [1] are expected to be used in biomedical and dental applications because they comprise nontoxic and allergy-free elements, and they exhibit low Young's modulus, excellent biocompatibility, good balance of strength and ductility, and excellent corrosion resistance. These alloys also exhibit multifunctional characteristics such as super-elasticity and shape memory effect [2] [3]. These functions enable the use of Ti–Nb–Ta–Zr system alloys in not only biomedical and dental implants but also orthodontic wires and stents. These multifunctional characteristics can be obtained through thermomechanical treatment in a considerably narrow range of the chemical composition change.

Further, these alloys are also required to possess low Young's modulus. Single crystals of a β-type Ti–Nb–Ta–Zr system alloy with a specific direction are expected to exhibit considerably low Young's modulus that is equivalent to that of the cortical bone for which the value lies between 10 and 30 GPa [4].

The development, multifunctional characteristics, improvement of mechanical properties, biological and mechanical biocompatibilities, and dental applicability of β-type Ti–Nb–Ta–Zr system alloys in addition to their Young's modulus will be described here.

2 Low-rigidity β-type titanium alloys

The research and development of titanium alloys for biomedical applications started fairly recently. The elements that are judged to be nontoxic and allergic-free through the reported data on cell viability for pure metals [5], polarization resistance (corrosion resistance) and tissue compatibility of pure metals and representative metallic biomaterials [5], and allergy-causing tendencies of pure metals [4] are selected as alloying elements for titanium. As a result, Nb, Ta and Zr are selected as the safest alloying elements for titanium.

It is desirable for Young's moduli of biomaterials to be equal to Young's modulus of the cortical bone (10–30 GPa) because if the former is considerably greater than the latter, bone resorption occurs. Young's modulus of the (α + β)-type titanium alloy, Ti-6Al-4 V, which is the most widely used titanium alloy for biomedical applications, is considerably lower than Young's moduli of stainless steels and Co-based alloys. However, its Young's modulus is still considerably greater than that of the cortical bone. Young's moduli of β-type titanium alloys are known to be smaller than those of α or (α + β)-type titanium alloys. Then the small Young's modulus leads to a low-rigidity in these alloys. In addition, β-type titanium alloys show excellent cold workability and high strength. Therefore, the research and development of low-rigidity β–type titanium alloys, which comprise nontoxic and allergy-free elements as stated above, are receiving considerable attention. The representative low-rigidity β-type titanium alloys that have been developed for use in biomedical applications are listed in Table 1 [6].

Table 1. Selected low modulus β-type titanium alloys for biomedical applications

Titanium and its alloys	ASTM standard	ISO standard	JIS standard	Type of alloy
Ti-13Nb-13Zr	ASTM F 1713			β
Ti-12Mo-6Zr-2Fe	ASTM F 1813	—	—	β
Ti-12Mo-5Zr-5Sn	—	—	—	β
Ti-15Mo	ASTM F 2066	—	—	β
Ti-16Nb-10Hf	—	—	—	β
Ti-15Mo-2.8Nb-0.2Si	—	—	—	β
Ti-15Mo-5Zr-3Al	—	—	JIS T 7401-6	β
Ti-30Ta	—	—	—	β
Ti-45Nb	AMS 4982	—	—	β
Ti-35Zr-10Nb	—	—	—	β
Ti-35Nb-7Zr-5Ta	Task Force F-04.12.33	—	—	β
Ti-29Nb-13Ta-4.6Zr	—	—	—	β
Ti-8Fe-8Ta	—	—	—	β
Ti-8Fe-8Ta-4Zr	—	—	—	β

3 Further lowering Young's modulus

The lowest Young's modulus reported to date in a bulk titanium alloy developed for biomedical applications seems to be 40 GPa for a Ti–Nb–Sn system alloy [7]. It is difficult to lower Young's modulus of bulk titanium alloys than this value. However, as it has been reported that the deformation characteristics of β-type Ti–Nb–Ta–Zr system titanium alloys are significantly anisotropic, i.e., they depend on the crystal direction [8]; Young's modulus is also considered to be dependent on the crystal direction. Therefore, a single crystal of a β-type titanium alloy with low Young's modulus and grown in a certain direction appears to exhibit Young's modulus less than 40 GPa. It has been reported [9] that Young's modulus less than 40 GPa (approximately 35 GPa) can be obtained in a single crystal of Ti-29Nb-13Ta-4.6Zr (TNTZ) [1, 2] with a crystal direction of [100]. Therefore, metallic single crystals of titanium alloys may be applicable as biomaterials; this may be referred to as "single crystal biometals".

Young's moduli of bulk β-type titanium alloys greater than Young's modulus of the cortical bone, which is approximately 10–30 GPa as stated above. In order to further reduce Young's moduli of titanium and its alloys, it is very effective to make titanium and its alloys porous; this is another way to drastically reduce Young's modulus of titanium. Young's modulus can be easily controlled by varying the ratio of porosity. It has been reported [10] that in the relationship between Young's modulus and the porosity of porous titanium samples made of titanium powders with different diameters and its comparison with Young's modulus of bulk titanium, at a porosity of approximately 30%, Young's modulus is nearly equal to that of the cortical bone. By using a titanium alloy with low Young's modulus, Young's modulus can be made to equal that of the cortical bone in a porosity lower than that of titanium. In addition, pores with appropriate sizes enhance bone conductivity. On the other hand, it has been also reported [10] that strength of the alloy decreases drastically with increasing porosity. At a porosity of approximately 30%, which leads to Young's modulus equal to that of the cortical bone, the 0.2% proof stress is below 100 MPa. The decrease in the strength of porous titanium can be effectively inhibited by combining it with a biocompatible polymer. One possible method to induce polymer penetration into the porous titanium is pressing. In this method [11], high molecular density polyethylene is pressed into porous titanium. Another method that has been proposed [12] involves the initial use of a monomer of poly(methyl methacrylate) (PMMA). Porous titanium is first immersed into the monomer of PMMA, thereby causing the monomer to penetrate the porous titanium. Subsequently, the monomer in the porous titanium is subjected to polymerization by heating. When combined with the PMMA, the strength of the porous titanium is increased as shown in Fig. 1. In this figure, for convenience, the data of the porous titanium and PMMA composites are plotted against the porosity, although the composites do not contain pores. The tensile strength of the porous titanium and PMMA composites is greater than that of porous titanium alone. Furthermore, it is considered that biofunctionalities are easily added to porous titanium

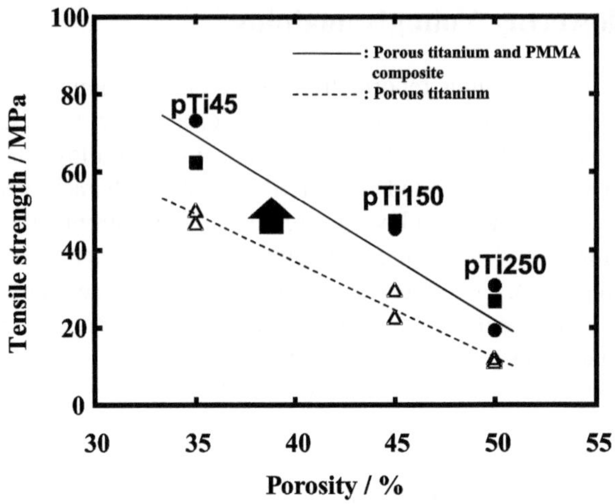

Fig. 1. Relationship between tensile strength and porosity of porous titanium and PMMA composites: pTi45, pTi150 and pTi250 are made of titanium powders with average diameters of 45, 150, and 250 mm

and polymer composites because the surface of porous titanium can be covered with polymer. Such a covering is also expected to reduce the porosity, which gives Young's modulus equal to that of the cortical bone, when the porous titanium alloy is made of the powder of a low-rigidity β-type titanium alloy such as TNTZ; the strength of the porous titanium alloy and polymer composites will be greater than that of the porous titanium and polymer composites.

It is important to understand what level of Young's modulus is effective in inhibiting bone absorption after implantation.

4 Super elastic and shape memory behaviors of low-rigidity β-type titanium alloys

Currently, Ni-free shape memory alloys are being developed for biomedical applications because the most popular practical shape memory alloy TiNi, which is applied to stents, catheters, etc., contains a large amount of Ni, which is an element that can cause allergic disease. There are many β-type alloys [13] such as Ti–Nb–Sn system alloys, Ti–Mo–Ga system alloys, Ti–Nb–Al system alloys, Ti–Mo–Al system alloys, Ti–Ta system alloys, Ti–Nb system alloys, Ti–Sc–Mo system alloys, Ti–Mo–Ag system alloys, Ti–Mo–Sn system alloys, Ti–Nb–Ta–Zr system alloys. Among these alloys, around 6% elastic strain has been reported to be obtained in one of the Ti–Nb–Al system alloys by controlling the texture. Very recently, over 6% total elastic strain has been reported to be obtained in Ti–Nb–Zr, Ti–Nb–Zr–Ta, Ti–Nb–Zr–Ta–O and Ti–Nb–Zr–Ta–N system alloys [14]. The shape memory effect of these alloys is associated with a deformation-induced martensite transformation and its reverse transformation.

The shape memory alloys mentioned above also show super-elastic behavior based on deformation-induced martensite transformation and its reverse transformation. One of the β-type Ti–Nb–Ta–Zr system alloys, TNTZ, also shows super-elastic behavior, although the shape memory effect is not observed in this alloy. Figure 2 [15] shows the tensile loading–unloading stress–strain curve of a drawn wire of TNTZ with a diameter of 0.3 mm. In the elastic region, the tensile loading–unloading stress–strain curve shows nonlinear elastic behavior. The behavior of deformation-induced martensite transformation is not recognized in the stress–strain curve shown in Fig. 3 [16] and in the deformed microstructure. The super-elastic behavior of this alloy cannot be explained by increasing the amount of deformation-induced martensite transformation and its reversion to the present state. In this case, the maximum elastic strain is around 3%.

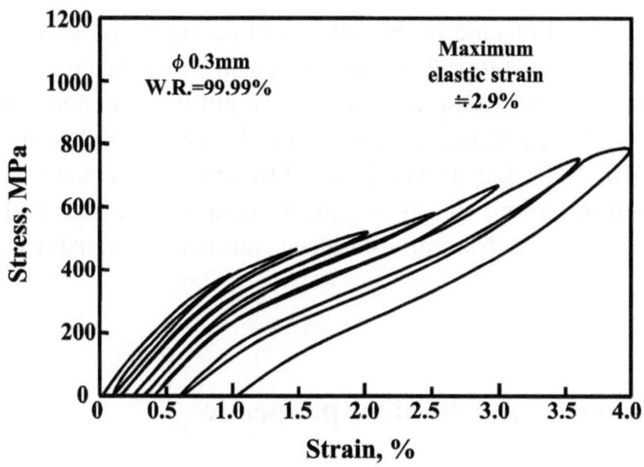

Fig. 2. Tensile loading–unloading Stress–strain curves of as-cold drawn TNTZ (ϕ0.3 mm). *W.R.* indicates work ratio

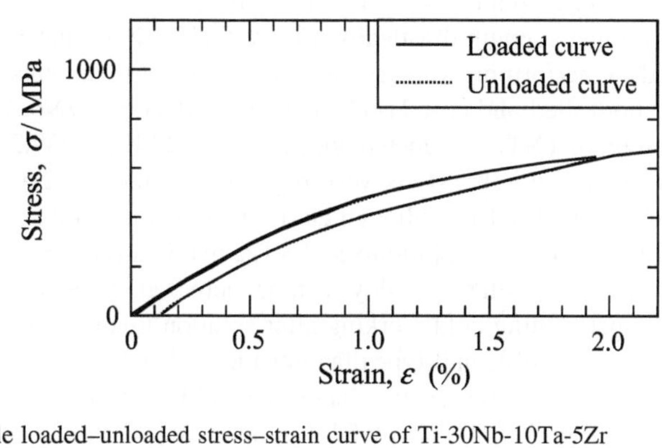

Fig. 3. Tensile loaded–unloaded stress–strain curve of Ti-30Nb-10Ta-5Zr

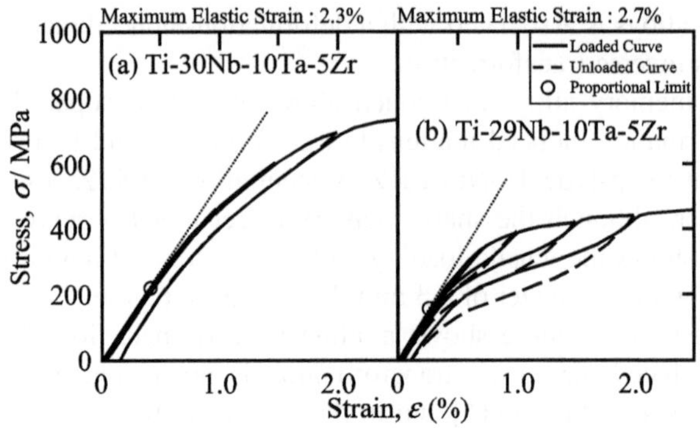

Fig. 4. Tensile loading–unloading stress–strain curves of (**a**) Ti-30Nb-10Ta-5Zr and (**b**) Ti-29Nb-10Ta-5Zr; stress–strain curve of (**b**) exhibits a shape memory character

Further, by a small change in the amount of elements, particularly Nb content in the Ti–Nb–Ta–Zr system alloys, the shape memory behavior caused by the deformation-induced martensite transformation and its reversion can be observed as shown in Fig. 4. The figure shows the tensile loading–unloading stress–strain curve of Ti-29Nb-10Ta-5Zr in which two different gradients are observed in the elastic deformation region; this is a typical characteristic of the shape memory effect caused by the deformation-induced martensite transformation and its reversion.

5 Improvement of mechanical properties

Among the mechanical properties of titanium alloys, the fatigue properties are significantly important for biomedical applications. In the case of β-type titanium alloys in biomedical applications, because the fatigue strength in as-solutionized conditions is low, heat treatment or thermomechanical treatment is effective in improving the fatigue strength drastically. Figure 5 [17] shows the S–N curves of TNTZ obtained by performing various thermomechanical treatments with fatigue limit ranges of hot rolled and cast Ti-6Al-4 V ELI and Ti-6Al-7Nb. For example, the fatigue strength of TNTZ conducted with aging at 723 K for 259.2 ks after cold rolling (CR) marked with B723K is very high, and in the highest range of the fatigue limit of Ti-6Al-4 V ELI although elastic modulus is a little greater, that is round 80 GPa, than that of as-solutionized TNTZ, that is around 60 GPa [18].

The strength of β–type titanium alloys can be increased while keeping Young's modulus low by performing cold working after solution treatment because a high cold working ratio is possible in β-type titanium alloys [15]. Strength such as tensile strength and 0.2% stress increases with increasing cold work ratio. The increase in the strength of cold worked TNTZ by 84% is nearly equal to that of Ti-6Al-4 V

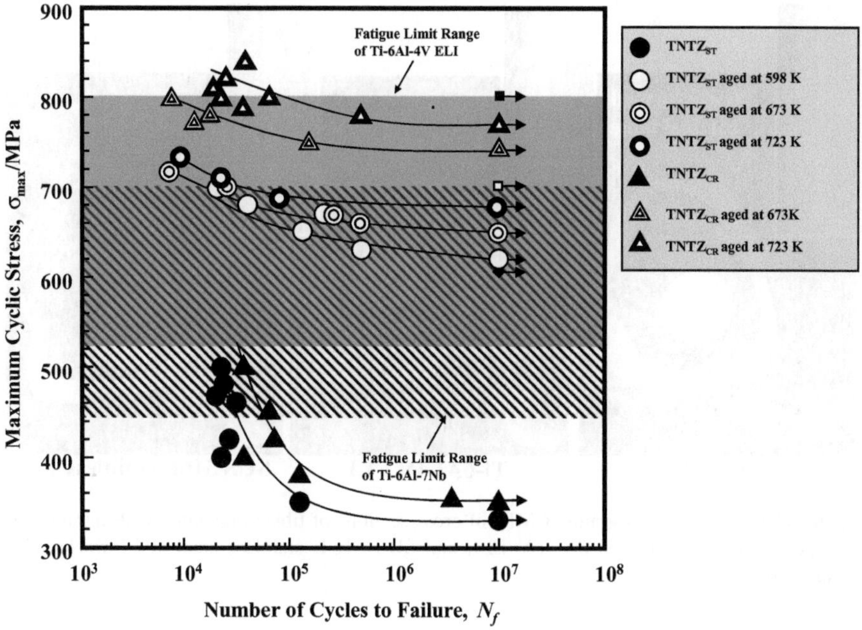

Fig. 5. S–N curves of Ti-29Nb-13Ta-4.6 in as-solutionized conditions (*$TNTZ_{ST}$*) and as-cold rolled conditions (*$TNTZ_{CR}$*), and $TNTZ_{ST}$ and $TNTZ_{CR}$ conducted with aging at 598 K, 673 K, and 723 K for 259.2 ks with fatigue limits range of Ti-6Al-4 V ELI and Ti-6Al-7Nb in air

ELI. On the other hand, Young's modulus is constant at a low value regardless of the cold work ratio. The ductility such as elongation and reduction of area is a little lowered by around 20% at low cold work ratio, but more than 20% at higher cold work ratio; ductility is nearly constant to be high.

6 Biological and mechanical biocompatibility

It has been reported [19] that the contact microradiograms (CMR) of the boundaries of bone and low-rigidity TNTZ (in this case, as-solutionized conditions), Ti-6Al-4 V ELI and SUS 316 L stainless steel implanted in the lateral femoral condyles of a rabbit show that each specimen is surrounded by newly formed bones, and the bone tissue shows a partial direct contact with the specimen. However, the extent of the direct contact is greater for TNTZ as compared with the extent for Ti-6Al-4 V ELI and SUS 316 L stainless steel. Therefore, the biocompatibility of TNTZ with bone is excellent.

Figure 6 [6] shows the CMR of the cross-section of a tibia implanted with each rod at 24 weeks after implantation. Among all the cases, the remodeling of bones is evidently the best in the case of TNTZ.

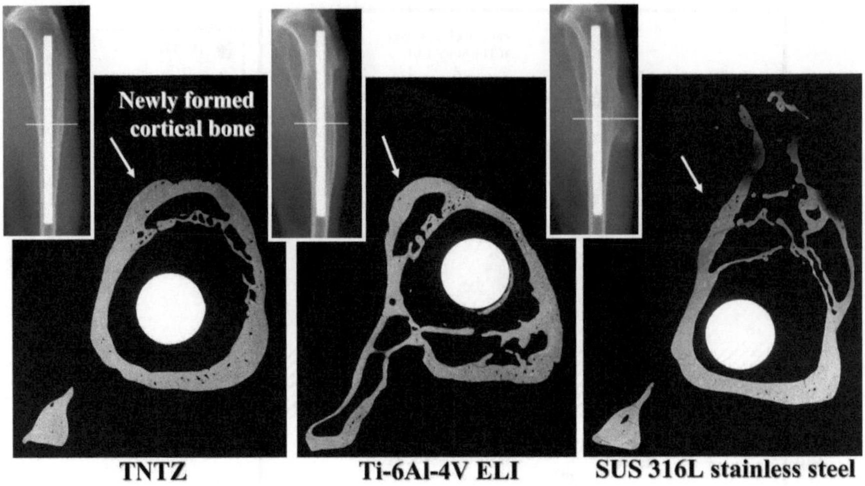

Fig. 6. Contact microradiograms (CMR) of cross-sections of tibias implanted with intramedullary rods made of TNTZ, Ti-6Al-4 V ELI, and SUS 316 L stainless steel at 24 weeks after implantation

Therefore, the biological and mechanical biocompatibility, which indicate good transmission between the bone and the implant are excellent for TNTZ.

7 Dental applications

It is well known that titanium is used in dental implants such as artificial tooth roots. These alloys are also receiving attention for application in dental prostheses such as crowns, inlays, and bridges, which are mainly fabricated by dental precision casting. Currently, β-type titanium alloys containing nontoxic and allergy-free elements are being developed for dental applications. For example, TNTZ is expected to be used for fabricating dental prostheses. The dental precision casting process is mainly used for this process. The melting point of TNTZ is considerably higher than that of pure Ti or Ti-6Al-4 V ELI that is generally used because the former contains large amounts of Nb and Ta, which have very high melting points. Therefore, a magnesia-based investment mold, which is generally used for the casting of pure Ti and Ti-6Al-4 V ELI, is not suitable. To solve this problem, a casting process using a calcia investment mold is currently being developed [20]. In this process, a duplex calcia coating method is used in which fine calcia is first coated on a pattern; this is followed by the coating of silica-fiber-reinforced fine calcia, as shown in Fig. 7. This method has been successfully applied for the casting of TNTZ.

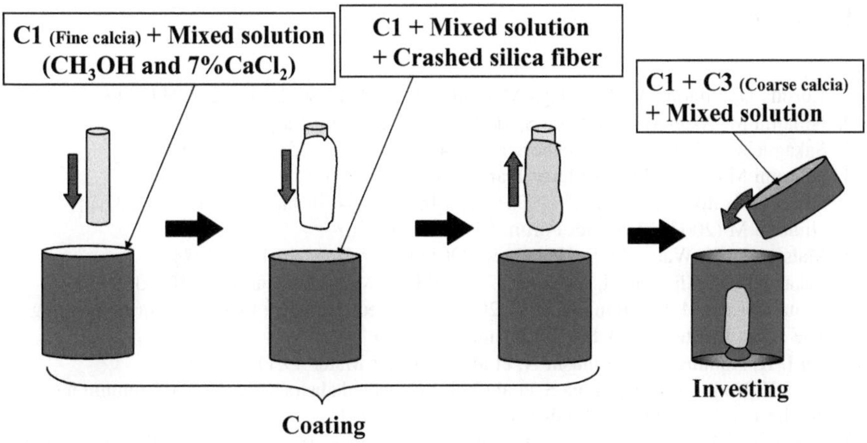

Fig. 7. Schematic drawing of duplex coating process for making calcia mold for dental precision casting of TNTZ

On the other hand, lowering the melting point of TNTZ by modifying the chemical composition using the alloy design method based on the electron theory in order to achieve the ideal properties of TNTZ for using conventional mold materials such as the magnesia-based mold material [21] have also been investigated. As a result, Ti-29Nb-13Zr-2Cr, Ti-29Nb-15Zr-1.5Fe, Ti-29Nb-10Zr-0.5Si, Ti-29Nb-10Zr-0.5Cr-0.5Fe, and Ti-29Nb-18Zr-2Cr-0.5Si alloys have been developed for dental applications [21]. The melting points of these alloys have been found to be considerably lower than the melting point of TNTZ. Among these alloys, the melting point of Ti-29Nb-13Zr-2Cr is the lowest approximately 2050 K. In comparison with TNTZ, it is easier to release the specimen of each of these designed alloys from the mold.

8 Summary

The low-rigidity β-type Ti–Nb–Ta–Zr system alloys exhibit super-elastic and shape memory characteristics according to their chemical compositions. TNTZ, which is one such alloy and exhibits the lowest Young's modulus among the Ti–Nb–Ta–Zr system alloys, is effective in inhibiting bone resorption and enhancing the remodeling of bones. The mechanical properties of TNTZ can be considerably improved by microstructural control through thermomechanical treatments. Dental prostheses of TNTZ can be fabricated by applying the calcia duplex coating method.

Multifunctional low-rigidity β-type Ti–Nb–Ta–Zr system alloys such as TNTZ are expected to be used in both biomedical and dental fields.

References

1. Kuroda D, Niinomi M, Morinaga M, et al (1998) Mater Sci Eng A 243:244–249
2. Niinomi M, Akahori T, Katsura S, et al (2007) Mater Sci Eng C 27:154–161
3. Sakaguchi N, Niinomi M, Akahori T (2004) Mater Trans 45:1113–1119
4. Niinomi M (2001). Metall Mater Trans A 32A:477–486
5. Okazaki Y, Ito Y, Ito A, et al (1993) J Jpn Inst Metals (in Japanese) 57:332–337
6. Niinomi M (2007) Mater Sci Forum 539–543:557–562
7. Matsumoto H, Watanabe S, Hanada S (2005) Mater Trans 46:1070–1078
8. Sakaguchi N, Niinomi M, Akahori T, et al (2005) Mater Sci Eng C 25:363–369
9. Akita S, Tane M, Nakajima H, et al (2007) Collected abstracts of the 2007 spring meeting of The Japan Institute of Metals (in Japanese), vol 89
10. Oh I. H, Nomura N, Masahashi N, et al (2003) Scr Mater 49:1197–1202
11. Baba Y, Nomura N, Fujinuma S, et al (2005) Collected abstracts of the 2005 autumn meeting of The Japan Institute of Metals (in Japanese), vol 441
12. Yamanoi H, Niinomi M, Akahori T, et al (2007) Collected abstracts of the 2007 spring meeting of The Japan Institute of Metals (in Japanese): 72
13. Niinomi M (2003) Sci Technol Adv Mater 4:445–454
14. Kim J, Hosoda H, Miyazaki S (2007) Collected abstracts of the 2007 spring meeting of The Japan Institute of Metals (in Japanese), vol 91
15. Niinomi M, Akahori T, Morikawa K (2005) J ASTM Int (on line publication), vol 2: Paper ID JAI12818
16. Niinomi M, Hanawa T, Narushima T (2005) JOM 57:18–24
17. Niinomi M (2003) Biomaterials 24:2673–2683
18. Akahori T, Niinomi M, Ishimizu K, et al (2003) J. Jpn Inst Metals (in Japanese) 67:652–660
19. Niinomi M, Hattori T, Morikawa K, et al (2002) Mater Trans 43:2970–2877
20. Niinomi M, Akahori T, Takeuchi T, et al (2005) Mater Sci Forum 475–479:2303–2308
21. Niinomi M, Akahori T, Takeuchi T, et al (2005) Mater Sci Eng C 25:417–425

Interface Oral Health Science 2007: 85–94

Study of in vivo bone tissue engineering

Chongyun Bao[1], Hongyu Zhou[1], Wei Li[1]*, Yunfeng Li[1], Hongsong Fan[2], Jinfeng Yao[1], Yunmao Liao[1], and Xingdong Zhang[2]
[1]*Key Laboratory of Oral Biomedical Engineering, Sichuan University, Chengdu, 610041;*
[2]*Research Center on Biomaterials, Sichuan University, Chengdu, 610065;*
People's Republic of China
*leewei2000@sina.com

Abstract. In vivo bone tissue engineering is an emerging field of regenerative medicine for bone defects, which differs from classical tissue engineering. In vivo bone tissue engineering uses the body as a "bioreactor" to construct bone graft with intrinsic osteoinductive biomaterials in the non-osseous or osseous sites. This technique relies on the body's own capacity to regenerate itself, and it does not rely on the delivery of exogenous growth factors or cells. This study has focused on the exploration of osteoinductive biomaterials, the construction of tissue engineering bone with osteoinductive biomaterials, and investigating possible applications in repairing mandibular bone defects in animals in order to develop a feasible technique for bone restoring. The results demonstrated that biomaterials without any exogenous growth factors or cells can induce bone formation in non-osseous sites, and in vivo tissue engineering bone formed with osteoinductive calcium phosphate (Ca-P) ceramics has similar histological characteristics with natural bone. It is possible to construct large bone graft with good blood supply and certain mechanical strength in vivo by using osteoinductive biomaterials. In vivo tissue engineering bone has been provided with a hope of being used to repair segmental bone defects of mandible and support dental implant.

Key words. osteoinductivity, calcium phosphate ceramic, in vivo bone tissue engineering, bone defect

1 Introduction

Bone defect is a common deformity caused by tumor, injury, and congenital malformation. More than 450,000 bone graft procedures are performed each year in the United States, at a cost of about $120 billion [1], and the numbers are expected to increase as life expectancy increases. To date, there is no ideal method for treating the segmental bone defect. Autograft bone was considered to be an ideal bone graft material because of its immunocompatibility, good biological properties, and compatible mechanical strength; however, several disadvantages accompany it: For example, a second surgical procedure is required for graft harvest, available bone quantity is limited, and potential possibility of morbidity at the donor-site probably exists. Allograft and xenograft, while less limited in supply, are accomplished by immunoexclusion and infection risk. The trend, therefore, will be to develop a kind of bone substitute with excellent biologic and mechanical properties for bone defect repair.

Since the 1980s, a new method called tissue engineering has developed in life science. Tissue engineering, one of the major parts of regenerative medicine, combined cell transplantation, materials science, and engineering to develop biological substitutes that can restore and maintain normal function [2]. Although tissue engineering science has made big progress in the past 20 years, several problems are encountered, such as: (1) Seed cell sources. The use of embryonic stem cells involves an ethical conflict, and adult stem cells present some disadvantages such as limited life-cycle and multiplication. (2) Lack of satisfactory scaffolds. The extracellular matrix not only provides support and nutrition, but also adjusts biologic molecules to promote cells migration, attachment, differentiation, and proliferation. (3) Limited dimension of grafts. After the cell-scaffold composites are implanted, most of the initial nutrition for cells are just directly penetrated from adjacent body fluid before new blood vessels grow into the scaffold. Because the distance of oxygen diffusion is limited in no more than several hundred microns, it is almost impossible to create a graft with enough dimension to fit the actual bone defect. (4) Stabilization of the seed cells. It is very important that the seed cells can differentiate, proliferate, and terminate growth as expected. So far, there is no report that answers the question of whether we have obtained an effective technique to certainly change these properties of in vitro cultivated cells [3, 4].

As commonly used bone substitutes, Ca-P ceramics have outstanding properties, such as similarity to bone mineral in composition; bioactivity, which means the ability to induce bone-like apatite (BLA) or carbonate hydroxyapatite (HA) formation on the surface and the ability to promote cellular function and expression leading to the formation of a uniquely strong bone Ca-P ceramics interface [5]; and osteoconductivity (ability to allow bone to expand along the inside pores or spaces of the appropriate scaffold or template). In addition, Ca-P ceramics with appropriate three-dimensional geometry are capable of absorbing and concentrating endogenous bone morphogenetic proteins (BMP) from surrounding circulation and becoming osteoinductive (capable of osteogenesis). In 1988, Heughebeart reported that bone-like deposits were found on the surface of Ca-P ceramics which were not carried with any growth factors or cells when implanted in non-osseous site of animals. In 1991, Xingdong Zhang and Ripamonti reported that new bone tissue formed after porous HA was placed in dogs' and baboons' non-osseous sites. As the theory that Ca-P ceramics were provided with osteoinductivity developed, combined with principles of tissue engineering, in the late 1990s, Zhang and Bao put forward the concept of in vivo bone tissue engineering and defined it by stating that "in vivo bone tissue engineering means to use the body as bioreactor to construct bone graft with intrinsic osteoinductive biomaterials in the non-osseous or osseous sites, then transplant the graft to repair bone defect without any exogenous growth factors or cells" [6]. This technique relies on the body's own capacity to regenerate itself and differs from general tissue engineering. In 2005, Stevens and his colleagues created a bioreactor between the surface of a long bone (tibia) and the covered periosteum, which is rich in pluripotent cells, by using a hydraulic elevation procedure which was recently developed in their laboratory. Once a space of desired dimension and shape was created, the space was filled with calcium–alginate gel. Their previous studies indicated that calcium within the matrix could favor the osteogenic differentiation of the appropriate progenitor cell population. As the researchers reported, a few weeks later, the space was filled with neo-osseous tissue [7]. In this animal model a small, fluid-filled cavity was created inside which they hoped new bone could grow, and when the formed bone was extracted and transplanted to a bone defect site in the same animal, it was observed that the new bone seamlessly integrated with the surrounding tissue [8].

Based on the previous studies, this study aimed to investigate the relationship between Ca-P ceramics' osteoinductivity and species of animals, as well as the possible material factors affecting osteoinductivity of Ca-P ceramics, and then tried to optimize Ca-P ceramics with high osteoinductivity to provide a sufficient theoretical and experimental basis for clinical applications. Simultaneously, in order to develop a feasible technique for bone repairing, we also investigated the construction of bone tissue engineering in vivo and its possible application in segmental bone defect repairing in load-bearing sites.

2 Experimental studies

The whole study consisted of the following three parts: study of osteoinductivity in vivo, constructing bone graft with osteoinductive scaffold in vivo, repairing mandibular bone defect, and supporting dental implants with in vivo tissue engineering bone.

2.1 Part one: study of osteoinductivity in vivo

Ca-P ceramics are frequently used as bone substitute materials in clinics because of their similarity to the mineral phase of bone, absence of antigenicity and excellent osteoconductivity [9]. Whether Ca-P biomaterials have an osteoinductivity has been debated for a long time, but extensive fundamental studies and clinical application have demonstrated that Ca-P ceramics with special structure may induce bone formation in soft tissue. Based on the previous work of the group, this study further explored the relationship between osteoinductivity of Ca-P materials and animal species, and the material factors affecting osteoinductivity of Ca-P ceramics. It also optimized and selected Ca-P materials with high osteoinductivity to provide theoretical and experimental evidence for the application of osteoinductive materials.

2.1.1 Materials and methods

Hydroxyapatite/tricalcium phosphate (TCP) was sintered at 1,100°C and 1,200°C, and HA was sintered at 1,200°C. The chemistry of the ceramics was analyzed by X-ray diffraction (XRD). The average pore size of the ceramics was tested by scanning electronic microscope (SEM). The porosity of the ceramics was determined by the Archimedes water displacement method. Each ceramic was prepared with three cylinder dimensions: (1) 5 mm in diameter and 8 mm in length, (2) 4 mm × 6 mm, (3) 2 mm × 3 mm. Numbers of samples were decided statistically with the formula:

$$N = \frac{(\mu_\alpha + \mu_\beta) 4\bar{\pi}(1-\bar{\pi})}{(\pi_1 - \pi_2)^2}$$

A total of 27 dogs, 42 New Zealand rabbits, and 54 SD rats were used in this experiment. Three kinds of ceramics were prepared and implanted in the muscles of the dogs, rabbits, and rats. Dogs and rabbits were implanted with type (1) and (2) materials, respectively, at the site of bilateral spine muscles; rats were implanted with type (3) in the inside thigh bilaterally.

Samples were harvested at 6, 12, and 24 weeks after implantation. The osteoinductivity and the quantity of bone/bone-like tissue formation in the three kinds of Ca-P materials and animals were compared by morphology observation, ^{99m}Tc-MDP absorption evaluation, and statistical analysis. The judgment standard of bone/bone-like tissue was determined by three pathologists (Fig. 1). Light micrograph images were programmed by IPP (Image-Pro-Plus) software, and the data was analyzed by SPSS11.5.

2.1.2 Results and discussion

The results tested by XRD, SEM, and the Archimedes water displacement methods showed that the three ceramics had different physicochemical characters (Table 1).

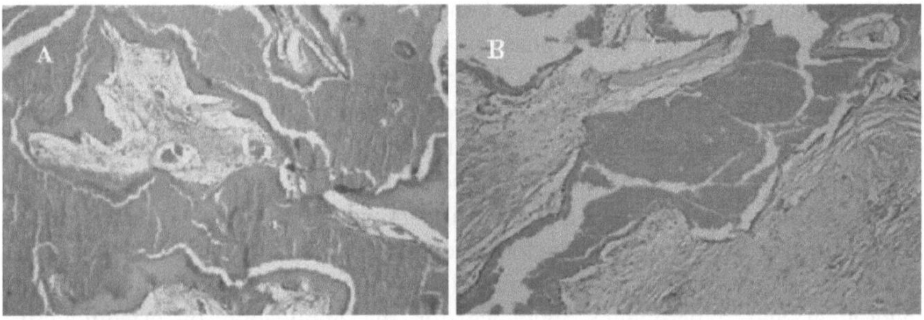

Fig. 1. Decalcified light micrographs after ceramics implanted in dorsal muscles of dogs. HE, ×100. **A** Bone tissue, **B** bone-like tissue

Table 1. Physicochemical characters of the three ceramics

	HA/TCP 1,100°C	HA/TCP 1,200°C	HA 1,200°C
Chemistry	6/4 HA/β – TCP	6/4 HA/α – TCP	HA
Pore size	350 ± 50 μm	400 ± 50 μm	400 ± 50 μm
Porosity (%)	70.8	55.2	62.9

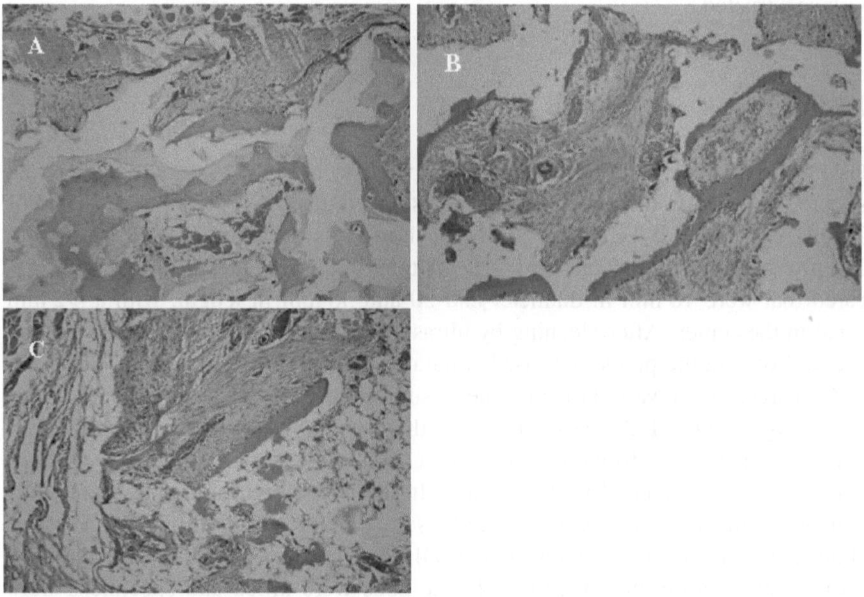

Fig. 2. Decalcified light micrographs after three kinds of ceramics had been implanted in the muscles of dogs for 12 weeks. HE, ×100. **A** HA/TCP sintered at 1,100°C; **B** HA/TCP sintered at 1,200°C; **C** HA sintered at 1,200°C

Thirty-six samples of each kind of ceramics implanted in animal species in different periods were decided statistically. Generally, the amount of osseous or osteoid tissue formation was greatest in HA/TCP sintered at 1,100°C and least in HA sintered at 1,200°C (Fig. 2). As for animal species, the quantity of bone formation was greatest in dogs and least in rats. Similar results were reported in the previous studies, but no reports provided statistical data. The quantity of newly formed bone of the same Ca-P materials increased with time, but the increase was not significant for the ceramics with high osteoinductivity. The morphology and composition of the ingrown tissue were related to the osteoinductivity of the ceramics, and cells differentiated directly toward osteoblasts in high osteoinductive materials, whereas in low osteoinductive materials, fibroblasts appeared first and inflammatory reaction could be found, and then bone formation took place [10].

2.2 Part two: constructing bone graft with osteoinductive scaffold in vivo

No tissue engineering bone composed of scaffold and cultured cells in vitro was permitted to be applied clinically. Ca-P ceramics with special structures can induce bone formation in soft tissue. This study was to explore a new method for bone substitute, by which vascular bone graft with osteoinductive Ca-P ceramics in vivo was constructed.

2.2.1 Materials and methods

Osteoinductive calcium phosphate was prepared by the H_2O_2 foaming method and sintered at 1,100°C for 3 h. XRD analysis showed that the precipitate consisted of 70% HA and 30% β-TCP. The SEM test showed its average pore size was about 395 μm and the porosity rate was 70%. A total of 32 ceramic cylinders were prepared that were 10 mm in diameter and 20 and 30 mm in length, with a Φ2 mm canal in the center. After cleaning by ultrasonication, a BLA layer was formed on the wall of ceramic pores in revised simulated body fluid (RSBF).

Construction of vascularized bone tissue engineering graft in vivo: 10 dogs (16–20 kg weight, 1–2 years old) were divided equally into two groups. Two muscle pockets were formed by separating tissue in one thigh, and two kinds of materials were implanted in the pockets. In the other thigh, materials were compounded with fresh marrow and cancellous bone before being implanted. Tetracycline and ^{99m}Tc-MDP were injected in all dogs intravenously, a week and 4 h, respectively, before they were killed. The dogs were killed and specimens were harvested at 6 and 12 weeks, respectively, after operation. The specimens were detected by blood vessel staining, histological observation, tetracycline fluorescence, ^{99m}Tc-MDP single photon electronic computer tomography (SPECT), and mechanical property testing.

2.2.2 Results and discussion

Bone formation was found in the ceramics implanted in dorsal muscles. Vascular pigmentation with 3% methylthioninium chloride showed that these bone grafts turned blue 6 weeks after operation. Histological observation and tetracycline labeling of cylinders mixed with marrow and porous bone showed obviously new bone formation (Fig. 3). SPECT showed that ^{99m}Tc-MDP concentrated at the sites of ceramic cylinders implantation.

Compared to not-implanted ceramics, two kinds of implanted materials during different periods were provided with higher compressing strength and bending strength as time passed. Compression stress of ceramic cylinders was from 4.15 to 8.12 MPa after 6 weeks and from 4.13 to 13.81 MPa after 12 weeks. Bending

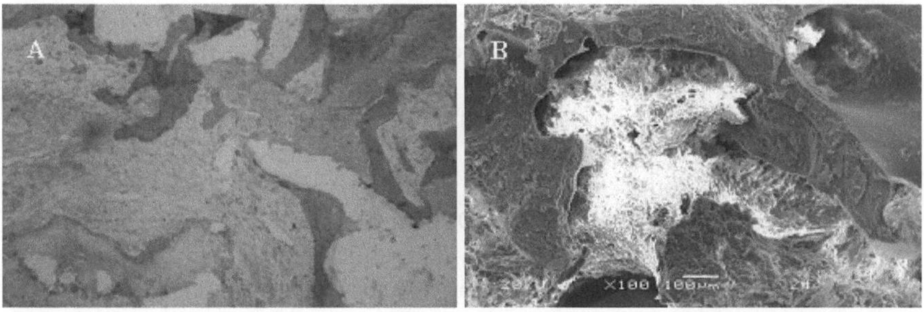

Fig. 3. Light micrograph (**A**) and SEM micrograph (**B**) of samples after implantation for 12 weeks. It was shown that new bone ingrowth and osseous matrix filled in pores of ceramics

stresses varied from the original 4.13 to 7.93 MPa after 6 weeks, and from 4.13 to 12.54 MPa after 12 weeks. Like natural bone, bone grafts did not crack under mechanical property testing. Of the two kinds of materials, the morphology of HA/TCP with BLA and marrow was better, but their biochemical properties were comparable.

2.3 Part three: repairing mandibular bone defect and supporting dental implant with in vivo tissue engineering bone

Bone defect is a frequently encountered disease clinically, and until now there has been no satisfactory method to treat massive or segmental bone defects [11]. Previous studies showed that in vivo tissue engineering bone could be constructed with osteoinductive Ca-P ceramics in vivo. The purpose of this study therefore was to explore feasible techniques of repairing massive bone defect with in vivo tissue engineering bone and to provide experimental evidence for the clinical application of this technique in the future.

2.3.1 Materials and methods

The osteoinductive Ca-P ceramics (HA/TCP sintered at 1,100°C as referred to in this article) were prepared and fabricated for 16 samples with a size of 20 mm × 10 mm × 8 mm. All the ceramics were soaked in RSBF for 72 h and sterilized by autoclave before being implanted.

Eight dogs were used in this study. After the animals were anesthetized with 3% sodium pentobarbital, two ceramics were implanted bilaterally in the femoral muscles of each dog, to obtain living ceramic bone grafts, namely, in vivo tissue engineering bone. Six weeks after implantation, operations were performed on the dogs again. The box-like bone defects 20 mm in length and 10 mm in depth were formed bilaterally in the mandibles of the dogs along the alveolar bone, where the teeth had been extracted. Meanwhile bone grafts were explanted from the femoral muscles and transplanted immediately to reconstruct the box-like bone defects of bilateral mandible. Simultaneously, one prefabricated dental implant was implanted in each graft as well as in alveolar bone, and then an internal fixation system was used to fix the bone grafts.

Tetracycline (50 mg/kg) and ^{99m}Tc-MDP (1 mCi/kg) were injected intravenously in all dogs 1 week and 4 h, respectively, before the animals were killed. Four of the dogs were killed at 8 weeks and four at 12 weeks after operation. Samples were harvested for gross observation, X-ray examination, tetracycline fluorescence labeling, ^{99m}Tc-MDP SPECT, and histological observation. The samples for histological observation were fixed, dehydrated, and embedded in methyl methacrylate; 15 μm undecalcified sections were made and stained with HE and Masson for light microscope observation.

2.3.2 Results and discussion

When the ceramics were explanted from the femoral muscles, it was found that they were enveloped with the vascular rete, and the tissue grew into the ceramics. After transplantation in the mandible, no infection or inflammation was observed. The bone graft fused with the host bone and restored the shape of the mandible.

Roentgenograms showed that the transplant was in close apposition with the host bone 8 weeks after transplantation. The density of bone graft was similar to that of the host bone. The boundaries between host bone and transplant had become indistinct (Fig. 4).

Of the SPECT results, the color distribution in implanted ceramics and in femoral diaphysis of nature bone coincided, which revealed that the density of implanted ceramic is equal to that of the natural bone.

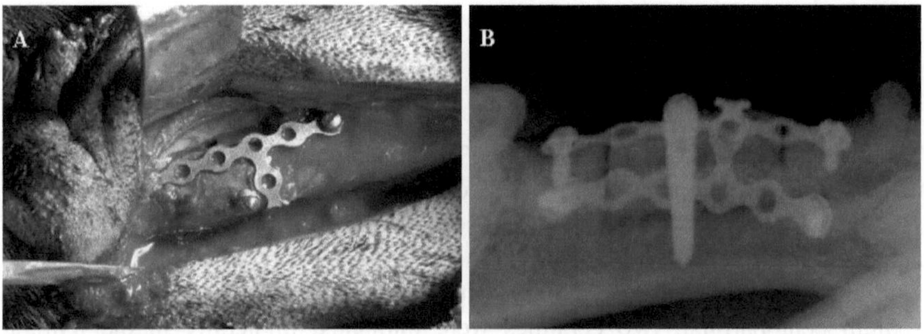

Fig. 4. The implant was fixed firmly in the graft and alveolar bone (**A**), and 2 months after transplantation, the boundaries between host bone and transplant from the X-ray became indistinct but could still be detected (**B**)

The fluorescence was detected in the transplant by tetracycline fluorescence labeling. It showed that fluorescence is strong in the pore walls of ceramics, while weak in the center of the pore.

Histological observation showed that active bone regeneration was found in the boundaries between the host bone and the transplant. On the interface of transplant and host bone, new bone formed into the pores of ceramics. Osteoblast rimming, even overlapping around the developing bone could be detected (Fig. 5a), but thin connective tissue was also seen. Inside the transplant, mineralized bone matrix with osteocytes filled in the pores of ceramics. The Haversian system was clearly observed (Fig. 5b). The matured trabeculae connected with one another. Macrophages gathered into a ball appeared in the ceramics.

In this study, bone graft was constructed directly by the technique of bone tissue engineering in vivo. Compared with the traditional methods, the superiority of bone tissue engineering in vivo is significant. First, the complicated procedures in vitro [12] are omitted and replaced by relatively simple methods in vivo. Second, it is well known that they are frequently used as a bone substitute because of their excellent properties such as biocompatibility, osteoconductivity, and osteoinductivity. In this study, in vivo tissue engineering bone is just based on Ca-P ceramics, so it has the same excellent properties as Ca-P ceramics. Moreover, as the bone grows into the pores of ceramics, the amount of organic tissue increases; as a result, the tissue engineering bone in vivo modified the poor biomechanical property of Ca-P ceramics which have high elastic modulus and low flexibility. Therefore, because of bone graft with living auto-cells and good blood supply, the biomechanical properties and bioactivity of in vivo tissue engineering bone were similar to the autograft. After being transplanted it participated in the bone metabolism of the host as early as possible. Thirdly, because it is difficult to treat the massive or segmental bone defect clinically nowadays, in vivo bone tissue engineering makes it possible to reconstruct the living bone graft in considerable size; so it is a satisfactory method to treat a massive or segmental bone defect.

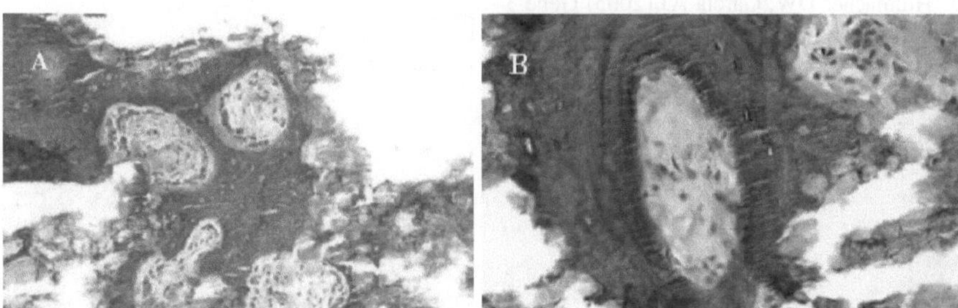

Fig. 5. Light micrographs for histological observation (12 weeks postoperatively). **A** New bone with osteoblast rimming, HE staining, ×200. **B** Haversian system with osteocytes, HE staining, ×400

3 Conclusion

Biomaterials without any exogenous growth factors or cells can induce bone formation in non-osseous sites. In vivo tissue engineering bone formed with osteoinductive Ca-P ceramics has histological characteristics similar to those of natural bone. The ratio of osseous or osteoid tissue formation and the quantity of bone formed are related to the physicochemical property and animal species. By the in vivo tissue engineering technique, it is possible to construct a large bone graft with good blood supply and a certain mechanical strength. The characteristics of the ingrown tissue and the pattern of bone formation are related to the osteoinductivity of the materials. In vivo tissue engineering bone has been provided with the hope that it will be used to repair segmental bone defects of the mandible and support dental implants in the future.

Acknowledgment. This research was financially supported by the National Natural and Science Foundation of China (Contract No. 30672337).

References

1. Robert F Service (2000) Science 289:1498–1500
2. George FM, Chizu N, Linda GG (2004) *J Bone Joint Surg* 86:1541–1558
3. Linda G, Griffith, Gail Naughton (1997) Science 276:181–185
4. Logeart AD, Anagnostou F, Bizios R, et al (2005) J Cell Mol Med 9(1):72–84
5. Ripamonti U (2006) Biomaterials 27(6):807
6. C Bao, H Fan, X Zhang, et al (2004) Key Eng Mater 255:801–804
7. Stevens MM, Marini RP, Schaefer D, et al (2005) Proc Natl Acad Sci 102(32):11450–11455
8. Robert F Service (2005) *Science* 309:683
9. Laurencin C, Khan Y, El-Amin SF (2006) Expert Rev Med Devices 3(1):49–57
10. Bao CY, Li P, Tan YF, et al (2005) Key Eng Mater 288–289:281–286
11. Schnurer SM, Gopp U, Kuhn KD, et al (2003) Orthopade 32(1):2–10
12. Hutmacher DW, Garcia AJ (2005) Gene 347(1):1–10

Interface Oral Health Science 2007: 95–104

Toughening of bioabsorbable polymer blend by microstructural modification

Mitsugu Todo[1]* **and Tetsuo Takayama**[2]*
[1]*Research Institute for Applied Mechanics;* [2]*Interdisciprinary Graduate School of Engineering Science, Kyushu University, Fukuoka 816-8580; Japan*
*todo@riam.kyushu-u.ac.jp, takayama@riam.kyushu-u.ac.jp

Abstract. In order to improve the phase morphology of bioabsorbable polymer blend of poly(lactic acid) (PLA) and poly (ε-caprolacton) (PCL), an additive with the isocyanate group, lysine tri-isocyanate (LTI), was used, and the effects of LTI addition on the fracture energy, J_{in}, and the related fracture micromechanism were investigated. The study showed that J_{in} effectively increases with an increase in LTI. Microscopic examination of the mode I fracture surfaces also exhibited that the size of the PCL phases dramatically decreases due to LTI addition, leading to the reduction of void formation and suppression of local stress concentration, and therefore resulting in the increase of the fracture energy. The improved miscibility also contributes to the ductility enhancement, which further increases the fracture energy. In order to improve the mechanical properties, such as the bending modulus and strength, the annealing process was conducted for PLA/PCL and PLA/PCL/LTI blends. The mechanical properties of both the blends effectively increased due to the strengthened structures by crystallization of PLA. J_{in} of PLA/PCL largely reduced by annealing; on the other hand, that of PLA/PCL/LTI effectively improved. The well-entangled structure of PLA/PCL/LTI results in the elongated ductile fracture of firmly connected fibrils; as a result, the energy dissipation during fracture initiation is largely increased.

Key words. poly(lactic acid), poly(ε-caprolacton), fracture energy, additive, crystallization

1 Introduction

Poly (lactic acid) (PLA) is one of the typical bioabsorbable polymers, which has been used as a biomaterial in medical fields such as orthopedics and oral surgery [1, 2]. PLA is, for example, utilized as bone fixation implants owing to its relatively high strength and stiffness. Its importance has led to many studies on its mechanical properties and fracture behavior [3–7] which found that the mode I fracture behavior of PLA is relatively brittle in nature. Therefore, blending with a ductile bioabsorbable polymer such as poly (ε-caprolacton) (PCL) has been adopted to improve the fracture energy of brittle PLA [8–13]. It was, however, also found that the immiscibility of PLA and PCL causes phase separation, and tends to lower the fracture energy especially when PCL content increases. It has recently been found that addition of lysine tri-isocyanate (LTI) to PLA/PCL blend effectively improves their immiscibility [14–18], and therefore the fracture energy [13–18].

Although PCL blending effectively improves the brittleness of PLA and furthermore LTI addition results in dramatic improvement of the fracture energy of PLA/PCL, the fundamental mechanical properties, such as the bending strength and modulus of PLA/PCL and PLA/PCL/LTI, tend to be lower than the base polymer PLA as a result of the blending of ductile soft polymer PCL. It is known that these mechanical properties of PLA can be improved by crystallization using the annealing method [19]. However, the effects of crystallization on the mechanical properties of PLA/PCL and PLA/PCL/LTI blends have not yet been fully understood.

In this article, the effects of LTI addition and its content on a mode I fracture property, the fracture energy at initiation J_{in}, are presented. Fracture micromechanisms are also discussed on the basis of the microscopic results obtained using a polarizing-light optical microscope (POM) and a field emission scanning electron microscope (FE-SEM). The effects of annealing on the mechanical properties, such as the bending strength and modulus of PLA/PCL and PLA/PCL/LTI, were examined in the present study. J_{in} values of the annealed PLA/PCL and PLA/PCL/LTI were also evaluated, and compared to that of the quenched blends to assess the effectiveness of crystallization. Fracture micro-mechanism was also characterized by observing fracture surfaces using the FE-SEM. The fracture mechanism was then correlated with the macroscopic fracture energy.

2 Experimental

2.1 Materials and specimens

PLA/PCL and PLA/PCL/LTI blends were fabricated using PLA pellets (Lacty #9030, Shimadzu Kyoto, Japan) and PCL pellets (CelgreenH7, Daicel Chemistry Industries, Osaka, Japan), and LTI (Kyowa Hakko Chemical, Tokyo, Japan) by melt-mixing in a conventional melt-mixer at 180°C for 20 min and at a rotor speed of 50 rpm. The mixing ratio of PLA and PCL was fixed at 85:15 in weight fraction. For PLA/PCL/LTI, LTI contents were chosen to be 0.5, 1, 1.5, and 2 phr. The blend mixtures were then press-processed using a conventional hot press at 180°C and 30 MPa. Single-edge-notch-bend (SENB) specimens of $70 \times 10 \times 2\ mm^3$ with 5 mm notch were prepared from these plates for mode I fracture testing.

Some of the fabricated plates of PLA/PCL and PLA/PCL/LTI were annealed at 100°C for 3 h using a forced convection oven and then naturally cooled down to room temperature. For the annealed PLA/PCL/LTI, LTI content was fixed at 1 phr. Bean specimens of $70 \times 10 \times 2\ mm^3$ were prepared from the annealed plates to evaluate bending mechanical properties such as the modulus and strength. SENB specimens were also processed from the annealed plates to measure the mode I fracture energy.

The microstructures of the prepared specimens were characterized by observing the fracture surfaces of cryo-fractured SENB specimens prepared under liquid nitrogen environment using FE-SEM.

2.2 Crystallinity and molecular weight measurement

The enthalpy of crystallization and melting, dH_c and dH_m, of the quenched and the annealed PLA/PCL and PLA/PCL/LTI were determined by a differential scanning calorimeter (DSC). Small pieces of the blends were heated under nitrogen gas flow at a rate of 10°C/min for DSC measurements. The crystallinity of PLA, $x_{c,PLA}$, in the blends having constant PLA content, X_{PLA}, was evaluated according to the following formula [19]:

$$x_{c,PLA}\,(\%) = \frac{100 \times (dH_m + dH_c)}{93 \times X_{PLA}} \tag{1}$$

where 93 (J/g of the polymer) is the enthalpy of melting of the PLA crystal having the infinite crystal thickness reported by Fischer et al. [20].

The weight-average molecular weight, M_w, of the prepared samples was evaluated in chloroform at 40°C by a GPC system with a spherical porous gel made of a styrene–divinylbenzene copolymer using polystyrene standards.

2.3 Mode I fracture and bending tests

Three-point-bend-type fracture tests of the SENB specimens were conducted using a servo-hydraulic testing machine at a loading rate of 1 mm/min. Time histories of load and displacement data were recorded using a digital recorder, and later, load–displacement relationship was evaluated from these data. The critical J-integral value, J_{in}, at crack initiation was obtained as the fracture energy using the following formula:

$$J_{in} = \frac{\eta U_{in}}{B(W-a)} \tag{2}$$

where U_{in} is the critical energy, B the specimen thickness, W the specimen width, a the initial crack length, and η the geometrical correction factor, which is equal to 2 for the standard SENB specimen. The critical point corresponding to crack initiation was defined as the point at which the specimen rigidity dropped suddenly due to the onset of crack propagation [15].

Three-point bending tests of the beam specimens were performed at a loading-rate of 10 mm/min using the servo-hydraulic testing machine, and the load–displacement curves were obtained as described above. Bending modulus, E, and bending strength, σ_f, were then evaluated from the linear portion of the load–displacement curves and the maximum load, respectively, using the following formulae:

$$E = \frac{L^3}{4bh^3}S \quad \text{and} \quad \sigma_f = \frac{3PL}{2bh^2} \tag{3}$$

where S is the initial slope of the load–displacement curve. L, b, and h are the span, the width, and the thickness, respectively, and P is the maximum load.

2.4 Microscopic observation

For each of PLA, PLA/PCL and PLA/PCL/LTI specimens, a thin section was prepared using the petro-graphic thin-sectioning technique from the region in the vicinity of the crack-tip where process zone was developed in the mode I fracture test. The thin-sectioned samples were then observed using POM to characterize the mechanism of process zone formation. Fracture surfaces of the mode I fracture specimens were also observed using FE-SEM to characterize the fracture mechanism and the effect of LTI addition and annealing on the fracture behavior.

3 Results and discussion

3.1 Effect of LTI addition on the fracture energy of PLA/PCL

3.1.1 Phase morphology

FE-SEM micrographs of cryo-fractured surfaces of PLA/PCL and PLA/PCL/LTI are shown in Fig. 1. Spherical features appeared on the micrograph are thought to

a

b

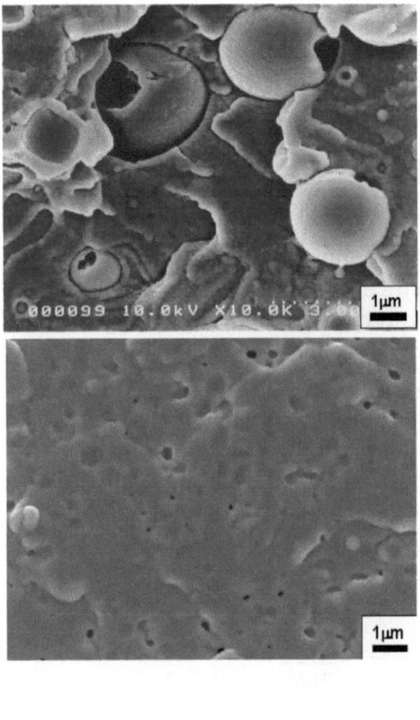

Fig. 1. FE-SEM micrographs of cryo-fracture surfaces of PLA/PCL and PLA/PCL/LTI **a** PLA/PCL; **b** PLA/PCL/LTI

be PCL-rich phases [9]. These micrographs clearly showed that the size of the PCL-rich phase dramatically decreases by LTI addition. It is thus presumed that LTI addition effectively improves the miscibility of PLA and PCL. This is thought to be related to the following chemical reaction, that is, the hydroxyl group of PLA and the isocyanate group of LTI creates a urethane bond:

$$\text{HO–R}' + \text{R–N=C=O} \rightarrow \text{R–NHCOO–R}'$$

3.1.2 Fracture energy and mechanisms

Dependence of Jin on LTI content is shown in Fig. 2. It is seen that J_{in} of PLA/PCL is a little larger than that of PLA, indicating the effectiveness of PCL blend on J_{in} is very low. J_{in} of PLA/PCL is effectively improved by LTI addition, and J_{in} increases with increase in LTI content up to 1.5 phr. There is no difference of J_{in} between 2 and 1 phr of LTI addition, suggesting that the improvement of J_{in} is saturated with about 1.5 phr of LTI.

POM micrographs of crack growth behaviors in PLA, PLA/PCL, and PLA/PCL/LTI are shown in Fig. 3. Craze-like features are clearly seen in front of the crack-tip of the pure PLA in Fig. 3a. This kind of crack-tip damage is broadened by PCL blending as shown in Fig. 3b. With LTI addition, the craze-like feature is no longer generated, and instead, the crack-tip region is plastically deformed, very similar to the crack-tip deformation in ductile plastics and metal. It is known that this kind of plastic deformation dissipates more energy than the craze-like damage, resulting in the greater fracture energy. It is therefore thought that LTI addition to PLA/PCL dramatically changes the crack-tip deformation mechanism; as a result, J_{in} is greatly improved.

FE-SEM micrographs of fracture surfaces in the vicinity of the initial notch-tip are shown in Fig. 4. The fracture surface of PLA is very smooth, corresponding

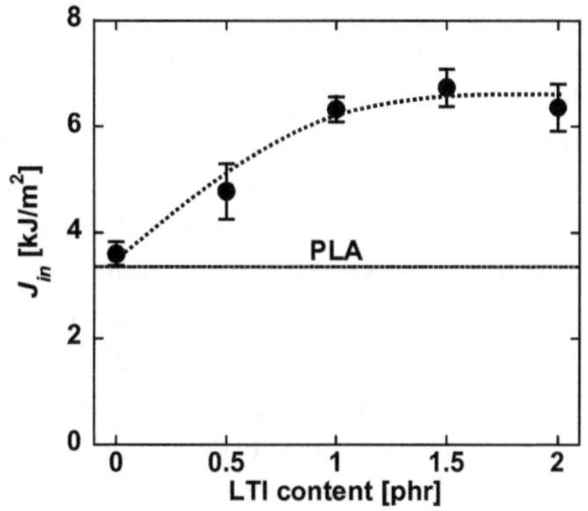

Fig. 2. Dependence of the fracture energy, J_{in}, on LTI content

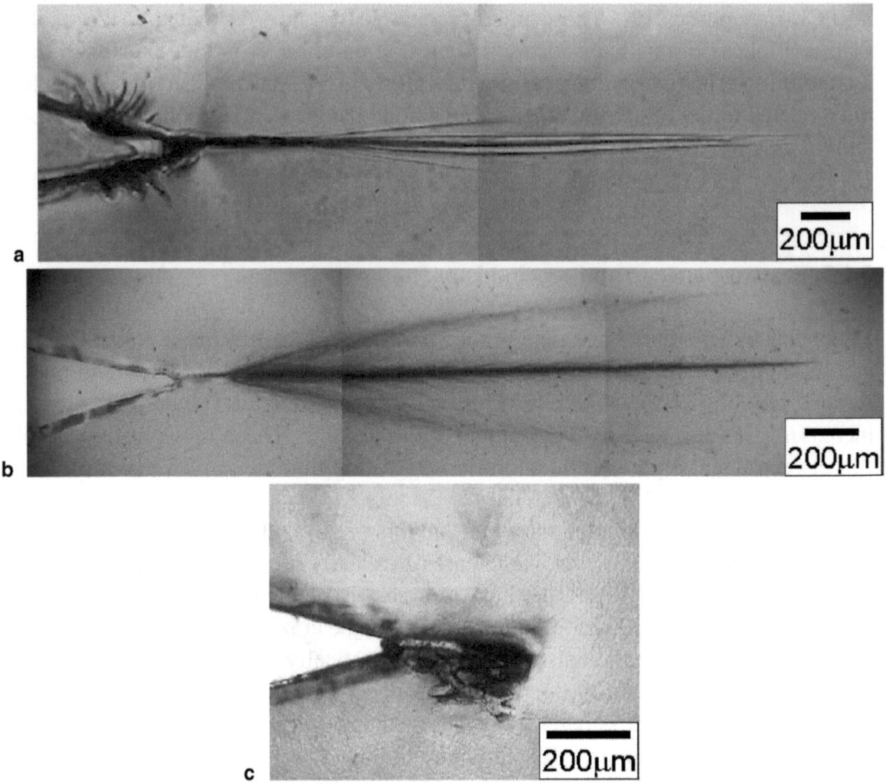

Fig. 3. POM micrographs of crack growth behaviors. **a** PLA; **b** PLA/PCL; **c** PLA/PCL/LTI

to a brittle fracture behavior with low fracture energy. The surface roughness increases with the existence of elongated PCL and cavities by PCL blending. These cavities are thought to be created by debonding of the PCL-rich phases from the surrounding PLA matrix phase and usually cause local stress concentration in the surrounding regions. Thus, this kind of cavitation tends to lower the fracture energy because of the local stress concentration, and compensates to the increase in fracture energy due to the ductile deformation of PCL. This is the reason for the slight improvement of J_{in} in PLA/PCL shown in Fig. 2. It is clearly seen from Fig. 4c that cavities do not exist on the fracture surface of PLA/PCL/LTI, indicating that the miscibility of PLA and PCL improves due to LTI addition. In addition, elongated structures are more on PLA/PCL/LTI than PLA/PCL. Thus, extensive ductile deformation associated with disappearance of cavitation is the primary mechanism of the dramatic improvement of J_{in}.

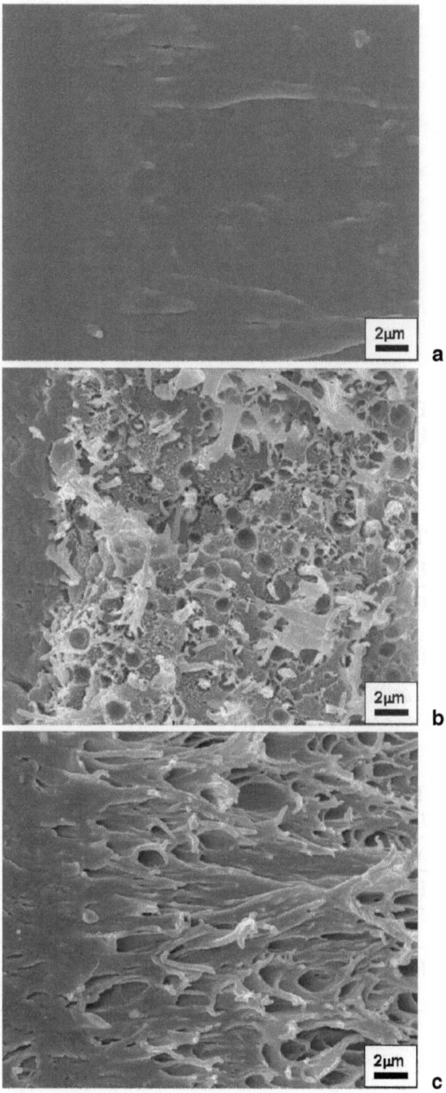

Fig. 4. FE-SEM micrographs of mode I fracture surfaces. **a** PLA; **b** PLA/PCL; **c** PLA/PCL/LTI

3.2 Effect of annealing on the mechanical properties and fracture energy of PLA/PCL and PLA/PCL/LTI

3.2.1 Microstructure and microstructural properties

FE-SEM micrographs of the cryo-fracture surfaces of the annealed PLA/PCL and PLA/PCL/LTI are shown in Fig. 5. By comparing Fig. 5 with Fig. 1, the effect of annealing on the micro-structural change is recognized as a cauliflower-like surface, indicating the formation of PLA spherulites. Crystallinity, $x_{c,PLA}$, and molecular

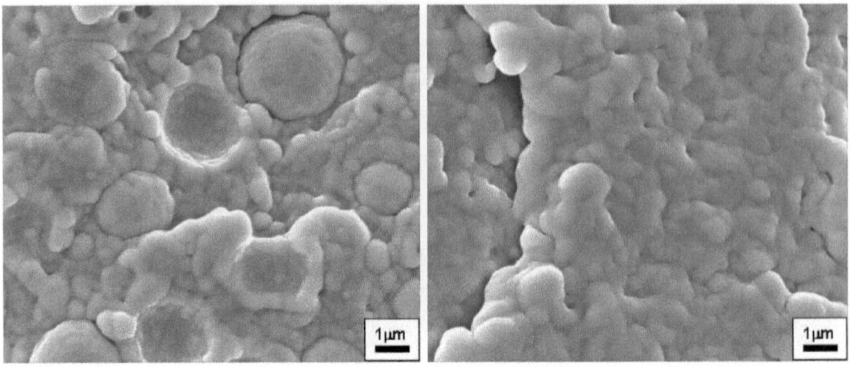

a b

Fig. 5. FE-SEM micrographs of cryo-fracture surfaces of annealed blends. **a** PLA/PCL; **b** PLA/PCL/LTI

Table 1. Effect of annealing on $x_{c,\,PLA}$ and M_w of PLA/PCL and PLA/PCL/LTI

Polymer blend	Process condition	$x_{c,PLA}$ (%)	M_w (g/mol)
PLA/PCL	Quenching	11.4	1.03×10^5
	Annealing	45.6	9.08×10^4
PLA/PCL/LTI	Quenching	4.8	1.13×10^5
	Annealing	36.6	1.52×10^5

Table 2. Effect of annealing on the bending and fracture properties of PLA/PCL and PLA/PCL/LTI

Material	Process condition	E (GPa)	σ_f (MPa)	J_{in} (kJ/m^2)
PLA		3.70	103.2	3.32
PLA/PCL	Quenching	3.26	93.5	3.60
	Annealing	3.74	95.7	0.61
PLA/PCL/LTI	Quenching	3.06	85.6	6.33
	Annealing	3.56	100.3	8.68

weight, M_w, are shown in Table 1. It is clearly seen that $x_{c,PLA}$ increases dramatically with annealing. M_w of PLA/PCL slightly decreases by annealing and this might be due to thermal degradation of molecules during annealing at 100°C for 3 h. On the other hand, M_w of PLA/PCL/LTI increases by annealing. This is thought to be attributable to the polymerization by urethane bond formation.

3.2.2 Bending and fracture properties

Effect of annealing on the bending mechanical properties, E and σ_f, of PLA/PCL and PLA/PCL/LTI are shown in Table 2. These values of the pure PLA are also shown in the table. For both the polymer blends, E and σ_f tend to increase by annealing. Annealing causes crystallization of PLA, as indicated by increase in $x_{c,PLA}$ (see Table 1), and therefore, likely to strengthen the structure of PLA and the blends, resulting in the increase of E and σ_f. It should be noted that E values of both the blends and σ_f of PLA/PCL/LTI become very close to those of PLA.

Effects of annealing on J_{in} are also shown in Table 1. J_{in} of PLA/PCL largely degrades by annealing; on the other hand, J_{in} of PLA/PCL/LTI effectively improves by 161% due to annealing.

3.2.3 Fracture mechanism

Field emission scanning electron microscope micrographs of mode I fracture surfaces of the annealed blends are shown in Fig. 6. For the annealed PLA/PCL, ruptured PLA fibrils are created on the fracture surface; on the other hand, elongated structures of the spherical PCL phases are observed in the quenched PLA/PCL as shown in Fig. 4b. As reported by Park et al. previously that annealing process tends to reduce the fracture energy of PLA [5–7] due to the embrittlement of the PLA structure by crystallization, for PLA/PCL, the PLA-rich phases are likely to become brittle by annealing and because of the phase separation caused by the immiscibility of PLA and PCL, J_{in} degrades. For the annealed PLA/PCL/LTI, elongated entangled fibril structures generated by the chemical bonding between PLA and PCL molecules through LTI addition are observed as also seen in the quenched PLA/PCL/LTI (Fig. 4c), and these fibrils are firmly connected to each other due to the crystallization of PLA. This kind of micro-structural change is thought to result in the improvement of J_{in}.

a 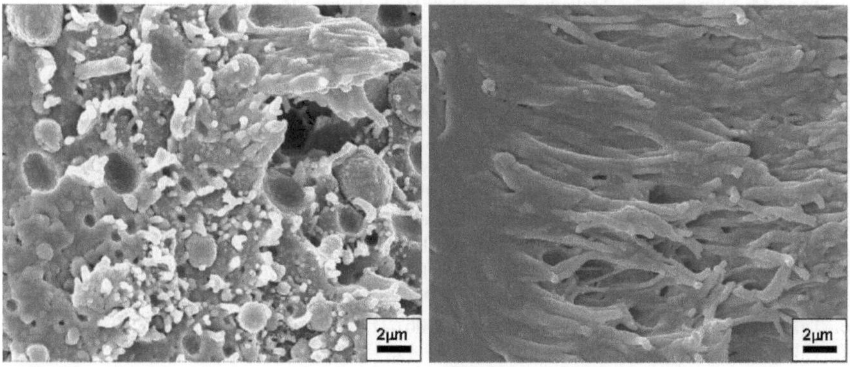 b

Fig. 6. FE-SEM micrographs of mode I fracture surfaces of annealed blends. **a** PLA/PCL; **b** PLA/PCL/LTI

4 Summary

Effects of LTI additive on the fracture energy, J_{in}, and the fracture micro-mechanism of PLA/PCL polymer blend were investigated, and furthermore, effects of annealing process on the bending mechanical properties, E and σ_f, and J_{in} of PLA/PCL and PLA/PCL/LTI blends were assessed and the micro-structural modification due to annealing was characterized. The results obtained are summarized as follows:

(1) The miscibility of PLA and PCL is effectively improved by LTI addition; as a result, the phase morphology of PLA/PCL is dramatically changed such that the size of the PCL-rich phases become small.
(2) J_{in} of PLA/PCL is effectively improved by LTI addition, and increases with increase in LTI up to 1 phr. The improvement is saturated more than 1 phr of LTI.
(3) E of both the polymer blends and σ_f of PLA/PCL/LTI are effectively improved by annealing. Especially, E of the blends becomes very close to that of the pure PLA. These improvements of the mechanical properties are thought to be related to the strengthened structures of the blends due to the crystallization of PLA phase by annealing.
(4) J_{in} of PLA/PCL largely degrades by annealing as a result of embrittlement of the PLA phases. On the contrary, J_{in} of PLA/PCL/LTI is effectively improved by annealing. The well-entangled structure of PLA/PCL/LTI results in the elongated ductile fracture of firmly connected fibrils; as a result, the energy dissipation during fracture initiation is largely increased.

References

1. Mohanty AK, Misra M, Hinrichsen G (2000) Macromol Mater Eng 276/277:1–24
2. Higashi S, Tamamoto T, Nakamura T, et al (1986) Biomaterials 7:183–187
3. Todo M, Shinohara N, Arakawa K (2002) J Mater Sci Lett 21:1203–1206
4. Todo M, Shinohara N, Arakawa K, et al (2003) Kobunshi Ronbunshu 60:644–651
5. Park SD, Todo M, Arakawa K (2004) J Mater Sci 39:1113–1116
6. Park SD, Todo M, Arakawa K (2004) Key Eng Mater 261/263:105–110
7. Park SD, Todo M, Arakawa K (2005) J Mater Sci 40:1055–1058
8. Tsuji H, Ikada Y (1996) J Appl Polym Sci 60:2367–2375
9. Todo M, Park SD, Takayama T, Arakawa K (2007) Eng Frac Mech 74:1872–1883
10. Wang L, Ma W, Gross RA, et al (1998) Polym Degrad Stab 59:161–168
11. Hiljanen M, Varpomaa P, Sppala J, et al (1996) Macromol Chem Phys 197:1503–1523
12. Meredith JC, Amis EJ (2000) Macromol Chem Phys 201:733–739
13. Tsuji H, Yamada T, Suzuki M, et al (2003) Polym Int 52:269–275
14. Dell' Erba R, Groeninckx G, Maglio G, et al (2001) Polymer 42:7831–7840
15. Harada M, Hayashi H, Iida K, et al (2003) Polym Prepr 52:965
16. Takayama T, Todo M, Arakawa K, et al (2006) Trans Jpn Soc Mech Eng 72:713–718
17. Takayama T, Todo M (2006) J Mater Sci 41:4989–4992
18. Takayama T, Todo M, Tsuji H, et al (2006) J Mater Sci 41:6501–6504
19. Tsuji H, Ikada Y (1995) Polymer 36:2709–2716
20. Fischer EW, Sterzel HJ, Wegner G (1973) Kolloid-Z u Z Polym 251:980

Interface Oral Health Science 2007: 105–114

Corrosion resistance and biocompatibility of a dental magnetic attachment

Osamu Okuno* and Yukyo Takada
Division of Dental Biomaterials, Tohoku University Graduate School of Dentistry, Sendai 980-8575, Japan
*okuno@mail.tains.tohoku.ac.jp

Abstract. A dental magnetic attachment is used for over-dentures, removable partial dentures, and orthodontic and maxillofacial prostheses. The dental magnetic attachment is composed of a magnetic assembly and a keeper. The magnetic attractive force between the magnetic assembly and the keeper is used as the retention. The magnetic assembly is composed of a small Nd–Fe–B magnet that is covered within a magnetic stainless steel yoke and a non-magnetic stainless steel spacer. The keeper is also made of magnetic stainless steel. The yoke of the magnet assembly and the keeper form a closed magnetic circuit, which is necessary to concentrate on the magnetic flux and make efficient use of it. The covering non-magnetic stainless steel and magnetic stainless steel yoke are welded seamlessly using micro-laser to protect the magnet from corrosion. In an oral cavity, the keepers and the magnetic assemblies are contacted with root caps made of dental precious alloys. It is important to examine the galvanic corrosion behavior of those stainless steels with dental precious alloys from the electrochemical properties and released ions. The dental magnetic attachment was implanted in rabbit tibia to investigate the influence of the static magnetic flux on hard tissue.

Key words. corrosion resistance, biocompatibility, magnetic attachments

1 Introduction

It is tempting to apply magnets to dentistry. However, the practical dental application of a magnet was difficult until rare earth magnets were invented in the later half of the 1960s (Fig. 1) [1–4]. The first dental application of the Sm–Co magnet to a retaining denture seems to have been that undertaken by Sasaki and Kinouchi [5–8]. Afterwards, worldwide research on magnetic attachments began [1–4, 9, 10]. In 1990s, introducing micro-laser welding technology, the magnetic attachment which perfectly covered and sealed the magnet by magnetic stainless steel and non-magnetic stainless steel from the oral fluid was developed in Japan [1–4, 11–14]. After that, the magnetic attachment was further improved, and more than 1,200,000 magnetic attachments have been produced since then. The dental magnetic attachment is used for over-dentures, removable partial dentures, and orthodontic and maxillofacial prostheses. The devices make it easier to remove and replace dentures and reduce lateral and rotational stresses to the abutment tooth or implant under function.

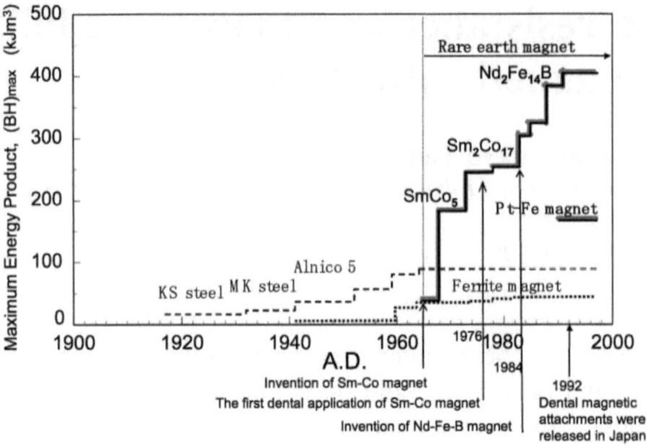

Fig. 1. Progress in permanent magnets

Table 1. Magnetic properties of magnets

Magnet	Residual flux density Br (T)	Coercive force HCB (kA/m)	Maximum energy product (BH) max (kJ/m^3)	Curie temperature (°C)
$SmCo_5$	0.85–0.95	636–756	143–176	710
Sm_2Co_{17}	1.04–1.14	732–852	198–247	770
$Nd_2Fe_{14}B$	1.37–1.43	1034–1114	351–390	310
Fe–Pt–Nb	0.90–1.10	358–438	143–185	460
Pt–Co	0.7	358	92	550
Alnico (Al–Ni–Co)	1.05	127	88	850

2 Magnets

The magnetic characteristics of magnets are evaluated by the magnetization curve, which is the intensity of magnetization of the magnet in the magnetic field (Fig. 2) [3]. Table 1 shows the magnetic properties of magnets [3]. The coercive force of the rare earth magnet is much larger than that of the Alnico magnet. The coercive force of the Nd–Fe–B magnet is more than nine times that of the Alnico magnet. The effect of the demagnetizing field for the Nd–Fe–B magnet is then less than that for the Alnico magnet, and the thin Nd–Fe–B magnet can achieve a much higher attractive force than that of the Alnico magnet. The maximum energy product of the Sm_2Co_{17} and Nd–Fe–B magnets is three times higher and five times higher than that of the Alnico magnet, respectively. (The magnetic attractive force of magnets can be compared by the maximum energy product of the magnet.)

Although rare earth magnets have excellent magnetic properties, with the progress of research, it became clear that the poor corrosion resistance and lack of retention force of rare earth magnets were obstacles to overcome.

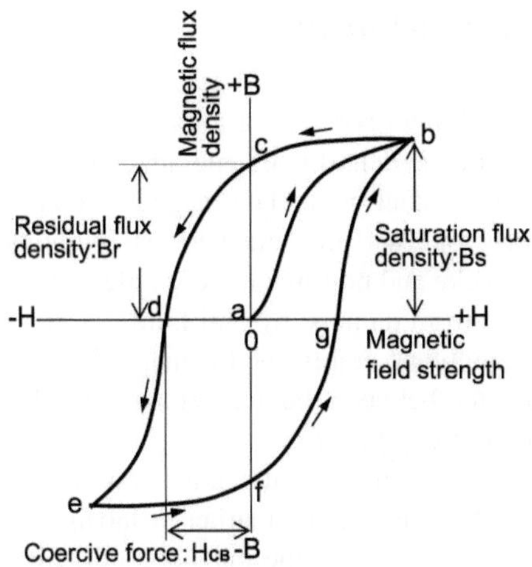

Fig. 2. Magnetization curve for ferromagnetic materials.

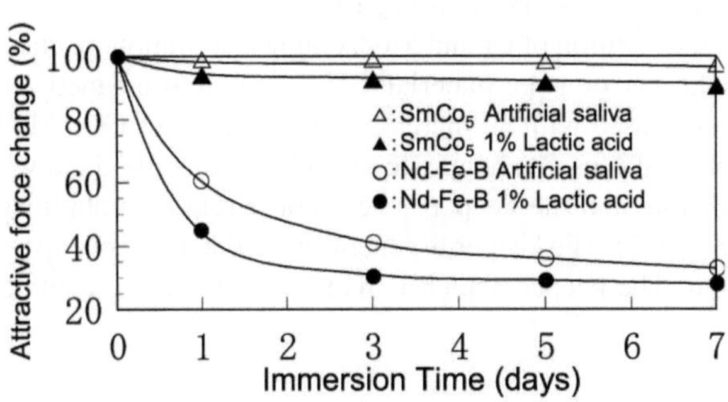

Fig. 3. Attractive force change of the magnets in artificial saliva and 1% lactic acid

Figure 3 shows an attractive force change of the $SmCo_5$ magnet and Nd–Fe–B magnet in artificial saliva and 1% lactic acid for 7 days, respectively [15]. In particular, the Nd–Fe–B magnet may corrode and deteriorate severely in contact with oral fluids.

The retention force of the small magnet, which can be placed into the denture, is insufficient for retaining the denture. The attractive force between the 4 mm × 1.5 mm Nd–Fe–B magnet and the magnetic stainless steel of SUS XM27 is 2.8 N at maximum. It is not exactly clear what the necessary force for the retaining denture is. It seemed to lack sufficient denture retention because the retaining forces of Konuskrone and precision attachment are 5–10 and 4–10 N, respectively [16, 17].

3 Dental magnetic attachment

The dental magnetic attachment consists of a magnet assembly and a keeper. The rare earth metal magnet of the dental magnetic attachment enhances the attractive force by magnetic stainless steel yoke. The magnet is completely covered with a magnetic stainless steel yoke and non-magnetic stainless steel. The joint of the magnetic stainless steel yoke and non-magnetic stainless steel is welded seamlessly by micro-laser to protect from damage by oral fluids. A closed magnetic circuit is formed between the magnetic assembly and the magnetic stainless keeper. A strong attractive force is generated between the magnet assembly fixed in the denture and the keeper fixed in the root cap [1–3].

The structure of the dental magnetic attachments is shown in Fig. 4. The dental magnetic attachments, which are on the market in Japan, are used in the cup-type or sandwich-type yoke for enhancing the magnetic attraction by concentrating of the magnetic flux density and closed circuit. The magnets are Nd–Fe–B. The magnetic stainless steels for the yoke and keeper are SUS 444, SUS XM27, or SUS447J1 ferritic stainless steels, as shown in Table 2. Non-magnetic stainless steel is austenite stainless steel of SUS316L [1–3].

Both magnetic stainless and non-magnetic stainless steels have good corrosion resistance with the addition of Cr and Mo. A higher saturation magnetic flux density is required for keeper or yoke materials. The saturation magnetic flux density of stainless steel decreased with an increase in the content of Cr + Mo [18, 19].

In general clinical uses, the magnetic assembly is directly fixed within a denture or a prosthetic base, and the keeper is set upon a retaining abutment tooth or an implant fixture (Fig. 5). The keeper is generally fixed in the root cap by cast bonding or dental cement. The keeper, which is fixed directly on the abutment tooth with

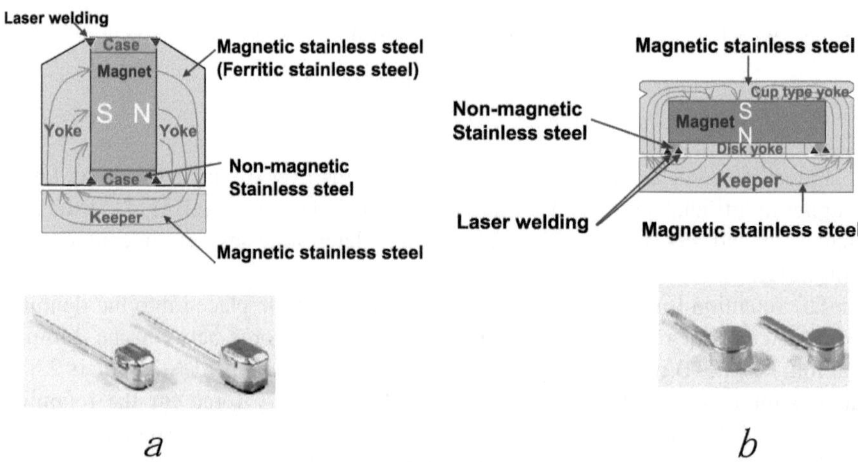

Fig. 4. Structure of the dental magnetic attachments. **a** Sandwich yoke-type; **b** cup yoke-type

Table 2. Chemical compositions of the stainless steels for dental magnetic attachments

Stainless steels	Compositions (mass%)							
	Cr	Mo	Ni	C	Si	Mn	P	Fe
444	18.56	1.97	0.20	0.006	0.27	0.17	0.029	Balance
XM27	26.0	1.0	0.17	0.002	0.34	0.09	0.019	Balance
447J1	30.0	2.0	0.15	0.003	0.15	0.04	0.015	Balance
316L	17.73	2.04	12.32	0.011	0.057	0.85	0.033	Balance

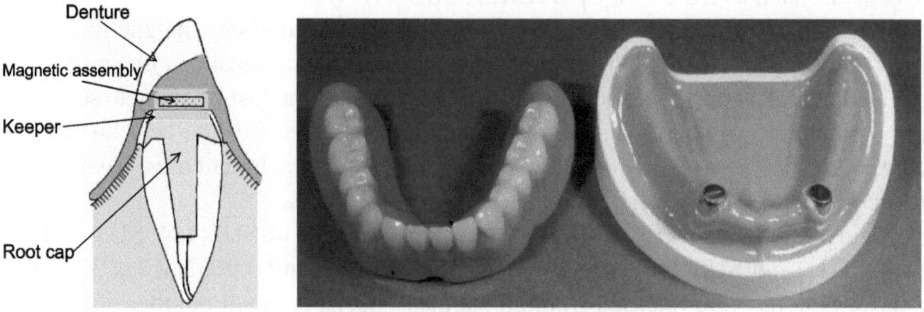

Fig. 5. Applications of a dental magnetic attachment to dentures

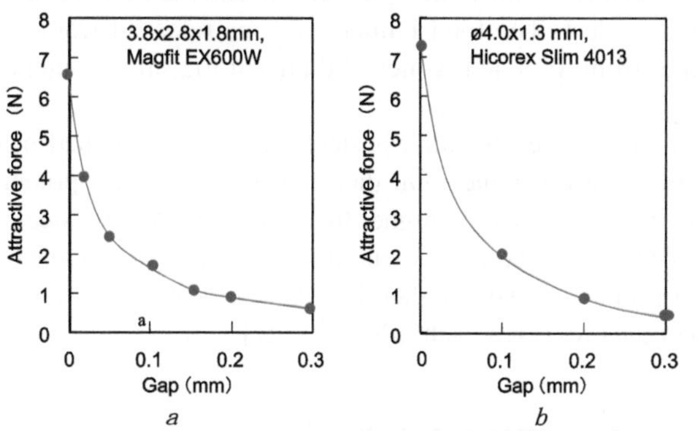

Fig. 6. Attractive force of a dental magnetic attachment for the gap between the magnetic assembly and the keeper. **a** Sandwich yoke-type (3.8 mm × 2.8 mm × 1.8 mm, Magfit EX600W, Aich Steel, Tokai); **b** cup yoke-type (ø4.0 mm × 1.3 mm, Hicorex Slim 4013, Hitachi Metals, Tokyo)

dental cement, is also on the market. In the case of an implant, the keeper is fixed on the abutment of the implant [1–3].

Figure 6 shows the attractive force of the cup yoke-type magnetic attachment and the sandwich yoke-type magnetic attachment. The retention forces of the 3.8 mm × 2.8 mm × 1.8 mm sandwich yoke-type magnetic attachment and the ø4.0 mm × 1.3 mm cup yoke-type magnetic attachment are 5.9N and 7.2N, respectively. Both the retention forces decrease considerably with an increase in the gap between the magnetic assembly and the keeper. Proprietary companies market attachment designs based upon reported retention force ranges of 2.4–11.8N for the magnetic assembly of Φ2.5 × t1.3–Φ5.5 × t1.3 mm in Japan [1–3].

4 Corrosion resistance

Dental magnetic attachments are usually composed of ferritic and austenitic stainless steels, such as SUS 444, SUS XM27, SUS 447J1, and SUS 316L, as described earlier. When dental magnetic attachments work in an oral cavity, magnetic assemblies and keepers contacted root caps made of dental precious alloys in a corrosive environment. Keepers are often heated with investment for cast bonding to sensitizing temperature of stainless steel for corrosion. The effects of the heat history on corrosion must be investigated [20]. A dental magnetic attachment must have excellent corrosion resistance when in contact with oral fluids.

Figure 7 shows ions released from the stainless steels of SUS 444, SUS XM27, SUS 447J1, and SUS 316L in a 1% lactic acid solution and 0.9% NaCl at 37°C for 7 days. Fe ions were released from all stainless steels. A small number of Ni ions were released only from SUS 316L in the lactic acid and NaCl solutions. The total number of released ions decreased in both solutions as the concentration of Cr contained in the ferritic stainless steels increased, and the number of ions from SUS 447J1 was the smallest. The total number of ions released from SUS 316L was smaller than that from SUS 444, which had almost the same content of Cr as SUS 316L [21–25].

The rest potentials of each stainless steel after 24 h remained near 0.17 V and showed no significant difference. On the other hand, the rest potentials of dental precious alloys were significantly higher than those of stainless steels, except those of the Au–Ag–Pd alloy. The rest potentials of the Au–Ag–Pd alloy for about 5 h from right after immersion were higher than those of stainless steels; however, after that, those potentials reversed each other [21, 22].

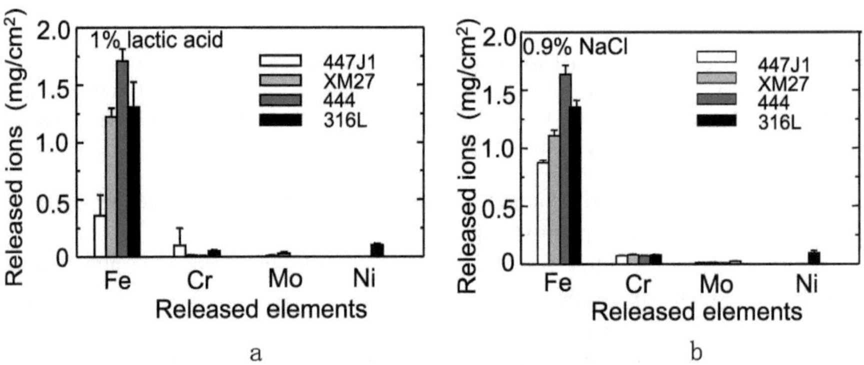

Fig. 7. Ions released from stainless steels in a 1% lactic acid solution and a 0.9% NaCl solution at 37°C for 7 days. **a** 1% lactic acid solution; **b** 0.9% NaCl solution

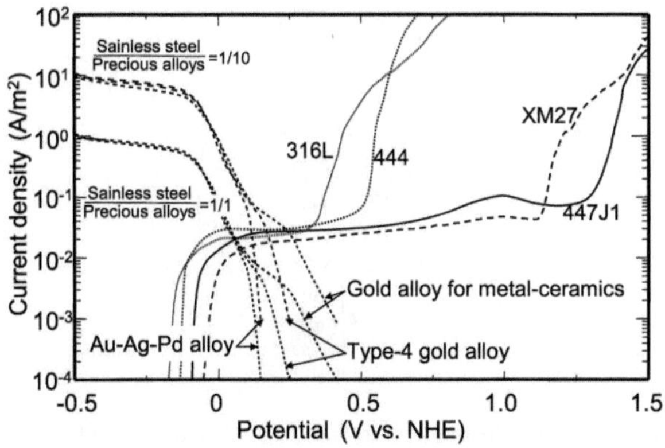

Fig. 8. Corrosion potentials of stainless steels in contact with precious alloys that are electrochemically obtained from their anodic and cathodic polarization curves in a 0.9% NaCl solution at 37°C

According to the anodic polarization curves (Fig. 8), the pitting potentials of the stainless steels were in proportion to their Cr concentrations [21, 22]. SUS 316L showed the lowest value of the pitting potential among the stainless steels in the NaCl solution. However, all rest potentials stayed within the passive regions of the stainless steels.

The cathodic polarization curves of the precious alloys (Au–Ag–Pd alloy, type-4 gold alloy, and gold alloy for metal-ceramics) and the anodic polarization curves of the stainless steels are summarized, including the current density multiplied by ten times the cathodic polarization curves, in Fig. 8 [21, 22]. The corrosion current and corrosion potential of each stainless steel in contact with the precious alloys can be electrochemically obtained from the intersection points of the cathodic and anodic polarization curves.

The corrosion potentials of each stainless steel are maintained within the passive region, and the value of the corrosion potential is sufficiently lower than that of the pitting potentials even when in contact with precious alloys at a surface area ratio of (stainless steel)/(precious alloy) = 1/1 [21, 22]. However, as the surface area of the precious alloys grows at a ratio of 1/10, the corrosion potential also rises.

According to the elution test, the ferritic stainless steels used for the yokes and keepers mainly released Fe ions, and their number was extremely small and similar to that of the Cu ions released from gold alloys. The corrosion potentials of each stainless steel remained within the passive region even when the steel was in contact with the precious alloys. Although the pitting potential of SUS316L used for the magnetic shield was the lowest of those of stainless steels (0.35 V), it was higher than the corrosion potential of stainless steels in contact with precious alloys. These findings indicate that the ferritic stainless steels used in dental magnetic attachments have superior corrosion resistance to SUS316L, which is agreed to have high reliability as biomedical stainless steel, and that they can conceptually overcome galvanic corrosion when they are in contact with dental precious alloys. Therefore,

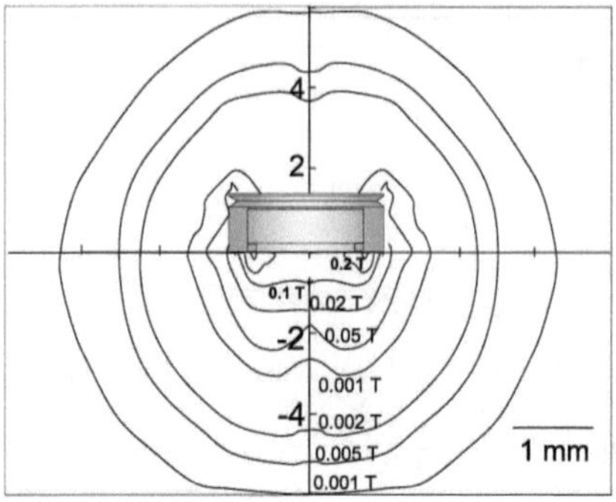

Fig. 9. Magnetic flux density distribution around the magnetic assembly of a dental magnetic attachment

these stainless steels could be safely used as magnetic yokes or keepers in the oral cavity even when in contact with precious alloys. If these stainless steels are in contact with precious alloys with a large surface area, however, the corrosion potential of the stainless steels also rises. It is best to choose dental alloys whose corrosion potential does not increase when the alloys are used for root caps with a large surface area [21, 22].

5 The effects of the static magnetic field on hard tissue

A dental magnetic attachment was implanted in rabbit tibia to investigate the influence of the static magnetic flux on hard tissue. Figure 9 shows the magnetic flux density distribution around the magnetic assembly of the dental magnetic attachment that was implanted in the rabbit tibia. The magnetic assembly has a locally high magnetic flux density of 0.5 T.

Osteogenesis around the magnetic assembly with magnetization was superior to that around the magnetic assembly without magnetization at 2 weeks of implantation. After 12 weeks of implantation, no differences were observed between the osteogenesis around the magnetic assembly with and without magnetization. Although further research is necessary, it appears that the osteoblast was promoted by the stimulation of the static magnetic field, and that the static magnetic field has good biocompatibility.

6 Conclusion

The corrosion resistance and biocompatibility of dental magnetic attachments mainly made in Japan have been described with regard to the structure, retention force, corrosion resistance, and biocompatibility. The ferritic stainless steels composing dental magnetic attachments showed excellent corrosion resistance in the elution test and electrochemical evaluations. Although further research is necessary, it appears that the static magnetic field stimulates and promotes the formation of osteoblasts and that the static magnetic field has good biocompatibility. Many kinds of dental magnetic attachments are produced and sold in Europe and the USA. Research on the dental magnetic attachments made in Europe and the USA will continue.

References

1. Okuno O (1994) Tohoku Univ Dent J (in Japanese) 13:1–10
2. Okuno O (1993) NKZKAU (in Japanese) 32(1):6–13
3. Okuno O (2007) J J Dent Mater (in Japanese) 26:291–300
4. Jackson TR (1986) JOMI 1:81–91
5. Sasaki H, Kinouchi Y (1976) Pract Prosthodent (in Japanese) 9:77–82
6. Sasaki H, Shiota M, Tsuda S, et al (1976) Pract Prosthodent (in Japanese) 9:229–234.
7. Tsutsui H, Kinouchi Y, Sasaki H, et al (1979) J Dent Res 58:1597–1606
8. Yohsuke K, Tomiyuki U, Tsutsui H, et al (1981) J Dent Res 60:50–58
9. Gillings B (1981) J Prosthet Dent 45:484–491
10. Caputo AA, Pezzoli M (1986) J Prosthet Dent 56:104–106
11. Okuno O, Ishikawa S, Iimuro FT, et al (1991) Dent Mater J 10:172–184
12. Tanaka Y, Honkura Y, Arai K, et al (1992) J J Mag Dent (in Japanese) 1:23–29
13. Tanaka Y, Honkura Y, Furushima Y, et al (1991) J Jpn Prosthodont Soc (in Japanese) 35:167–177
14. Okuno O, Takada Y, Nakamura K, et al (1993) J J Magn Dent (in Japanese) 2:1–10
15. Kitsugi A, Okuno O, Nakano T, et al (1992) Dent Mater J 11:119–129.
16. Stewart BL, Edwards AO (1983) J Prosthet Dent 49:28–34
17. Körber KH (1986) Ishiyaku Publishers Inc (translated in Japanese), pp 58–59
18. Takada Y, Okuno O (1995) J J Mag Dent (in Japanese) 4:10–18
19. Takada Y, Okuno Y (1994) J J Magn Dent (in Japanese) 3:14–22
20. Takada Y, Okuno O (2005) Dent Mater J 24:391–397
21. Okuno O, Takada Y, Kikuchi M, et al (2002) In: Proceedings of the 1st International Conference. J J Magn Dent 11:100–104
22. Takada Y, Takahashi N, Okuno O (2007) J J Magn Dent 16(in press)
23. Okuno O, Iimuro FT, Nakano T, et al (1992) J J Magn Dent (in Japanese) 1:14–22
24. Iimuro FT, Yoneyama T, Okuno O (1992) Dent Mater J 12:136–144
25. Haoka K, Kanno T, Takada Y, et al (2000) Dent Mater J 19:116–124

Symposium III: Biomaterials: Scaffolds for oral tissue regeneration

Interface Oral Health Science 2007: 117–128

Developmental genetics of the dentition

Wei-Yuan Yu and Paul Sharpe*
Department of Craniofacial Development, Dental Institute, King's College London, Guy's Hospital, London SE1 9RT, UK
*paul.sharpe@kcl.ac.uk

Abstract. Much is now known of the molecular signalling and genetic determinants for tooth development. Here, we discuss the current understanding of the cellular origin of teeth, the early patterning that determines where a tooth is positioned in the mouth, how the tooth initiates, and the number of the teeth that will form in mouse. The genes that are important for regulating the tooth shape are also discussed.

Key words. tooth development, morphogenesis, dentition

Introduction

Teeth are complex organs that develop by an increasingly well-understood temporal series of epithelial/mesenchymal interactions. Developmentally, teeth are unusual in two significant aspects: (1) Although considered as a single organ, teeth can have completely different shapes, even in the same species. (2) Teeth do not exist or function as single entities; they are organized as "dentitions" where shapes occupy precise relative locations. The position, number, and shape of teeth in a dentition are thus interlinked. This is dramatically illustrated by the clinical observation that despite the existence and reports of numerous dental anomalies, there is only one documented case of an individual with an inappropriate tooth shape [1]. The genetic regulation of position and shape determination must therefore be highly controlled and integrated during development.

Tooth position—where in the embryo?

In all animals, teeth only develop at two sites, the oral cavity and the pharynx. The existence of teeth in regions outside the jaw margins such as the palate, tongue, and pharynx is restricted to certain fishes and amphibians. In mammals, teeth are exclusively found on the margins of the jaws. All teeth comprise cells from at least two distinct origins, epithelium and mesenchyme. As far as we know the mesenchyme is always of neural crest origin, whereas the epithelium can be of ectodermal or endodermal origin. Epithelial cells are in contact with the neural crest-derived mesenchyme in many locations in the embryo so there must be some mechanism that either activates odontogenesis at specific sites or represses odontogenesis everywhere except the oral cavity (and pharynx). In mammals for example, the 2nd branchial arch is composed of the same cell derivatives as the 1st branchial arch, ectoderm and cranial neural crest-derived mesenchyme (Fig. 1A). Teeth, however,

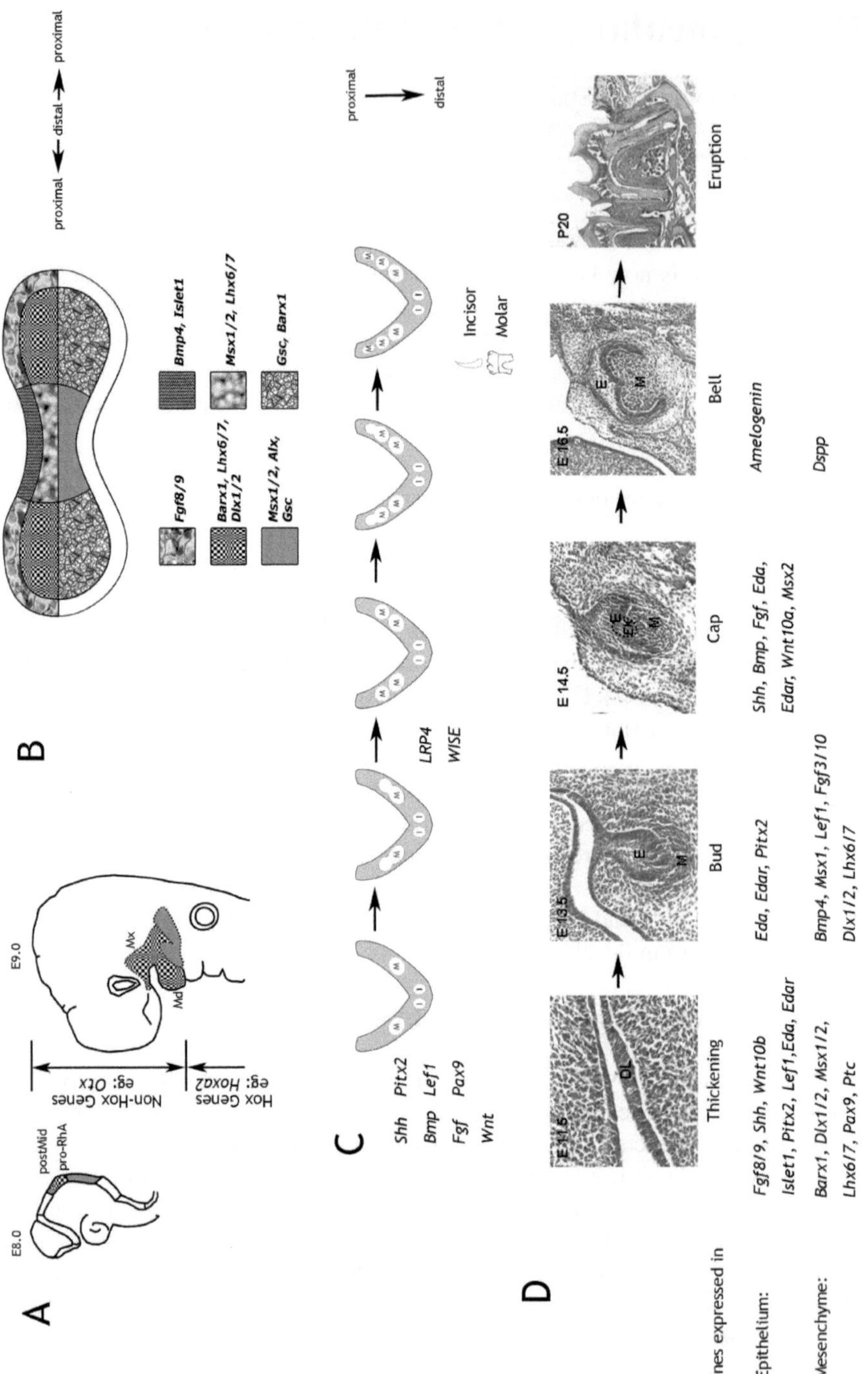
A
E8.0
postMid
pro-RhA
E9.0
Non-Hox Genes
eg: Otx
Hox Genes
eg: Hoxa2
Mx
Md
B
proximal ← distal → proximal
Fgf8/9
Barx1, Lhx6/7, Dlx1/2
Msx1/2, Alx, Gsc
Bmp4, Islet1
Msx1/2, Lhx6/7
Gsc, Barx1
C
Shh Pitx2
Bmp Lef1
Fgf Pax9
Wnt
LRP4
WISE
Incisor
Molar
proximal
distal
D
E11.5
DL
E13.5
E
M
E14.5
E
EK
M
E16.5
E
M
P20
Thickening
Bud
Cap
Bell
Eruption
Genes expressed in
Epithelium:
Fgf8/9, Shh, Wnt10b
Islet1, Pitx2, Lef1, Eda, Edar
Eda, Edar, Pitx2
Shh, Bmp, Fgf, Eda,
Edar, Wnt10a, Msx2
Amelogenin
Mesenchyme:
Barx1, Dlx1/2, Msx1/2,
Lhx6/7, Pax9, Ptc
Bmp4, Msx1, Lef1, Fgf3/10
Dlx1/2, Lhx6/7
Dspp

Fig. 1. The embryonic development of the mammalian tooth. **A** The tooth mesenchyme is derived from the cranial neural crest. In mouse, at *E8.0*, neural crest cells that migrate from the posterior midbrain (*postMid*) and pro-rhombomere A (*pro-RhoA*) will later form the ectomesenchyme of the 1st branchial arch including the maxillary and mandibular tooth mesenchyme. Stem cells can be used experimentally to replace the cranial neural crest-derived mesenchyme. Hox genes such as Hoxa2 are expressed in branchial arches except in the 1st branchial arch. Hox genes are not needed in tooth morphogenesis. **B** Proximo-distal patterning of the 1st branchial arch gives tooth its identities. From E10.0, signals from the oral epithelium, such as Fgfs and Bmp, pattern the 1st branchial arch into different domains; these signals will induce different mesenchymal homeobox genes and further mark the branchial arch into proximal and distal regions. The combined expression of the homeobox transcription factors in the ectomesenchyme cells constitutes the "Odontogenic Homeobox Code". The proximo-distal information determines the complex (*proximal*) and simple (*distal*) tooth crown shape. **C** Factors that determine the tooth position and tooth numbers. *Shh*, *Bmp4*, *Fgf8*, *Wnt10b*, *Pitx2* and *Lef1* are expressed in dental epithelium and *Pax9* in the dental mesenchyme at the sites of tooth initiation. The tooth initiation is only partly responsible for controlling the tooth number. The mouse 1st molar (*M*) is developed from more than one single epithelial bud, and the 2nd and 3rd molars are formed by the outgrowth of the primary molar buds. The low-density lipoprotein LRP4 and Ectodin/WISE are involved in controlling the breaking of the expanding tooth epithelium into separate teeth [Sharpe lab, unpublished results]. **D** Morphology and genes expressed in the epithelium (*E*) and mesenchyme (*M*) at different stages of the molar tooth development. The thickening of the dental lamina (*DL*) at the oral epithelium can be seen at *E11.5*. Later at bud stage, the tooth epithelium invaginates into the neural crest derived ectomesenchyme, and the mesenchyme cells condensed around the epithelium to form the dental papilla. Enamel knot (*EK*) forms at around *E14.5* cap stage and later, secondary EKs forms at bell stage. EKs are known to determine the tooth crown shape. The odontoblasts and ameoloblasts at bell stage interact within the adjacent mesenchymal and epithelial cell layers and secrete dentin and enamel matrix proteins, respectively. The 1st molar of a mouse erupts at around 20 days after birth. Several genes of each stage expressed in the epithelium and mesenchyme are listed only develop on the 1st branchial arch in mammals, and never on the 2nd. This restriction to certain facial tissues (1st arch and frontonasal process) is, however, not linked to the same process that control craniofacial skeletal development. Unlike the development of the cranial skeleton that is spatially regulated by Hox genes, teeth develop by local interactions that are independent of the Hox status of the mesenchyme cells. Thus when 2nd arch mesenchyme is recombined with 1st arch epithelium (at the appropriate developmental state), teeth can form [2]. Similarly, whereas ectopic expression of Hoxa2 in 1st arch mesenchyme disrupts skeletal morphogenesis, teeth develop normally [2]. These results illustrate that it is the orofacial epithelium that contains the instructions to initiate odontogenesis. The developing oral and dental epithelial cells express a large variety of signalling proteins, representative of all the major groups of cell signals: Shh, BMP's, FGF's, Wnt's, and TNF's (Fig. 1b–d). Clearly there is some temporal/spatial combination of these that gives this epithelium its unique odontogenic inducing capacity. Significantly, however this capacity is only transient, existing for a very short time window in development (E10–11 in mouse embryos). Once the epithelium has induced an odontogenic response in the mesenchyme, the epithelium loses its inductive properties and becomes "passive". The mesenchyme that has been exposed to the inductive influence of the oral/dental epithelium itself becomes the inducer and is capable of inducing odontogenesis in non-oral epithelium [3]. This is the basis of the first exchange of reciprocal epithelial/mesenchymal interactions.

Early, now classic, tissue recombination experiments, established that although neural crest-derived mesenchymal cells can form teeth, mesodermal cells cannot [4, 5]. This has long been interpreted as showing that the combination of oral epithelium and cranial neural crest-derived mesenchyme cells that is unique to the oral cavity is what determines where in the embryo teeth will develop. More recently however this idea has been challenged and what appears to make cranial neural crest cells "special" in this context and different to mesoderm is their stem cell-like properties. This has been dramatically demonstrated by showing that the cranial neural crest-derived mesenchyme can be replaced in tooth development with stem cells from other sources, including adult bone marrow for examples [6].

The phenomenon of reciprocal signalling in early tooth development and the identification of adult stem cell sources that can participate in tooth development have potentially major consequences for tooth tissue engineering. The developmental biology shows us that at particular times in early tooth development the epithelium and mesenchyme cells need not be derived from the embryonic oral cavity in order to participate in tooth development. Thus if teeth can form from non-oral cell sources experimentally, this could provide a way of providing teeth for replacement. As long as the mesenchymal cells have stem cell-like properties and can respond to the unique signal combinations from the epithelium, tooth development can be initiated. Clues to the identities of the inducing signals in tooth development have come from experiments where recombinant signalling proteins are added exogenously to explant cultures of developing 2nd branchial arches from mouse embryos. Such cultures cannot form dental tissues, but in the presence of just one exogenous protein, BMP4, differentiation of odontogenic cells can be induced [7].

Tooth position—where in the mouth?

Animals have what appears to be an infinite variety of dentitions, reflecting the importance of tooth patterns in feeding and evolution. These do not simply involve different tooth shapes but also edentulous areas where the dental pattern requires no teeth. Rodents and mice in particular are prime examples of this. The positions in the oral cavity where tooth formation is initiated must therefore be genetically determined. A number of signalling proteins (Shh, BMP4, FGF8, Wnt10b) and transcription factors (Pitx2, Lef1) are expressed in dental epithelium at the sites of tooth initiation (Fig. 1c,d). These signalling pathways clearly interact in some way because overactivation of any of the pathways can result in the formation of extra teeth [8–22]. If localized expression of these molecules determines where a tooth primordium will form, it follows that mechanisms must exist to limit their expression to the sites of tooth formation. The best studied of these pathways in this respect is Shh. Shh expression is restricted to dental epithelium from E11.5 in mice. Ptc receptors are present in both the epithelium and mesenchyme at this time, indicating a dual role for Shh signalling. In the epithelium, Shh stimulates local proliferation of dental cells to form a tooth bud [23]. Inhibition of Shh at this time prevents bud formation and ectopic Shh produces ectopic buds [8, 17, 24]. Shh signalling to the mesenchyme is likely to form part of the epithelial–mesenchymal interactions at this time. The mechanisms that restrict Shh to the sites of tooth formation are not fully understood. However, the maintenance of these expression domains is regulated by the repressive action of Wnt7b in non-dental epithelium [25].

Lef1 expression is localized to dental epithelium at the initiation stage and is required for tooth development. However, in the absence of Lef1, tooth buds do form [18, 26–28]. Similarly, Pax9 expression in the dental mesenchyme at the sites of tooth initiation has been suggested to play a role in determining tooth position. However, in the absence of Pax9, tooth buds still form at the correct position [29, 30].

Tooth number—initiation

The control of tooth position and tooth number are in part of the same process. The mechanisms described earlier that determine the sites of tooth initiation are in turn partly responsible for tooth number. Simplistically, tooth number equates to the sites of initiation. However, this view is a gross over simplification because formation of the dentition is complicated by two additional features: (1) Teeth often develop from multiple buds. (2) Formation of many teeth such as much of human dentition and mouse 2nd and 3rd molars does not involve an independent "initiation" event (see below and Fig. 1c). Detailed 3D reconstruction of early tooth bud epithelium has revealed a surprisingly complex pattern of bud formation [31, 32]. Mouse 1st molars for example develop from four or five epithelial "swellings", the most mesial of which is excluded from the final tooth bud and rapidly undergoes apoptosis. This mesial swelling is believed to be a vestigial bud that is a surviving remnant of the tooth (premolar) that has been lost in mouse evolution [33–35].

Tooth number—outgrowth of the dental epithelium

In mice, initiation of tooth buds as described earlier produces primordia for the incisors and 1st molars only. The 2nd molars develop slightly later as outgrowths of the 1st molar thickenings and the 3rd molars similarly develop as the outgrowth of the 2nd molar buds (Fig. 1c) [36–38]. Tooth number is thus also controlled after the primary initiation process. Developmentally, little attention has been paid to understanding the mechanisms of secondary outgrowth but this is an important topic because much of the human dentition develops via this process. There are two key processes of secondary outgrowth that need to be understood. What determines the position of the outgrowth and what ensures the breakdown of the epithelium to create two separate tooth primordial? At present nothing is known about how the outgrowth position is determined. The physical separation of the tooth germs has however begun to be understood from the identification of mutant mice with fused molars. Three mouse mutants have been reported to have fused molars; hypomorphic alleles of the low density lipoprotein receptor, Lrp4, the BMP/Wnt antagonist Ectodin/Wise, and K14 Cre-mediated deletion of Smo [24, 39, 40]. In the absence of Ectodin/Wise, BMP signalling is increased in the regions connected to the 1st and 2nd molars. As a consequence of this, Shh expression is decreased, and apoptosis that is normally activated to breakdown the epithelial connections is repressed. This same series of molecular and phenotypic changes also occurs in Lrp4 mutant mice, and sequence evidence shows that Ectodin/Wise binds to the extracellular domain of Lrp4 (Unpublished results, Sharpe's lab). This pathway has a general conserved role in epithelial organization because similar phenotypes and molecular changes are also observed in the palate in these mutants.

The canonical Wnt signalling pathway also plays an important role in tooth number. Expression of a stabilized form of β-catenin in the epithelium of mouse tooth buds results in the development of extra teeth [41].

Tooth morphogenesis—determination

The earliest event in tooth development is the determination of tooth crown shape. Thus by the time a bud starts to develop, the eventual shape the tooth crown will form is irreversibly specified. The ectomesenchyme of the early jaw primordial expresses a number of homeobox genes in different spatially restricted domains (Fig. 1b). Thus Dlx and Barx1 genes are expressed proximally, whereas Msx1/2 is expressed more distally [42]. The Lim-domain genes Lhx6/7 are co-expressed in oral ectomesenchyme but not in more aboral mesenchyme [11]. Ectomesenchyme cells at different positions thus express different combinations of homeobox and other transcription factors. These different complements of genes provide the cells with positional information which they interpret and differentiate accordingly to the "Odontogenic Homeobox Code" [43–45]. For mammalian dentitions, the proximo-distal information equates to complex (proximal) and simple (distal) tooth crown shapes. Thus, loss of function of Dlx (1&2) or Barx1 affects molar (proximal) morphogenesis. Ectopic expression of Barx1 in distal ectomesenchyme cells results in incisors developing with a molar crown shape [12]. These proximo-distal expression patterns are transient, and by E11.5–12 in mice the expression has disappeared, by which time tooth crown shape has been determined. From E12, the expression of most of these genes is re-initiated in the mesenchyme condensing around the tooth buds and for all genes except Barx1, this expression is independent of tooth type. Barx1 expression remains restricted to molar tooth mesenchyme throughout development.

The proximo-distal expression domains are regulated by signals from the oral epithelium. For example, the restriction of Barx1 expression to proximal mesenchyme is established by antagonistic signals from the oral epithelium. FGF8, expressed proximally, positively regulates Barx1; whereas BMP4 expressed more distally negatively regulates Barx1 expression [12]. Thus a simple proximo-distal prepattern of signals present in the oral epithelium is transferred to the mesenchyme in the form of different domains of homeobox gene expression (Fig. 1b).

Tooth crown morphogenesis-epithelial folding

Prior to the bud stage, tooth primordial are histologically indistinguishable from each other. From the bud stage onwards, the morphogenetic information present in the mesenchyme cells as a result of the combination of homeobox genes they expressed previously manifests itself by the folding of the tooth epithelium to produce cusps. The molecular links between early homeobox gene expression and the later control of epithelial folding are not understood. However, the NF-κB pathway has been identified as having an important role in molar tooth epithelial folding.

The first clues to the identification of this specific developmental role for NF-κB came from the identification of mutations in patients with hypohidrotic ectodermal dysplasia (HED), a disorder that affects ectodermal organ development. In HED, teeth are highly abnormal with molars in particular often having very little cusp formation. The first gene to be identified in HED patients was EDA (ectodysplasin), a secreted protein belonging to the TNF-family of ligands. Mutations in Eda were subsequently identified in a spontaneous mouse mutant, Tabby [46, 47]. Other mutated genes in HED patients and in two other mouse mutants, Downless and Crinkled, were found to be the Eda receptor, Edar, and an intracellular adaptor protein, Edaradd [48–52]. This ligand–receptor interaction occurs in the epithelial cells of bud stage tooth germs and is required for correct epithelial folding to form cusps. Moreover, mis-activation of the pathway in mice transgenic for a ligand-independent, activated form of Edar, produce molars with extra cusps [52].

Since TNFs are well-established regulators of NF-κB, it was not surprising to observe a severe loss of cusps in mice with suppressed NF-κB activity. Moreover, known intracellular adaptors of the pathway such as Traf's could also be identified in this tooth-specific pathway [53]. Mutations in Traf6 for example produce a molar cusp phenotype, similarly but more severe than loss of NF-κB [54]. The level of NF-κB activity, regulated by Eda signalling in bud stage epithelium, appears to be direct epithelial folding to produce cusps. Rather how this activity is connected to the earlier determination events by genes such as Barx1 remains to be established.

References

1. Kantaputra PN, Gorlin, RJ (1992) Double dens invaginatus of molarized maxillary central incisors, premolarization of maxillary lateral incisors, multituberculism of the mandibular incisors, canines and first premolar, and sensorineural hearing loss. Clin Dysmorphol 1:128–136
2. James CT, Ohazama A, Tucker AS, et al (2002) Tooth development is independent of a Hox patterning programme. Dev Dyn 225:332–335
3. Ferguson CA, Tucker AS, Sharpe PT (2000) Temporospatial cell interactions regulating mandibular and maxillary arch patterning. Development 127:403–412
4. Lumsden AG (1988) Spatial organization of the epithelium and the role of neural crest cells in the initiation of the mammalian tooth germ. Development 103(Suppl):155–169
5. Mina M, Kollar EJ (1987) The induction of odontogenesis in non-dental mesenchyme combined with early murine mandibular arch epithelium. Arch Oral Biol 32:123–127
6. Ohazama A, Modino SA, Miletich I, et al (2004) Stem-cell-based tissue engineering of murine teeth. J Dent Res 83:518–522
7. Ohazama A, Tucker A, Sharpe PT (2005) Organized tooth-specific cellular differentiation stimulated by BMP4. J Dent Res 84:603–606
8. Hardcastle Z, Mo R, Hui CC, et al (1998) The Shh signalling pathway in tooth development: defects in Gli2 and Gli3 mutants. Development 125:2803–2811
9. Vainio S, Karavanova I, Jowett A, et al (1993) Identification of BMP-4 as a signal mediating secondary induction between epithelial and mesenchymal tissues during early tooth development. Cell 75:45–58

10. Bei M, Maas R (1998) FGFs and BMP4 induce both Msx1-independent and Msx1-dependent signaling pathways in early tooth development. Development 125:4325–4333
11. Grigoriou M, Tucker AS, Sharpe PT, et al (1998) Expression and regulation of Lhx6 and Lhx7, a novel subfamily of LIM homeodomain encoding genes, suggests a role in mammalian head development. Development 125:2063–2074
12. Tucker AS, Matthews KL, Sharpe PT (1998) Transformation of tooth type induced by inhibition of BMP signaling. Science 282:1136–1138
13. Tucker AS, Yamada G, Grigoriou M, et al (1999) Fgf-8 determines rostral-caudal polarity in the first branchial arch. Development 126:51–61
14. Neubuser A, Peters H, Balling R, et al (1997) Antagonistic interactions between FGF and BMP signaling pathways: a mechanism for positioning the sites of tooth formation. Cell 90:247–255
15. Sarkar L, Sharpe PT (1999) Expression of Wnt signalling pathway genes during tooth development. Mech Dev 85:197–200
16. Dassule HR, McMahon AP (1998) Analysis of epithelial-mesenchymal interactions in the initial morphogenesis of the mammalian tooth. Dev Biol 202:215–227
17. Dassule HR, Lewis P, Bei M, Maas R, et al (2000) Sonic hedgehog regulates growth and morphogenesis of the tooth. Development 127:4775–4785
18. Kratochwil K, Dull M, Farinas I, et al (1996) Lef1 expression is activated by BMP-4 and regulates inductive tissue interactions in tooth and hair development. Genes Dev 10: 1382–1394
19. Vadlamudi U, Espinoza HM, Ganga M, et al (2005) PITX2, beta-catenin and LEF-1 interact to synergistically regulate the LEF-1 promoter. J Cell Sci 118:1129–1137
20. Lu MF, Pressman C, Dyer R, et al (1999) Function of Rieger syndrome gene in left-right asymmetry and craniofacial development. Nature 401:276–278
21. St Amand TR, Zhang Y, Semina EV, et al (2000) Antagonistic signals between BMP4 and FGF8 define the expression of Pitx1 and Pitx2 in mouse tooth-forming anlage. Dev Biol 217:323–332
22. Yamashiro T, Zheng L, Shitaku Y, et al (2007) Wnt10a regulates dentin sialophosphoprotein mRNA expression and possibly links odontoblast differentiation and tooth morphogenesis. Differentiation 75:452–462
23. Cobourne MT, Hardcastle Z, Sharpe PT (2001) Sonic hedgehog regulates epithelial proliferation and cell survival in the developing tooth germ. J Dent Res 80:1974–1979
24. Gritli-Linde A, Bei M, Maas R, et al (2002) Shh signaling within the dental epithelium is necessary for cell proliferation, growth and polarization. Development 129:5323–5337
25. Sarkar L, Cobourne M, Naylor S, et al (2000) Wnt/Shh interactions regulate ectodermal boundary formation during mammalian tooth development. Proc Natl Acad Sci USA 97: 4520–4524
26. Sasaki T, Ito Y, Xu X, et al (2005) LEF1 is a critical epithelial survival factor during tooth morphogenesis. Dev Biol 278:130–143
27. Zhou P, Byrne C, Jacobs J, et al (1995) Lymphoid enhancer factor 1 directs hair follicle patterning and epithelial cell fate. Genes Dev 9:700–713
28. van Genderen C, Okamura RM, Farinas I, et al (1994) Development of several organs that require inductive epithelial-mesenchymal interactions is impaired in LEF-1-deficient mice. Genes Dev 8:2691–2703
29. Peters H, Neubuser A, Kratochwil K, et al (1998) Pax9-deficient mice lack pharyngeal pouch derivatives and teeth and exhibit craniofacial and limb abnormalities. Genes Dev 12: 2735–2747
30. Mostowska A, Kobielak A, Trzeciak WH (2003) Molecular basis of non-syndromic tooth agenesis: mutations of MSX1 and PAX9 reflect their role in patterning human dentition. Eur J Oral Sci 111:365–370
31. Peterkova R, Peterka M, Vonesch JL, et al (1995) Contribution of 3-D computer-assisted reconstructions to the study of the initial steps of mouse odontogenesis. Int J Dev Biol 39:239–247

32. Lesot H, Peterkova R, Viriot L, et al (1998) Early stages of tooth morphogenesis in mouse analyzed by 3D reconstructions. Eur J Oral Sci 106(Suppl 1):64–70
33. Viriot L, Peterkova R, Peterka M, et al (2002) Evolutionary implications of the occurrence of two vestigial tooth germs during early odontogenesis in the mouse lower jaw. Connect Tissue Res 43:129–133
34. Keranen SV, Aberg T, Kettunen P, et al (1998) Association of developmental regulatory genes with the development of different molar tooth shapes in two species of rodents. Dev Genes Evol 208:477–486
35. Keranen SV, Kettunen P, Aberg T, et al (1999) Gene expression patterns associated with suppression of odontogenesis in mouse and vole diastema regions. Dev Genes Evol 209: 495–506
36. Lesot H, Vonesch JL, Peterka M, et al (1996) Mouse molar morphogenesis revisited by three-dimensional reconstruction. II. Spatial distribution of mitoses and apoptosis in cap to bell staged first and second upper molar teeth. Int J Dev Biol 40:1017–1031
37. Peterkova R, Lesot H, Vonesch JL, et al (1996) Mouse molar morphogenesis revisited by three dimensional reconstruction. I. Analysis of initial stages of the first upper molar development revealed two transient buds. Int J Dev Biol 40:1009–1016
38. Viriot L, Peterkova R, Vonesch JL, et al (1997) Mouse molar morphogenesis revisited by three-dimensional reconstruction. III. Spatial distribution of mitoses and apoptoses up to bell-staged first lower molar teeth. Int J Dev Biol 41:679–690
39. Johnson EB, Hammer RE, Herz J (2005) Abnormal development of the apical ectodermal ridge and polysyndactyly in Megf7-deficient mice. Hum Mol Genet 14:3523–3538
40. Kassai Y, Munne P, Hotta Y, et al (2005) Regulation of mammalian tooth cusp patterning by ectodin. Science 309:2067–2070
41. Jarvinen E, Salazar-Ciudad I, Birchmeier W, et al (2006) Continuous tooth generation in mouse is induced by activated epithelial Wnt/beta-catenin signaling. Proc Natl Acad Sci USA 103:18627–18632
42. Tucker A, Sharpe P (2004) The cutting-edge of mammalian development; how the embryo makes teeth. Nat Rev Genet 5:499–508
43. Thomas BL, Sharpe PT (1998) Patterning of the murine dentition by homeobox genes. Eur J Oral Sci 106(Suppl 1):48–54
44. Thomas BL, Tucker AS, Ferguson C, et al (1998) Molecular control of odontogenic patterning: positional dependent initiation and morphogenesis. Eur J Oral Sci 106(Suppl 1): 44–47
45. Tucker AS, Sharpe PT (1999) Molecular genetics of tooth morphogenesis and patterning: the right shape in the right place. J Dent Res 78:826–834
46. Kere J, Srivastava AK, Montonen O, et al (1996) X-linked anhidrotic (hypohidrotic) ectodermal dysplasia is caused by mutation in a novel transmembrane protein. Nat Genet 13: 409–416
47. Srivastava AK, Pispa J, Hartung AJ, et al (1997) The Tabby phenotype is caused by mutation in a mouse homologue of the EDA gene that reveals novel mouse and human exons and encodes a protein (ectodysplasin-A) with collagenous domains. Proc Natl Acad Sci USA 94:13069–13074
48. Pispa J, Mustonen T, Mikkola ML, et al (2004) Tooth patterning and enamel formation can be manipulated by misexpression of TNF receptor Edar. Dev Dyn 231:432–440
49. Laurikkala J, Mikkola M, Mustonen T, et al (2001) TNF signaling via the ligand-receptor pair ectodysplasin and edar controls the function of epithelial signaling centers and is regulated by Wnt and activin during tooth organogenesis. Dev Biol 229:443–455
50. Tucker AS, Headon DJ, Schneider P, et al (2000) Edar/Eda interactions regulate enamel knot formation in tooth morphogenesis. Development 127:4691–4700
51. Courtney JM, Blackburn J, Sharpe PT (2005) The Ectodysplasin and NFkappaB signalling pathways in odontogenesis. Arch Oral Biol 50:159–163

52. Tucker AS, Headon DJ, Courtney JM, et al (2004) The activation level of the TNF family receptor, Edar, determines cusp number and tooth number during tooth development. Dev Biol 268:185–194
53. Ohazama A, Courtney JM, Sharpe PT (2003) Expression of TNF-receptor-associated factor genes in murine tooth development. Gene Expr Patterns 3:127–129
54. Ohazama A, Courtney JM, Tucker AS, et al (2004) Traf6 is essential for murine tooth cusp morphogenesis. Dev Dyn 229:131–135

Interface Oral Health Science 2007: 129–138

Involvement of PRIP, a new signaling molecule, in neuroscience and beyond oral health science

Masato Hirata*, Takashi Kanematsu, and Akiko Mizokami
Laboratory of Molecular and Cellular Biochemistry, Faculty of Dental Science, Kyushu University, Fukuoka 812-8582, Japan
*hirata1@dent.kyushu-u.ac.jp

Abstract. Investigation of chemically synthesized inositol 1,4,5-trisphosphate analogs has led to the isolation of a novel protein with a molecular size of 130 kDa, characterized as a molecule with domain organization similar to phospholipase C (PLC)-δ1 but lacking enzymatic activity; therefore the molecule was named PRIP (PLC-related, but catalytically inactive protein). Yeast two-hybrid screening of a brain cDNA library identified $GABA_A$ receptor-associated protein (GABARAP) and protein phosphatase 1(PP1), which led us to examine the possible involvement of PRIP in neuroscience, particularly in $GABA_A$ receptor signaling. PRIP knock-out (KO) mice were analyzed for $GABA_A$ receptor function with special reference to the action of benzodiazepines whose target is the γ subunit of the receptors; sensitivity to benzodiazepine was reduced as assessed by biochemical, electrophysiological, and behavioral analyses of KO mice, suggesting the dysfunction of γ2 subunit-containing $GABA_A$ receptors. The mesencephalic trigeminal nucleus, which mediates perceptions from periodontal mechanoreceptors and jaw-closer muscle spindles, receives many synaptic inputs, including those from $GABA_A$ receptors, indicating that PRIP might indirectly be involved in rhythmical jaw movement. In the present article, we summarize our current reach and the perspective of the functional significance of PRIP.

Key words. $GABA_A$ receptor, GABARAP, $Ins(1,4,5)P_3$, mesencephalic trigeminal nucleus

1 Introduction

Isolation of a novel signaling molecule, PRIP

Inositol 1,4,5-trisphosphate [$Ins(1,4,5)P_3$] emerged onto the stage of life science research in 1983 from a long background that had accumulated extensive information regarding techniques and knowledge about phosphoinositides, and thus had been well-prepared by British research [1]. We soon joined this research field and described for the first time the chemical modification of Ins(1,4,5)P_3 in 1985 [2]. The synthesized analog has an azidobenzoyl group at the C-2 position for photoaffinity labeling and causes irreversible inactivation of the receptor protein for Ca^{2+} release, following photolysis. As an extension of this research project, we also synthesized Ins(1,4,5)P_3 affinity matrices which led us to isolate two binding proteins: one was phospholipase C-δ (PLC-δ) and the other was a new molecule with a molecular size of 130 kDa (p130) [3]. Subsequent studies to isolate the cDNA encoding p130 revealed its considerable similarity to the PLC-δ family, but lacking the enzymatic activity, so we designated this molecule as PRIP (PLC-related, but catalytically inactive protein) [4].

2 Binding partners of PRIP

We have extended the project toward finding binding partners besides Ins(1,4,5)P_3 because the cellular signaling pathways are mediated by the relay of molecular interactions of proteins to proteins, nucleotides or lipids. We applied the yeast two-hybrid system to identify proteins that interact with PRIP in order to explore further biological functions. With the unique NH_2-terminal region of PRIP as bait for screening the human brain cDNA library, we isolated two positive clones, one of which was shown to encode the catalytic subunit of protein phosphatase 1α (PP1) [5]. Another clone was found to be GABARAP ($GABA_A$ receptor-associated protein) that was identified as a molecule capable of binding the γ2 subunit of $GABA_A$ receptor and tubulin [6]. In the process of experiments initiated by finding the above-mentioned binding partners, we also noticed that PRIP directly interacts with the β subunit of $GABA_A$ receptors [7] and protein phosphatase 2A (PP2A)

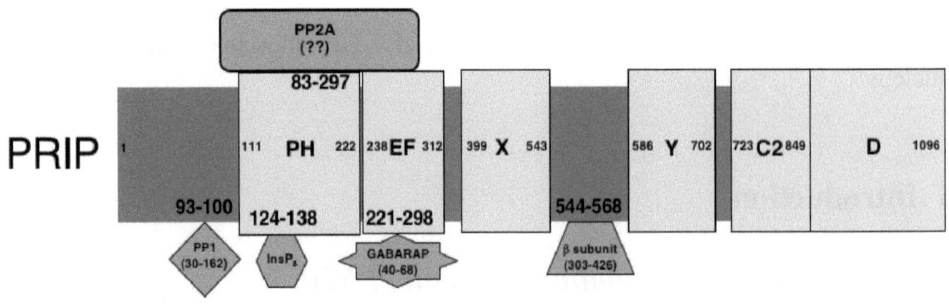

Fig. 1. Binding partners of phospholipase C-related, but catalytically inactive protein (*PRIP*). Molecules binding to PRIP are depicted together with the residues responsible for interaction (adapted with permission from Ref. [27])

[8]. Figure 1 depicts these interactions along with the amino acid residues responsible for the association.

3 Regulation of protein phosphatase activities by PRIP, depending on its phosphorylation

Analysis of the interaction of various regulatory subunits with PP1α has led to the identification of a consensus sequence for binding: K/R–K/R–V/I–X–F [9]. PRIP shares the consensus sequence KKTVSF (residues 92 to 97 in PRIP) although T is inserted, which was shown to bind to PP1α since mutations of 95V or 97F in PRIP to alanine resulted in no interaction [5]. It is possible that PRIP associates with PP1 because it is a substrate for the phosphatase activity of this enzyme. Indeed, PRIP itself was phosphorylated by the activation of cyclic AMP-dependent protein kinase (PKA); however, this explanation for the interaction between PRIP and PP1α is unlikely because phosphorylated PRIP could no longer associate with PP1α [5].

Matrix-assisted laser desorption/ionization time-of-flight mass spectrometry analysis of the phosphopeptides purified by Fe(III) affinity chromatography suggested the peptide (94-TVSFSSMPSEK-104) as a possible phosphopeptide in PRIP, which interestingly overlaps with the PP1α binding site (92-KKTVSF-97) in PRIP. Subsequent mutagenesis and binding studies revealed that residue 94T is critical for regulation of the association/dissociation with PP1 upon phosphorylation [7]. PP1 activity is inhibited by unphosphorylated PRIP, but not by phosphorylated PRIP [5]. On the other hand, the activity of PP2A was not modified by interacting with PRIP [8].

4 Involvement in $GABA_A$ receptor signaling

$GABA_A$ receptor-associated protein was first identified as a candidate molecule for the clustering of $GABA_A$ receptors at the post-synaptic membrane on the basis of its ability to bind to the γ2 subunit of the receptors and tubulin (microtubulus) [6]. However, the original proposal for GABARAP action has been revised based on the findings that GABARAP also binds to *N*-ethylmaleimide sensitive factor (NSF), an essential factor for vesicular fusion. GABARAP has a positive effect on the trafficking of $GABA_A$ receptors to the surface membrane of neurons, in relation to the ability to bind the specific molecules described above [10, 11]. The region of GABARAP responsible for binding to PRIP was clarified to be residues 40–68 [12], which are the same as the intracellular loop of the γ2 subunit of the $GABA_A$ receptor [6], explaining the competition. Therefore, it is expected that PRIP plays an important role in trafficking and/or stabilization of surface $GABA_A$ receptors.

4.1 Involvement in trafficking of $GABA_A$ receptors

To examine the possible role of PRIP interacting with GABARAP in a competitive manner with the γ subunit of $GABA_A$ receptors in the regulation of the receptor transport, we analyzed PRIP knock-out (KO) mice [12, 13] from electrophysiological and behavioral aspects in combination with the pharmacological effects of benzodiazepine (BZ) whose target is the α/γ subunit interface of $GABA_A$ receptors [14]. To examine the alteration of the surface expressed receptor number and composition, ligand-binding assays using [^{3}H]muscimol, a GABA agonist, and [^{3}H]flumazenil, a benzodiazepine antagonist, were performed. [^{3}H]muscimol saturation binding studies revealed about 20% elevation in binding sites (*B*max) with unaltered affinity (*K*d value) in KO mice. Benzodiazepine binding using [^{3}H]flumazenil revealed that maximal binding (*B*max) decreased by about 40% without alteration of the affinity. The interface of α/β subunits or α/γ subunits provides [^{3}H]muscimol- or [^{3}H]flumazenil-binding sites, respectively, suggesting that KO hippocampal neurons express more $GABA_A$ receptors with fewer γ2 subunits on their cell surface. Electrophysiological and behavioral analyses using a BZ agonist, diazepam, also supported the results obtained by ligand-binding analysis.

Most of the physiological pentameric structures of $GABA_A$ receptors in brain are made of three subunits (see Fig. 2a) [15]; therefore, the possibility must be considered that substitution of γ2 subunit occurs with another subunit, most probably δ subunit in KO neurons. Electrophysiological experiments using several chemicals specific to the δ subunit indicated no evidence for the substitution, excluding the possibility that δ subunit is a substitute for γ2 subunit in KO mice. In the light of a very recent report describing $GABA_A$ receptors composed of only α/β subunits lacking a third subunit in rat hippocampal pyramidal neurons by Mortensen and Smart [16], it is plausible that α/β pentamers lacking a third subunit are more expressed in KO hippocampal neurons.

As described above, this study was initiated by the finding that PRIP interacts with GABARAP in a competitive manner with the γ2 subunit of $GABA_A$ receptors. On the basis of the finding that GABARAP has a positive effect on the trafficking of $GABA_A$ receptors to the surface membrane of neurons, in relation to the ability of binding to γ subunits [10, 11], we predicted that gene knockout of PRIP would increase the cell-surface expression of $GABA_A$ receptors containing the γ2 subunit because of the deletion of competition. However, the results did not meet the expectation as described above [12, 13], rather indicating that PRIP is involved in supporting the trafficking and/or insertion into the surface membrane of the γ2 subunit-containing receptors, the processes of which probably require GABARAP to bind to the γ2 subunit [10, 11]. However, it is difficult to readily identify how PRIP molecules are implicated in the GABARAP-mediated transport of the γ2 subunits, again because PRIP molecules bind to GABARAP in a competitive manner with γ2 subunits [12].

Taking advantage of the fact that PRIP could form a ternary complex with the β subunit of the receptors and GABARAP, we assumed that the direct binding of

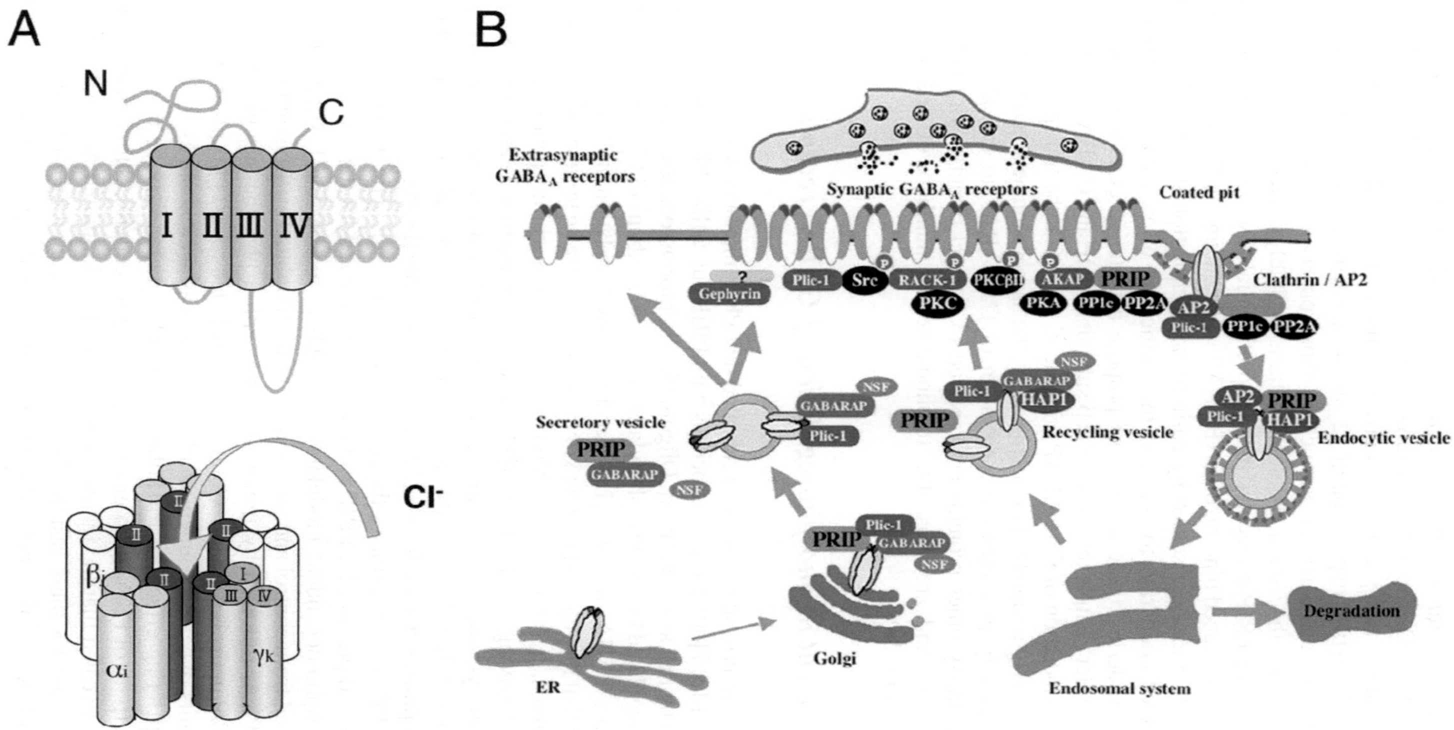

Fig. 2. **A** Schematic illustration of $GABA_A$ receptor structure. *Upper* Subunit structure of $GABA_A$ receptors with four transmembrane domains. *Lower* Heteropentameric $GABA_A$ receptors forming Cl-channel. **B** The life cycle of $GABA_A$ receptors and molecules including PRIP involved in the processes. For details, see Ref. [27] (adapted with permission from Ref. [27])

PRIP with the β subunit might be somehow concerned with the results observed with KO mice [13]. Disruption of the direct interaction between PRIP and $GABA_A$ receptor β subunits using a peptide corresponding to the PRIP binding site [8] reduced the cell surface expression of γ2 subunit-containing $GABA_A$ receptors in cultured cell lines and neurons, mimicking those observed with KO mice. The peptide did not alter the subunit assembly pattern, probably excluding the possibility that deletion of PRIP molecules induces preferential assembly between α and β subunits only. Reduced binding of GABARAP with pentameric $GABA_A$ receptors observed in KO mice as assessed by co-immunoprecipitation suggests that PRIP and GABARAP cooperate to promote the trafficking of γ2 subunit-containing receptors. Considering these results, we propose a tentative conclusion that the formation of triplet complexes among the β subunits, PRIP and GABARAP, would facilitate the association of GABARAP with the γ2 subunit to be transported to the right place at the right time [13] or alternatively, the association between β subunits and PRIP primarily promotes the association of GABARAP to the γ2 subunit. Furthermore, many other proteins are reported to interact directly with $GABA_A$ receptor subunits and are thus able to modulate the numbers of cell surface receptors (see Fig. 2b).

4.2 Involvement in phosphoregulation of $GABA_A$ receptors

Our yeast two-hybrid screening also revealed that PRIP interacts with the catalytic subunit of PP1 [5, 7]. PP2A also associates with PRIP but, unlike PP1, its activity is not modified by this association [8]. Oligomeric receptor-associated ion channels in neurons are regulated by the phosphorylation of various subunits [17]; this is indeed the case for the $GABA_A$ receptor [18]. Receptor phosphorylation or dephosphorylation requires that the relevant protein kinase or phosphatase be targeted to the receptor. For example, it is thought that Yotiao [19] and spinophilin (neurabin-II) [20] mediate the targeting of PP1 to NMDA (*N*-methyl-D-aspartate)- and AMPA (α-amino-3-hydroxy-5-methyl-4-isoxazole propionic acid)-sensitive glutamate receptors, respectively. It is thus possible that PRIP performs an analogous function for $GABA_A$ receptors, since PRIP associates directly with the β subunit as well as with the PP1 and PP2A, and regulates the activity of PP1 depending on its phosphorylation as described previously. PRIP KO mice were analyzed for the involvement of PRIP in phospho-dependent modulation of $GABA_A$ receptors [7]. Hippocampal slices prepared from control and mutant mice were stimulated by forskolin for 5 min to activate PKA, followed by immunoblotting with a phospho-specific antibody against S408/9 in the β3 subunit. Control mice exhibited robust phosphorylation of the β3 subunits, whereas a much smaller increase was seen in slices from mutant mice. The differing effects of forskolin on phosphorylation may be due either to decreased activity of PKA, or to modified phosphatase activity. Subsequent extensive studies revealed that the increased activity of PP1 in mutant mice is the most likely explanation for these results [7]. A recent observation by Kittler et al. [21] provides evidence of the molecular mechanisms that the phosphorylation of β subunit of $GABA_A$ receptors is also implicated in receptor internalization.

5 Possible involvement in oral health science

Human beings and most mammals ingest food by biting it into manageable pieces, followed by mastication and swallowing. During mastication, food is mechanically broken and mixed with saliva to create a slurry of small particles that can be easily swallowed. Mastication is brought about by the coordinated action of the muscles of mastication, the tongue, lips, and cheeks. There is a great deal of evidence from animal studies that variations in the parameters of rhythmical movements like mastication occur because sensory information generated during the movements alters the output of motor structures of the brain and spinal cord [22]. Thus, neural networks transmitting sensory information to the central masticatory area of the cerebral cortex and the brainstem central pattern generator (CPG) are implicated in rhythmical masticatory movements. The mesencephalic trigeminal nucleus, which mediates perceptions from the periodontal mechanoreceptors and jaw-closer muscle spindles to the motor nucleus, CPG and/or directly or indirectly to the cerebral cortex [22], receives many synaptic inputs, including those from $GABA_A$ receptors [23] (see Fig. 3). Furthermore, it is reported that there are extensive interconnections, both excitatory and inhibitory, between all of the lateral subgroups of the CPG, and commissural fibers are responsible for right–left coordination. These networks must be important in this organization, and perhaps also in generating the rhythm. Most of the interactions within these circuits involve amino acid neurotransmitters, including GABA. Thus, the evidence accumulated so far

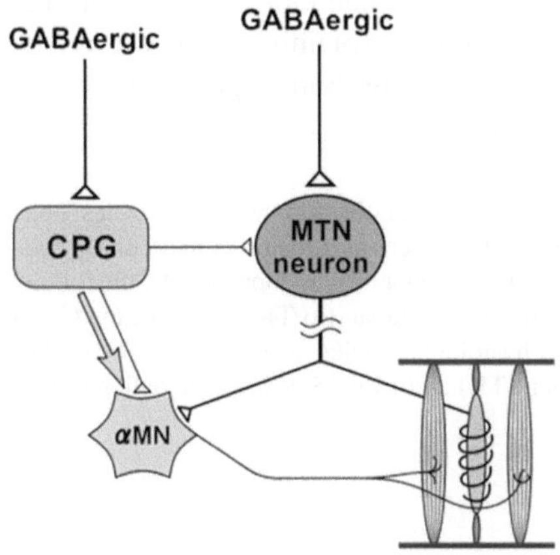

Fig. 3. Neuronal circuits around the mesencephalic trigeminal nucleus (modified picture taken from Ref. [23]). The mesencephalic trigeminal nucleus (*MTN*) mediates perceptions from jaw-closer muscle spindles and periodontal mechanoreceptors to the motor nucleus, central pattern generator (*CPG*) and/or directly or indirectly to the cerebral cortex. CPG for masticatory movements sends synaptic inputs not only to motoneurons but also to mesencephalic trigeminal nucleus neurons, thereby using mesencephalic trigeminal nucleus neurons as interneurons [23]. MTN and CPG receive many synaptic inputs including those from $GABA_A$ receptors, indicating that the characteristics of the receptors modulate masticatory movements

suggests that the characteristics (i.e., receptor numbers as well as physiological and pharmacological properties) of $GABA_A$ receptors in these areas are implicated in generating rhythmical jaw movement, indicating that PRIP might indirectly be involved in this movement.

6 Perspectives of PRIP functions

A clue to its other functions of PRIP are that GABARAP is a similar molecule to GATE-16, which plays an important role in vesicular transport through the trans-Golgi network in combination with binding to the *N*-ethylmaleimide-sensitive factor [24]. PRIP is also found to associate with GATE-16, albeit with weaker affinity. Therefore, we are currently examining KO mice regarding endocrine systems in the pancreas and pituitary glands, indicating some malfunction in secretion [25]. Thus, it appears that PRIP is involved in common intracellular vesicular transport systems in processes related to the actions of GATE-16 and/or GABARAP. PRIP might also be implicated in exocrine systems, including the salivary glands, bringing PRIP to the oral health science.

The question naturally arises as to why PRIP needs to be structurally similar to PLC-δ for functioning as described above. The PH domain is absolutely required for functioning in Ins(1,4,5)P_3-mediated Ca^{2+} signaling, whereas the necessity for non-catalytic X and Y barrels in PRIP is questionable. Recently, we noticed that it exhibits specific affinity for several phosphoinositides as ligands, but not as substrates, which would help PRIP remain in specific membranous compartments for functioning. We should also pay attention to this issue, in addition to exploring further functions of PRIP and to explaining the observations described above at a more molecular level. Furthermore, transcriptional regulation of PRIP is the next issue to be examined [26].

Acknowledgements. The Iwadare Scholarship, which provides financial support to graduate students taking basic oral health sciences as a major, is particularly acknowledged for the support to AM. The work performed in our laboratory is supported by the following grants: a Grant-in-Aid for Scientific Research from MEXT, Japan (to TK and MH), the Cooperative Study Program of the National Institute for Physiological Sciences (to TK and MH), The Naito Foundation (TK), Takeda Science Foundation (TK) and Brain Science Foundation (TK).

References

1. Streb H, Irvine RF, Berridge MJ, et al (1983) Nature 306:67–69
2. Hirata M, Sasaguri T, Hamachi T, et al (1985) Nature 317:723–725
3. Kanematsu T, Takeya H, Watanabe Y, et al (1992) J Biol Chem 267:6518–6525
4. Kanematsu T, Misumi Y, Watanabe Y, et al (1996) Biochem J 313:319–325
5. Yoshimura K, Takeuchi H, Sato O, et al (2001) J Biol Chem 276:17908–17913
6. Wang H, Bedford FK, Brandon NJ, et al (1999) Nature 397:69–72

7. Terunuma M, Jang I-S, Ha SH, et al (2004) J Neurosci 24:7074–7084
8. Kanematsu T, Yasunaga A, Mizoguchi Y, et al (2006) J Biol Chem 281:22180–22189
9. Egloff M-P, Johnson DF, Moorhead G, et al (1997) EMBO J 16:1876–1887
10. Leil TA, Chen ZW, Chang CSS, et al (2004) J Neurosci 24:11429–11439
11. Chen ZW, Chang CSS, Leil TA, et al (2005) Mol Pharmacol 68:152–159
12. Kanematsu T, Jang IS, Yamaguchi T, et al (2002) EMBO J 21:1004–1011
13. Mizokami A, Kanematsu T, Ishibashi H, et al (2007) J Neuorosci 27:1692–1701
14. Günther U, Benson J, Benke D, et al (1995) Proc Natl Acad Sci USA 92:7749–7753
15. Fritschy JM, Möhler H (1995) J Comp Neurol 359:154–194
16. Mortensen M, Smart TG (2006) J Physiol 577:841–856
17. Swope SL, Moss SJ, Raymond LA, et al (1999) Adv Second Messenger Phosphoprotein Res 33:49–78
18. Moss SJ, Smart TG (2001) Nature Rev Neurosci 2:240–250
19. Westphal RS, Tavalin SJ, Lin JW, et al (1999) Science 285:93–96
20. Yan Z, Wilson LH, Feng J, et al (1999) Nat Neurosci 2:13–17
21. Kittler JT, Chen G, Honing S, et al (2005) Proc Natl Acad Sci USA 102:14871–14876
22. Lund JP, Kolta A (2005) Int Congr Ser 1284:11–20
23. Saito M, Murai Y, Sato H, et al (2006) J Neurophysiol 96:1887–1901
24. Sagiv Y, Legesse-Miller A, Porat A, et al (2000) EMBO J 19:1494–1504
25. Doira N, Kanematsu T, Matsuda M, et al (2001) Biomed Res 22:157–165
26. Murakami A, Matsuda M, Nakashima A, et al (2006) Gene 382:129–139
27. Kanematsu T, Takeuchi H, Terunuma M, et al (2005) Mol Cells 20:305–314

Interface Oral Health Science 2007: 139–148

Conversion of functions by nanosizing—from osteoconductivity to bone substitutional properties in apatite

Fumio Watari[1]*, Atsuro Yokoyama[1], Michael Gelinsky[2], Wolfgang Pompe[2]
[1]*Graduate School of Dental Medicine, Hokkaido University, Sapporo 060-8586, Japan;*
[2]*Max Bergmann Center of Biomaterials, Institute of Materials Science, Technical University, 01069 Dresden, Germany*
*watari@den.hokudai.ac.jp

Abstract. Synthetic hydroxyapatite, in the usual case, of a macroscopic size, exhibits excellent osteoconductivity. However, it is not substituted with natural bone and remains permanently in the body; therefore it is suitable for using as an implant. It is well known that natural bone is composed of collagen and nanocrystallites of apatite with the size of approximately 50 nm. When the composite with collagen and nanoapatite synthesized in the biomimetic aspects is implanted, phagocytosis and inflammation are induced. Osteoclasts and osteoblasts are then differentiated and activated. The bone resemblant material and its phagocytizable nanometer size provide the conditions that composite is biologically degradable through phagocytosis by osteoclasts, and new bone formation by osteoblasts is simultaneously activated and proceeded. As a result, nanocomposite leads to the bone substitutional properties. Thus the conversion of functions is attained for apatite by nanosizing—from osteoconductivity in macroscopic size to bone substitutional properties in nano/micro scale. This tendency is more enhanced for carbonated hydroxyapatite. The mineralization surrounding collagen fibrils determines the crystallization of apatite for their size and orientations. Nanoparticles cause the reaction of cells/tissue and stimulate the occurrence of inflammation, which works as a stimulus in most cases or pronounces the conversion of functions leading to the bioactive properties for some cases, depending on the situation. Nano structure is essential for these stages to be processed.

Key words. nanosizing, apatite, tissue regeneration, inflammation, nanotoxicology

Introduction: non-resorbable and resorbable apatite

Synthetic hydroxyapatite in the usual case, that is, in a macroscopic size, exhibits excellent osteoconductivity. However, it is not substituted for natural bone and remains permanently in the body; therefore it is suitable for using as an implant.

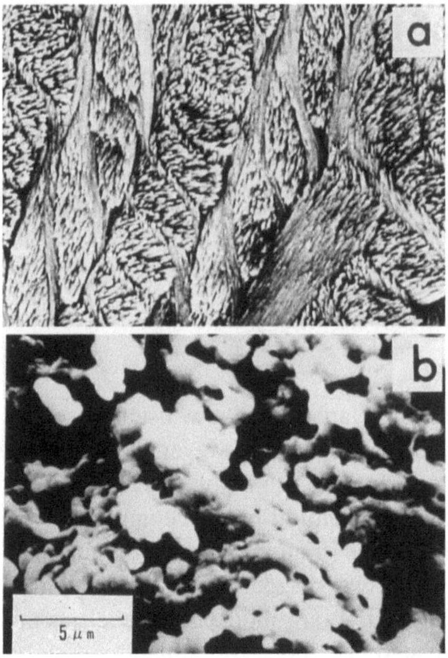

Fig. 1. Difference of morphology of hydroxyapatite. **a** Enamel of molar of rat, **b** sintered synthetic apatite

It is well known that natural bone is composed of collagen and nanocrystallites of apatite with the size of approximately 50 nm. In Fig. 1 the SEM photographs compare the difference in morphology of hydroxyapatite for natural hard tissue, in this case, enamel of molar of rat (a) and sintered synthetic apatite (b). In synthetic apatite the size of particles is a few microns, and they agglomerate in random, while in enamel, enamel prism of about 5 μm is composed of a bunch of apatite crystallites of about 50 nm. It is known that apatite crystallites are grown in their *c*-axis along collagen fibrils. Thus natural hard tissue is regarded as a kind of composite with the preferably oriented structure of nanocrystallites.

There is the difference in behavior between synthetic apatite and bone. Bone is continuously remodeled by resorption and new bone formation. Thus there exist apatites with different behaviors, non-resorbable and resorbable apatites. The problem arises: what is their difference and its cause? We will first see the nanosizing effect in general, and then the case of apatite and its mechanism.

Materials and methods

Both biochemical cell functional tests and animal implantation tests were done to investigate the reaction to fine particles of 99.9% pure Ti, Fe, Ni, and TiO_2 for the various sizes from 300 nm to 150 μm [1]. Human neutrophils were used as probe cells for various cell toxicity tests, after mixed with particles in Hank's balanced salt solution (HBSS) at 37°C. Histological investigations were done after implanting in the subcutaneous connective tissue of rats.

Hydroxyapatite-collagen composites were synthesized biomimetically on mineralized collagen type I. They have the three-dimensional scaffold structures with the interconnecting pores. They were implanted into the subcutaneous tissue, and bone defects made in the femur of rats for 1–12 weeks and observed histopathologically [2].

Results

Micro/nanosizing effect onto cell/tissue reaction

Figure 2 shows the dependence of TNF-α release from human neutrophils on the size of Ti particles. TNF-α was increased with the decrease in particle size. The increase was pronounced for 0.5 and 3 μm. The release of LDH, superoxide and cytokine Il-1β showed the similar behavior as TNF-α, while cell survival rate showed the inverse decreasing tendency. Under these conditions ICP elemental analysis indicated that the dissolution from Ti particles was negligible below detection limit [1].

Figure 3 shows the SEM image of human neutrophils of control (a) and the one exposed to 0.5 μm Ti particles (b) where a neutrophil extends its pseudopod to phagocytize Ti particles for the size less than 10 μm [3]. For the particles larger

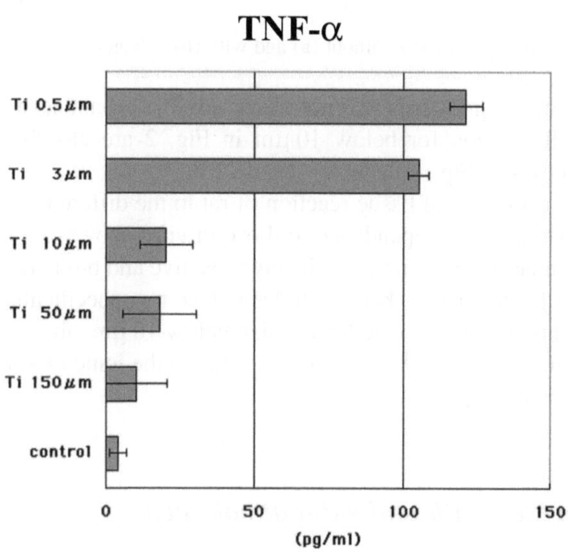

Fig. 2. Dependence of TNF-α release from human neutrophils on Ti particle size [1]

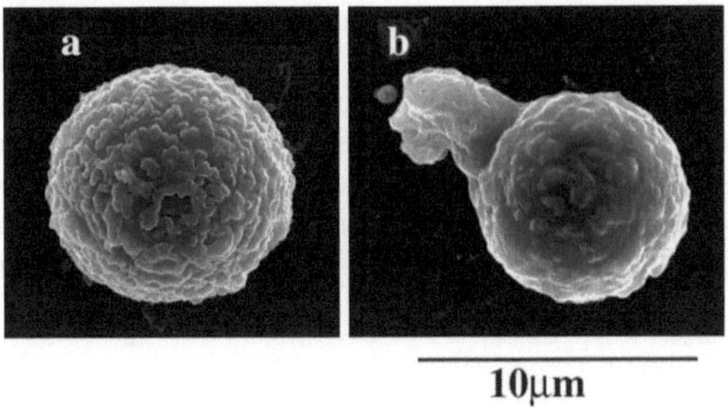

Fig. 3. SEM images of human neutrophils. **a** Control, **b** exposed to particles of Ti (500 nm) [3]

Fig. 4. Hydroxyapatite synthesized without (**a**) and with (**b**) collagen

than about 10 μm, phagocytosis was not observed. The pronounced phenomena of biochemical cell reaction for below 10 μm in Fig. 2 are closely related to the phagocytosis shown in Fig. 3.

The histological image of tissue reaction of rat to the different sizes of Ti particles showed a similar size dependence to those in vitro shown in Figs. 2 and 3.

These phenomena occur commonly in any bioactive and bioinert materials other than Ti, such as Fe and TiO_2 where particles induce non-specifically phagocytosis to cells and inflammation to tissue for the size below 10 μm, about the cell size. It is different from the usually observed toxicity due to the ionic dissolution effect in the macroscopic size [4].

Apatite formation with and without collagen

Figure 4 shows the comparison of morphology of hydroxyapatite synthesized without (a) and with (b) collagen by SEM observation. The particle size of apatite is mostly a few microns for without-collagen, while under the coexistence of collagen the product becomes the agglomerate of apatite crystallites of less than 100 nm with the lower crystallinity, revealed from X-ray diffraction analysis.

Failure of dental implants by bone desorption

In clinical cases of dental implants, failure sometimes occurs. Figure 5 shows the example of hydroxyapatite-coated titanium implant: before (a) and after (b) implantation. Failure occurs through inflammation and the resorption of apatite and the surrounding alveolar bone. Inflammation is induced by various reasons. The breakage of apatite-coating film or release of fine dust of apatite powders is one of the causes.

Resorption of nanoapatite and simultaneous osteogenesis in bone circumstances

When the biomimetic nanocomposites of apatite and collagen fibrils were implanted in the subcutaneous tissue, they were covered with fibrous connective tissue and then resorbed mostly at 8 weeks by phagocytosis.

Figure 6 shows the histopathological image when they were implanted in the bone marrow of rat for 8 weeks [2]. The area of nanocomposites (asterisks) was decreased and covered with new bone (white asterisks) of lamellar structures. Resorption of the nanocomposites and replacement by new bone proceeded. This tendency was progressed with time by 12 weeks. As shown in Fig. 6, phagocytosis

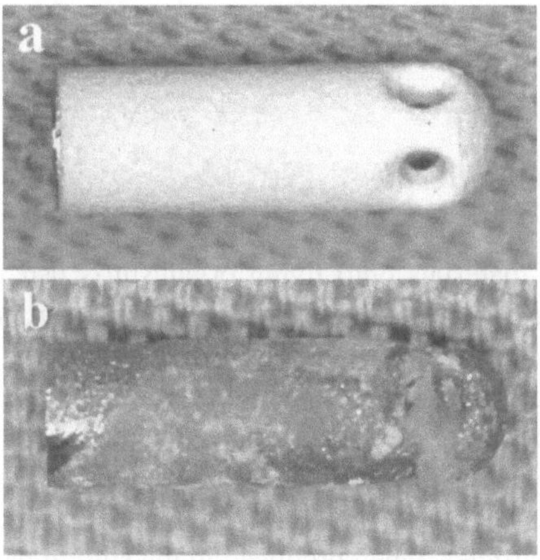

Fig. 5. Example of failure of dental implant of apatite-coated titanium: before (**a**) and after (**b**) implantation

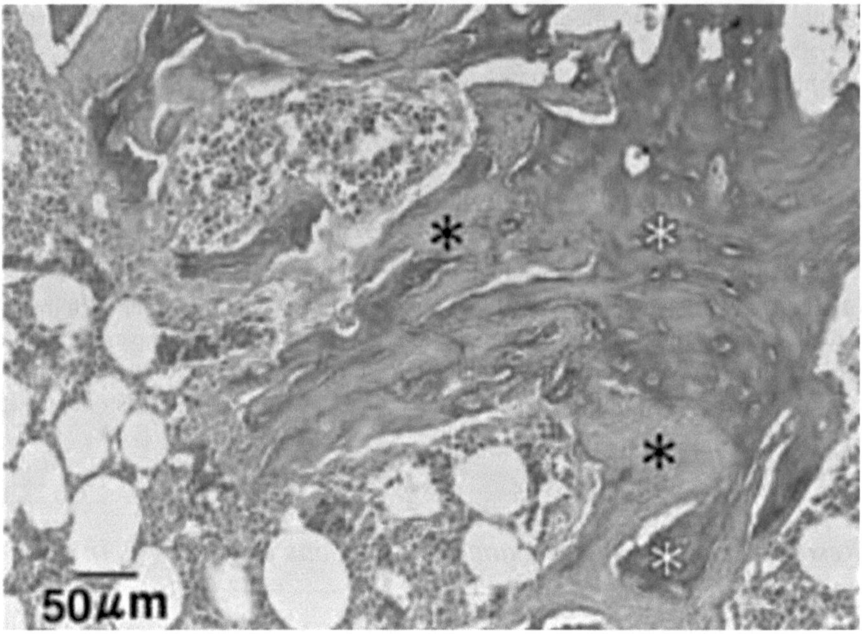

Fig. 6. Histology at 8 weeks after implantation in the bone marrow of rat. Materials (*asterisks*) were decreased and covered with new bone (*white asterisks*) with lamellar structures. AZ stain [2]

of nanoapatite by osteoclasts and osteogenesis by osteoblasts occurred adjacently to each other. Resorption and remodeling were similar to the case of autologous bone graft. As a result nanoapatite composites work as bone substitute materials for hard-tissue reconstruction.

Discussion

Nanosizing effect (general)

Nanosizing effect of materials onto living organism is usually interpreted as the aspects of the increase in specific surface area, which pronounces the chemical reactivity with the decrease in particle size. Effects related to the ionic dissolution correspond to this category, such as the acceleration of toxicity observed in Ni where tumor was generated in the long-term implantation for 0.5 μm particles [4], compared with necrosis that occurred in short term for macroscopic size [5]. There are, however, other kinds of effects [4]. Biocompatible titanium causes inflammation in abraded fine particles, when produced in the sliding parts of artificial joints, and asbestos [6], a kind of clay mineral, induces mesothelioma after a long term, large quantity of exposure. They can be understood as the physical particle effect, apart from the chemical material properties of either toxicity or biocompatibility.

Figure 2 showed clearly the cytotoxicity due to fine particles and its size dependence. Cytotoxicity and inflammation were pronounced when the particle size was smaller than 10 μm, about the cell size, where phagocytosis was induced.

Bioactive properties induced by nanosizing

Specific surface area effect is based solely on the material properties, and material-dependent, whereas the physical particle size effect has the origin in the relative size relationship between particles and cell/tissue and independent of materials. Stimulus arises as non-specific events to any bioinert, bioactive materials of metals [7], ceramics, polymers by biological process, which induces the occurrence of functionality of body defense system.

The term "biocompatible" may be classified into two categories: "bioinert" and "bioactive". "Bioinert" may be used for the materials which give neither harmful effects nor positive functional effects. Alumina, carbon, and Ti may also be included in this category. "Bioactive" is used for the materials which induce the intrinsic functional effects of the living organism, usually in a positive sense, for example, apatite inducing osteoconductivity.

The judgment of positive or negative is based on the evaluation system in the application for human beings whether they work usefully or obstructively, and indifferent from their generation mechanism. If we enlarge the definition of "bioactive" as the potential properties to induce the intrinsic functional effects of the living organism, including both the positive and negative sense, nanosizing effect can be classified as bioactive whether it generates inflammation or osteogenesis.

Nanosizing induces the non-specific phagocytosis of particles, which gives rise to the superoxide production, cytokine emission and differentiation/activation of cells which lead to inflammation in tissue.

Nanosizing effect in apatite

In the case of apatite, nanosizing effect induces phagocytosis and leads to the apparent inflammation, which causes bone resorption in some cases like Fig. 5 and bone formation in other cases like Fig. 6, depending on the bone circumstances of these events. The former causes the failure of dental implants where the hydroxyapatite-coating film on titanium implant and the surrounding newly formed bone were resorbed. The breakage of apatite-coating film and release of fine dust of nanoapatite powders could be one reason to activate osteoclasts and other phagocytizing cells. The similar phenomena are also well known for other materials. For example, abrasion particles produced from the sliding parts of artificial joints cause inflammation, whether material is polymer (polyethylene, etc.), metal (Ti, Co–Cr) or ceramics (alumina), and lead to osteolysis in the surrounding bone tissue, which determines the lifetime of using artificial joints.

When nanoapatite-collagen composites [2] or their derivatives reinforced with PLA or PLGA [8] were implanted into the bone defects of hard tissue, it leads apparently to the inflammation where cytokine emission and the differentiation or activation of osteoclasts and osteoblasts occur. Then the phagocytosis of nanoapatites by osteoclasts and osteogenesis by osteoblasts occur adjacently to each other, and the resorption of nanoapatite composites and new bone formation proceed simultaneously with time as shown in Fig. 6. As a result nanoapatite composites are substituted with new bone. Thus nanoapatite induces bioactive functions and works as bone substitutional. This tendency is more enhanced for carbonated hydroxyapatite. These phenomena are very similar to the bone remodeling process which occurs in natural bone.

Conversion of functions by nanosizing

Apatite in macroscopic size works as osteo-conductive but non-bone substitutional, while nano-size apatite works as bone substitutional.

Here, there is a conversion of functions of materials by nanosizing—from osteo-conductivity to bone substitutional properties in apatite.

Stimulus and bioactive properties of nanomaterials induced by biological process

Nanosizing causes the reaction of cells/tissue and stimulates to the occurrence of inflammation, which works as the stimulus in most cases. This toxicity is very weak compared with endotoxin [4]. Inflammation generates the conversion of functions leading to the bioactive functions for some cases, depending on the situation. These stimuli are different from those by specific surface effect where origin is solely from materials.

Conclusions

Synthesized hydroxyapatite, usually in macroscopic size, is osteoconductive but non-bone substitutional. Nanosizing of apatite induces bioactive reactivity to tissue where bone resorption or bone substitutional functions arise through the expression of inflammation, depending on the circumstances. Adjacent occurrence of resorption of nanoapatite composite by osteoclasts and simultaneous new bone formation by osteoblasts is very similar to the remodeling process of natural bone. Nanosizing works as a bioactive and causes inflammation, which leads to the conversion of functions through biological process such as from biocompatible to stimulative or from osteoconductive but non-bone substitute to bone substitute. Thus nanosizing of apatite is essential for hard tissue reconstruction and bone remodeling in the living organism.

Acknowledgments. The present study was performed under the support of Health and Labour Sciences Research Grants in Research on Chemical Substance Assessment from the Ministry of Health, Labour and Welfare of Japan (H18-Chemistry-General-006).

References

1. Tamura K, Takashi N, Kumazawa R, et al (2002) Effects of particle size on cell function and morphology in titanium and nickel. Mater Trans 43:3052–3057
2. Yokoyama A, Gelinsky M, Kawasaki T, et al (2005) Biomimetic porous scaffolds with high elasticity made from mineralized collagen—an animal study. J Biomed Mater Res Part B Appl Biomater 75B:464–472
3. Kumazawa R, Watari F, Takashi N, et al (2002) Effects of Ti ions and particles on cellular function and morphology of neutrophils. Biomaterials 23:3757–3764
4. Watari F, Tamura K, Yokoyama A, et al (2007) Biochemical and pathological responses of cells and tissue to micro- and nanoparticles from titanium and other materials. In: Bauerlein E (ed) Handbook of biomineralization, vol 3. Wiley-VCH, Weinheim, pp 127–144
5. Uo M, Watari F, Yokoyama A, et al (1999) Dissolution of nickel and tissue response observed by X-ray analytical microscopy. Biomaterials 20:747–755
6. Watari F, Inoue M, Akasaka T, et al (2006) Proceedings of the 6th Asian bioceramics symposium 2005, pp142–145
7. Matsuno H, Yokoyama A, Watari F, et al (2001) Biocompatibility and osteogenesis of refractory metal implants, titanium, hafnium, niobium, tantalum and rhenium, Biomaterials 22:1253–1262
8. Liao S, Wang W, Uo M, et al (2005) A three-layered nano-carbonated hydroxyapatite/collagen/PLGA composite membrane for guided tissue regeneration. Biomaterials 26:7564–7571

Interface Oral Health Science 2007: 149–158

Less response of osteocyte than osteoblast to mechanical force: implication of different focal adhesion formation

Teruko Takano-Yamamoto[1]*, Hiroshi Kamioka[2], and Yasuyo Sugawara[2]
[1]*Division of Orthodontics and Dentofacial Orthopedics, Tohoku University Graduate School of Dentistry, Sendai 980-8575;* [2]*Department of Orthodontics and Dentofacial Orthopedics, Okayama University Graduate School of Medicine, Dentistry and Pharmaceutical Sciences, Okayama 700-8525; Japan*
*t-yamamo@mail.tains.tohoku.ac.jp

Abstract. The structure of eukaryotic cells is controlled by a dynamic balance of mechanical forces. The mechanical force is exerted by the cytoskeleton on the extracellular matrix attachments, neighboring cells, and the substratum. Living bone is continually undergoing remodeling processes, which allow for continuous fine-tuning of the amount and spatial organization of the tissue in response to mechanical strain. The properties of osteocytes have been difficult to study and analyze because they are surrounded by mineralized tissue. Our successful isolation and maintenance of bone cells, especially osteocytes, facilitated the real-time analysis of osteocytes and osteoblasts in isolated culture. To identify the phenotype of osteocytes in the same field as that analyzed by calcium imaging, we used OB7.3, a chicken osteocyte-specific monoclonal antibody. We examined flow-induced calcium transients in primary chick osteocytes and osteoblasts, and analyzed the micro-elasticity of osteocyte and osteoblast to examine their stiffness measured with atomic force microscope (AFM) which is a powerful tool for measuring the micro-mechanics of the cells. We also examined the relationship among focal adhesion formation, calcium transients, and the micro-elasticity. It was concluded that focal adhesion formation and micro-elasticity are very important for bone cells to interact with cells and/or extracellular matrix for remodeling in response to mechanical stress.

Key words. osteocyte, mechanical stress, focal adhesion, calcium transient, micro-elasticity

1 Introduction

A great majority of bone cells are osteocytes derived from osteoblasts. There are approximately ten times as many osteocytes as osteoblasts in the normal human bone. Several functions have been proposed for osteocytes such as a calcium-sensor, a regulator for osteoid matrix maturation and mineralization, and a mechanosensor.

Living bone is continually undergoing the remodeling processes, which allow for continuous fine-tuning of the amount and spatial organization of the tissue in response to mechanical strain. Extensive in vivo reports by Lanyon et al. have identified osteocytes as load-sensitive cells [1–4]. We also demonstrated that mechanical stress induces osteopontin expression in almost all osteocytes, and in some osteoblasts and bone-lining cells at the resorption site during experimental tooth movement [5]. These studies suggest that osteocytes are sensors and/or transducers that respond to load-induced strain in the bone matrix. However, this up-regulation of biological response in osteocytes is always accompanied by the biological response of osteoblasts and bone-lining cells. Therefore, it is significant to compare the real-time response to mechanical stress among osteocytes and other bone cells in order to examine whether osteocytes are the primary mechanosensory cells in bone.

The mechanical property of the cell is controlled by a dynamic balance of mechanical forces of the cell. This balance of mechanical forces is exerted and contributed by the cytoskeleton, the neighboring cells, and the substratum [6, 7]. The mechanical properties are related to physiologically important processes, such as growth, cell cycle progression, gene expression, and other cell behaviors that are sensitive to changes in the mechanical force balance [8–12]. Therefore, measurements of the mechanical properties, such as elastic modulus of the living cell, would provide valuable insights into these processes of the living cells.

Focal adhesions are the assemblage of molecules that form a structural linkage between the extracellular matrix, the adhesion receptors, and the cytoskeleton [13]. At the cellular level, many biological responses to external forces originate in focal adhesions [14, 15]. Furthermore, the maturation and disassembly of focal adhesions control intracellular signaling cascades that ultimately alter cellular behavior [16]. Therefore, it is important to examine the focal adhesion formation of bone cells in vitro.

Actin filaments, that are the major elements of cytoskeleton, provide mechanical property of the cells and directly associate with the focal adhesions. Therefore, the relationship among focal adhesion formation, calcium transients, and elastic modulus of osteoblasts and osteocytes was examined by using Arg–Gly–Glu-containing peptides (GRGDS) which inhibit cell attachment through focal adhesion.

2 Lower calcium response of a single osteocyte to fluid shear stress

Regarding the exposure of bone to mechanical loading, it is thought that physical activities that produce various loads induce fluid flow in the porous spaces in bone tissue [17–19]. The fluid flow might be an important physical trigger that influences bone cell metabolism and bone adaptation to mechanical loading [20].

One of the earliest events in bone-cell mechano-transduction is intracellular calcium signaling, which is an important mechanism in many cellular processes e.g., differentiation, proliferation, and gene transcription [16]. It can also influence downstream signaling events. Therefore, the measurement of flow-induced intracellular calcium transients seems to be a useful indicator for evaluating the short-term mechanosensory response in bone cells. The successful isolation and maintenance of bone cells including osteocytes facilitated the real-time analysis of osteocytes and osteoblasts in isolated culture. Thus, we examined flow-induced calcium transients in primary chick osteocytes and osteoblasts. To identify the phenotype of osteocytes in the same field as calcium imaging, we used OB7.3, a chicken osteocyte-specific monoclonal antibody [21]. In addition, osteoblasts were identified with alkaline phosphate activity (ALP) (Fig. 1).

In flow-induced primary rat osteoblasts [21] and the MC3T3-E1 osteoblastic cell line [22], a calcium transient peak was seen within 15 s after the onset of flow. We showed a similar acute increment of calcium in the osteoblast population as well

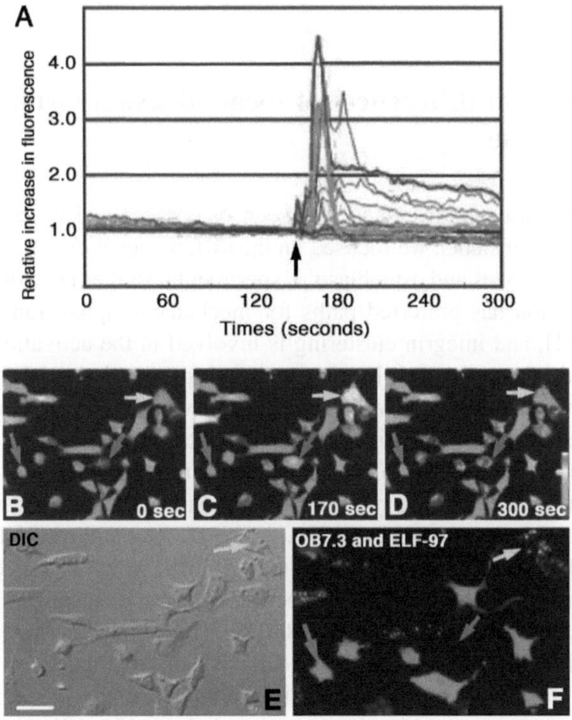

Fig. 1. **A** Representative intracellular calcium fluorescence profiles. The *arrow* indicates when flow (1.2 Pa) was initiated. *Red, blue,* and *green* lines represent the cells indicated by *arrowheads* of the respective colors in **B** and **C**. **B–D** Serial pseudo-color images of the cells. **E, F** Images of cells in **B** by differential interference contrast microscopy (**E**) and fluorescent microscopy (**F**). Fluorescent image shows the cells stained with OB 7.3 antibody (*red*) and ELF-97 (*green*)

as in the other cell populations; however, calcium response was mostly seen in the osteoblast population and OB7.3/ELF(ALP)-negative cell populations, and hardly seen in the osteocyte population (Fig. 1). In addition, elevated shear stress significantly increased the percentage of the responding osteoblast population and OB7.3/ELF-negative cell population; however, the percentage of the responding osteocyte population was not affected by elevated shear stress (Fig. 1). The percentages of responding cells in the osteoblast population, OB7.3/ELF-negative cell population, and standard cell line, MC3T3-E1, were close to the percentage in the previous study by Donahue et al. [22]; however, the percentages of responding cell in the osteocyte population were 5.5% at 1.2 Pa and 14% at 2.4 Pa. Taken together, our data suggest that the osteocyte population in vitro is less sensitive to shear stress.

Interestingly, hypotonicity-induced deformation of osteocytes caused calcium transients only in the cell processes [23], whereas calcium transients were almost identical across all cell types with or without processes in our study. We therefore think that a different calcium response between direct deformation and shear stress should be proven in the future.

3 Involvement of differences of focal adhesion formation in calcium response

Regarding the cause of differences between osteocytes and other cells in flow-induced calcium transients, we focused on the differences in focal adhesion formation between osteocytes and osteoblasts. Experimental studies previously confirmed that focal adhesion has preferred paths for mechanical signal transfer across the cell surface [14], and integrin clustering is involved in the activation of signaling pathways inside the cells [24]. Moreover, it is reported that calcium signaling is mediated by beta 1-integrin in MDCK cells [25], while both chick osteocytes and osteoblasts possess beta 1-integrin mediated cell attachment [26]. For these reasons, first, we observed focal-adhesion formation both in osteocytes and osteoblasts in primary culture and found the prominent expression of focal adhesion formation in osteoblasts but not in osteocytes (Fig. 2). Next, we employed GRGDS, which blocks integrin-mediated cell signaling, and GRGES, which is the control peptide of GRGDS. In Fig. 2, a significant reduction of calcium response in osteoblasts was confirmed by pretreatment with GRGDS. These findings suggested that the different calcium response among the cells is partially mediated via integrin-mediated cell signaling. In addition, it is reported that the calcium response is controlled by tension created by the actin cytoskeleton [27, 28]. As to differences in the actin cytoskeleton, we previously showed thick actin filaments in osteoblasts and fine actin filaments in osteocytes [29, 30]. Therefore, the inhibition of integrin-mediated cell attachment might indirectly affect the calcium response, due to differences in the actin cytoskeleton.

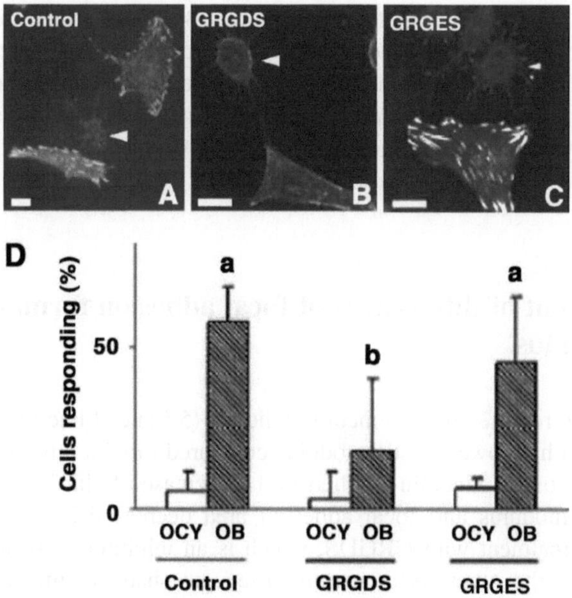

Fig. 2. Immunofluorescent images of focal adhesions labeled with anti-vinculin antibody. The cells were pretreated with 0.5 mM GRGDS (**B**), 0.5 mM GRGES (**C**), or medium (**A**, Control) for 10 min before flow experiments. *Bar* = 10 μm. **D** Influence of GRGDS or GRGES on the percentage of cells showing calcium increase during the fluid-flow periods. The *open column* shows the percentage of responding osteocytes at 1.2 Pa of shear stress (*OCY*), whereas the *closed column* shows the percentage of responding osteoblasts at 1.2 Pa of shear stress (*OB*). Values are expressed as the mean ± SE of eight separate experiments. **A** Significant difference between osteocytes and osteoblasts ($P < 0.01$), **B** significant difference between osteoblasts in the control and GRGDS pretreated culture ($P < 0.05$)

4 Lower elastic modulus of osteoctyes

Measurements of elastic modulus were performed by AFM. The elastic modulus decreased in osteoblasts, osteoid osteocytes, and mature osteocytes in the given order. However, it was reported that the elastic modulus of muscle cells was significantly increased depending on cell differentiation [31]. The level of elastic modulus was positively related to the expression level of actin-binding proteins, such as actinin and myosin. In our previous study, bone cells showed less expression of α-actinin and myosin in osteocytes [29]. Therefore, it was suggested that the different expression of actin-binding protein is related to the elastic modulus. Furthermore, we examined site-dependent elastic modulus in osteoblasts and osteocytes. The elastic modulus of the peripheral region was significantly higher than that of the nucleus region. Similarly, mouse fibroblasts (NIH3T3) and human umbilical vein endothelial cells (HUVEC) had higher elastic modulus of the peripheral region than that of nucleus region [32, 33]. Furthermore, the difference of elastic modulus depended on the amount of actin filaments in the site-dependent of the cells [32]. In the three groups, the elastic modulus of osteoblasts is the highest in both regions. As to comparison between osteoid osteocytes and mature osteocytes, the elastic modulus of osteoid osteocytes in peripheral region was significantly higher than that of mature osteocytes.

5 Involvement of differences of focal adhesion formation in elastic modulus

Wolfgang et al. reported that a vinculin-deficient (5.51) cell line of mouse embryonic carcinoma had lower elastic modulus compared to wild type (F9) [35]. They suggested that loss of vinculin resulted in the decrease of elastic modulus. In our study, elastic modulus and focal adhesion area decreased in a time-dependent manner after treatment with GRGDS, which is an inhibitor of focal adhesion in osteoblasts. On the other hand, mature osteocytes had no influence on elastic modulus by the treatment of GRGDS. Osteoid osteocytes have the same shape as mature osteocytes, but the difference in response between osteoid osteocytes and mature osteocytes might be one of the factors that osteoid osteocytes had the possibility of different response to extra-stimuli from that of mature osteocytes.

6 Focal adhesion formation in osteoblasts and osteocytes in vivo

Although we showed a significant difference in focal adhesion formation in osteocytes and osteoblasts in vitro, there was no information indicating whether both osteocytes and osteoblasts in chick calvaria form focal adhesion to the surrounding extracellular matrix. Recently, we developed a method of observing osteocytes and osteoblasts in chick calvaria by labeling the cells with fluorescence and using both a combination of confocal laser scanning (CLS) and differential interference contrast (DIC) microscopy [35]. Therefore, we applied this technique to visualize vinculin localization in calvaria to examine focal adhesion formation as seen in osteocytes and osteoblasts in vitro. There are several cell types that develop vinculin plaque in vivo that is structurally similar to focal adhesions [36, 37]. We also demonstrated the accumulation of vinculin in osteoblasts on the bone surface as well as in osteocytes within the bone (Fig. 3). It is not certain that vinculin plaque in bone possesses the same function of mechanical response as focal adhesion in 2D culture; however, vinculin accumulation is one of the characteristics of the focal adhesions, and we could see a similar distribution of vinculin both in vivo and in vitro.

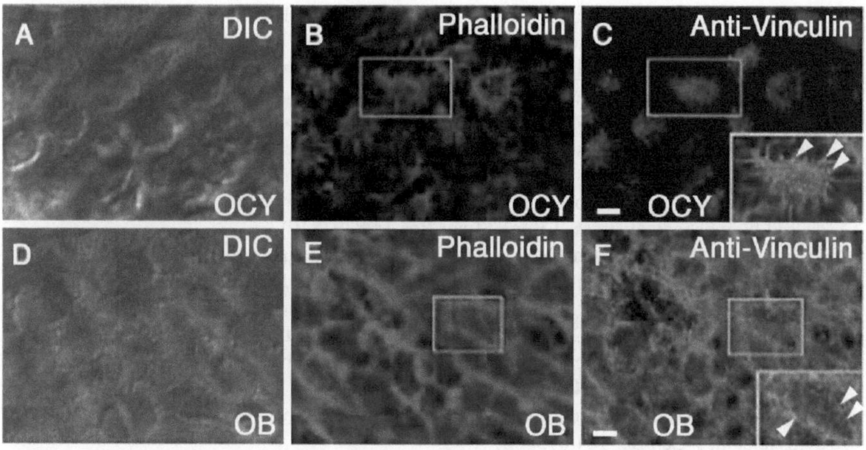

Fig. 3. Differential interference contrast (*DIC*) and fluorescent images of osteocytes (*OCY*) and osteoblasts (*OB*) in calvaria. *Red staining* represents Alexa 594-phalloidin, showing actin, and *green staining* represents anti-vinculin labeling. (**A–C**) Osteocytes in calvaria 5 μm below the osteoblast layer. (**D–F**) Mineral-facing side of the osteoblast layer. The large *inset* in **c** is a merged image comprising the small *insets* in **B** and **C**. The large *inset* in **F** is a merged image of the small *insets* in **E** and **F**. *Arrowheads* indicate vinculin plaque in an osteocyte and *arrows* indicate vinculin plaque in an osteoblast. *Bars* = 10 μm

7 Conclusion

Osteocytes isolated from chick calvaria are less sensitive to fluid shear stress in terms of the short-term calcium response. Transient intracellular calcium was mainly seen in the flow-treated osteoblast population. The difference in calcium response to mechanical stress was related to the difference in the formation of focal adhesion. Furthermore, it was suggested that there was a difference in the mechanical property related to focal adhesion formation among osteoblasts, osteoid osteocytes, and mature osteocytes. It is concluded that focal adhesion formation and micro-elasticity are very important for bone cells to interact with cells and/or extracellular matrix for remodeling in response to mechanical stress.

Acknowledgment. This study was supported by grant-in-aids from the Ministry of Education, Culture, Sports, Science and Technology of Japan.

References

1. Pead MJ, Suswilo R, Skerry TM, et al (1988) Increased ^{3}H uridine levels in osteocytes following a single short period of dynamic loading *in vivo*. Calcif Tissue Int 43:92–96

2. Skerry TM, Bitensky L, Chayen J, et al (1989) Early strain related changes in enzyme activity in osteocytes following bone loading *in vivo*. J Bone Miner Res 4:783–788
3. el Haj AJ, Minter SL, Rawlinson SC, et al (1990) Cellular responses to mechanical loading in vitro. J Bone Miner Res 5:923–932
4. Dallas SL, Zaman G, Pead MJ, et al (1993) Early strain-related changes in cultured embryonic chick tibiotarsi parallel those associated with adaptive modeling in vivo. J Bone Miner Res 8:251–259
5. Terai K, Takano-Yamamoto T, Ohba Y, et al (1999) Role of osteopontin in bone remodeling caused by mechanical stress. J Bone Miner Res 14:839–849
6. Ingber DE, Dike L, Hansen L, et al (1994) Cellular tensegrity: exploring how mechanical changes in the cytoskeleton regulate cell growth, migration, and tissue pattern during morphogenesis. Int Rev Cytol 150:173–224
7. Li ML, Aggeler J, Farson DA (1987) Influence of a reconstituted basement membrane and its components on casein gene expression and secretion in mouse mammary epithelial cells. Proc Natl Acad Sci USA 84:136–140
8. Ingber DE (1993) Cellular tensegrity: defining new rules of biological design that govern the cytoskeleton. J Cell Sci 104:613–627
9. Ingber DE, Prusty D, Sun Z (1995) Cell shape, cytoskeletal mechanics, and cell cycle control in angiogenesis. J Biomech 28:1471–1484
10. Mooney D, Hansen L, Vacanti J, et al (1992) Switching from differentiation to growth in hepatocytes: control by extracellular matrix. J Cell Physiol 151:497–505
11. Singhvi R, Kumar A, Lopez GP, et al (1994) Engineering cell shape and function. Science 264:696–698
12. Mitchison TJ (1995) Evolution of a dynamic cytoskeleton. Phil Trans R Soc Lond B 349:299–304
13. Burger EH, Klein-Nulend J (1999) Mechanotransduction in bone—role of the lacuno-canalicular network. FASEB J 13:S101–112
14. Owan I, Burr DB, Turner CH, et al (1997) Mechanotransduction in bone: osteoblasts are more responsive to fluid forces than mechanical strain. Am J Physiol 273:C810–C815
15. Smalt R, Mitchell FT, Howard RL, et al (1997) Induction of NO and prostaglandin E_2 in osteoblasts by wall-shear stress but not mechanical strain. Am J Physiol 273:E751–E758
16. Berridge MJ, Bootman MD, Lipp P (1998) Calcium—a life and death signal. Nature 395:645–648
17. Cowin SC, Weinbaum S, Zeng Y (1995) A case for bone canaliculi as the anatomical site of strain generated potentials. J Biomech 28:1281–1297
18. Duncan RL, Turner CH (1995) Mechanotransduction and the functional response of bone to mechanical strain. Calcif Tissue Int 57:344–358
19. You L, Stephen C, Cowin SC, et al (2000) A model for strain amplification in the actin cytoskeleton of osteocytes due to fluid drag on pericellular matrix. J Biomech 34:1375–1386
20. Nijweide PJ, Mulder RJ (1986) Identification of osteocytes in osteoblast-like cell cultures using a monoclonal antibody specifically directed against osteocytes. Histochemistry 84:342–347
21. Donahue SW, Jacobs CR, Donahue HJ (2001) Flow-induced calcium oscillations in rat osteoblasts are age, loading frequency, and shear stress dependent. Am J Physiol Cell Physiol 281: C1635–C1641
22. Chen NX, Ryder KD, Pavalko FM, et al (2000) Ca^{2+} regulates fluid shear-induced cytoskeletal reorganization and gene expression in osteoblasts. Am J Physiol Cell Physiol 278: C989–C997
23. Miyauchi A, Notoya K, Mikuni-Takagaki Y, et al (2000) Parathyroid hormone-activated volume-sensitive calcium influx pathway in mechanically loaded osteocytes. J Biol Chem 275:3335–3342
24. Schwartz MA, Schaller MD, Ginsberg MH (1995) Integrins: emerging paradigms of signal transduction. Annu Rev Cell Dev Biol 11:549–599

25. Praetorius HA, Praetorius J, Nielsen S, et al (2004) Beta1-integrins in the primary cilium of MDCK cells potentiate fibronectin-induced Ca^{2+} signaling. Am J Physiol Renal Physiol. 287: F969–F978
26. Aarden EM, Nijweide PJ, van der Plas A, et al (1996) Adhesive properties of isolated chick osteocytes in vitro. Bone 18:305–313
27. Sadoshima J, Takahashi T, Jahn L, et al (1992) Roles of mechano-sensitive ion channels, cytoskeleton, and contractile activity in stretch-induced immediate-early gene expression and hypertrophy of cardiac myocytes. Proc Natl Acad Sci USA 89:9905–9909
28. Wu Z, Wong K, Glogauer M, et al (1999) Regulation of stretch-activated intracellular calcium transients by actin filaments. Biochem Biophys Res Commun 261:419–425
29. Kamioka H, Sugawara Y, Honjo T, et al (2004) Terminal differentiation of osteoblasts to osteocytes is accompanied by dramatic changes in the distribution of actin-binding proteins. J Bone Miner Res 19:471–478
30. Tanaka-Kamioka K, Kamioka H, Ris H, et al (1998) Osteocyte shape is dependent on actin filaments and osteocyte processes are unique actin-rich projections. J Bone Miner Res 13:1555–1568
31. Collinsworth AM, Zhang S, Kraus WE, et al (2002) Apparent elastic modulus and hysteresis of skeletal muscle cells throughout differentiation. Am J Physiol Cell Physiol. 283: C1219–C1227
32. Haga H, Sasaki S, Kaswabata K, et al (2000) Elasticity mapping of living fibroblasts by AFM and immunofluorescence observation of the cytoskeleton. Ultramicroscopy 82:253–258
33. Kataoka N, Iwaki K, Hashimoto K, et al (2002) Measurements of endothelial cell-to-cell and cell-to-substrate gaps and micromechanical properties of endothelial cells during monocyte adhesion. Proc Natl Acad Sci USA 99:15638–15643
34. Wolfgang H, Galneder R, Ludwig M, et al (1998) Differences in F9 and 5.51 cell elasticity determined by cell poking and atomic force microscopy. FEBS Lett 424:139–142
35. Kamioka H, Honjo T, Takano-Yamamoto T (2001) A three-dimensional distribution of osteocyte processes revealed by the combination of confocal laser scanning microscopy and differential interference contrast microscopy. Bone 28:145–149
36. Small JV (1985) Geometry of actin-membrane attachments in the smooth muscle cell: the localisations of vinculin and alpha-actinin. EMBO J 4:45–49
37. Shear CR, Bloch RJ (1985) Vinculin in subsarcolemmal densities in chicken skeletal muscle: localization and relationship to intracellular and extracellular structures. J Cell Biol 101:240–256

Section I:
Biomechanical-biological interface

Interface Oral Health Science 2007: 161–166

Mechanical stretch inhibits chondrogenesis through ERK-1/2 phosphorylation in micromass culture

Ichiro Takahashi[1], Fumie Terao[1], Taisuke Masuda[3], Yasuyuki Sasano[2], Osamu Suzuki[3], and Teruko Takano-Yamamoto[1]*
[1]*Division of Orthodontics and Dentofacial Orthopedics;* [2]*Division of Craniofacial Development and Regeneration;* [3]*Division of Craniofacial Function Engineering, Tohoku University Graduate School of Dentistry, Sendai 980-8575, Japan*
*t-yamamo@mail.tains.tohoku.ac.jp

Abstract. Chondrocyte differentiation has been known to be affected by shearing stress, compressive or expansive force in vivo and in vitro. In the present study, we investigated the transient phosphorylation of extracellular signal-regulated kinase (ERK)-1/2 under stretch stimulation, and this ERK-1/2 phosphorylation mediates mechanical stretch signaling during chondrogenesis. Dissociated embryonic E12 rat limb bud cells were assembled to micromass culture on a silicon bottom plate. After 4 days, micromass cultures were stretched prior to the isolation of protein samples. Stretch stimulation was also loaded with or without MEK-1 or MEK-1/2 inhibitor. Western blot analysis revealed that phosphorylation of ERK-1/2 increased and peaked at 1.0 h after stretch loaded. Alcian blue staining and semi-quantitative reverse transcription-polymerase chain reaction analysis for type II collagen gene expression revealed that inhibited chondrogenesis by mechanical stretch stimulation was rescued by inhibiting MEK-1 and MEK-2 activities. It was concluded that signaling through ERK-1/2 was activated by stretch stimulation in micromass cultures and was involved in the inhibition of chondrogenesis by stretch stimulation.

Key words. chondrogenesis, mechanical stress stretch, ERK-1/2, MEK-1/2

1 Introduction

Cartilage, bone, muscle, and tendon form supporting elements for mechanical movement of mammalian body. These tissues consisted of tissue-specific cells, such as chondrocytes, osteoblasts, osteocytes, myocytes, and fibroblasts. Metabolism and formation of these tissues are regulated by soluble factors, such as growth factors, cytokines, and hormones. Recently, mechanical stimulation to the cells is considered as one of the key regulators of cellular metabolism and differentiation similar to those soluble factors.

Mechanical compression stimulates the gene expression and metabolism of cartilage-specific extracellular matrix (ECM) aggrecan and type II collagen. Physiological level of compressive force stimulates the differentiation of chondrocytes, whereas tension force inhibits [1–3]. Thus, mechanical stimulation plays a role in the regulation of cellular phenotypic changes of chondrocytes. During these processes, it is considered that the mechanical stimulation is translated into biochemical signals. Indeed, stretch-activated Ca^{2+} channel is activated and c-fos, a transcriptional factor, and mitogen-activated protein kinases (MAPKs) are phosphorylated by mechanical stimulation, in fibroblasts or myofibroblasts as demonstrated in a previous study [4].

We have previously demonstrated the mechanobiological response of chondrocytes for mechanical stretch stimulation where differentiation of chondrocytes is inhibited. Here, we describe the intracellular mechanobiological response of differentiating chondrocytes.

2 Materials and methods

Micromass culture

Fore- and hind-limb buds of embryonic day-12 embryos of Sprague-Dawley rats were microdissected, dissociated into the cells by 50 mg/ml collagenase and 50 mg/ml trypsin, and suspended to be 5×10^6 cells/ml in culture media. Dalbecco's Modified Essential Medium (DMEM) supplemented with 10% fetal bovine serum and antibiotics (100 U/ml of penicillin and 100 mg/ml of streptomycin) was used as the culture media. Three 50 µl drops of 5×10^6 cells/ml of cell suspension were assembled as spot micromass culture on a silicon bottom dish (Iwaki glass, Japan). Three hours after dropping the cell suspension, 2 ml of the culture media was added.

Application of mechanicals stress

Chondrogenic differentiation of the cells was assayed by alcian blue staining in the culture. Cultures were primarily maintained for 3 days prior to mechanical stimulation. For silicone bottom plate culture system, thereafter, stepwise stretch was applied every 24 h for 4 days. First stimulation gave 12% strain to the culture and 8% additional strain was loaded to the culture at every activation of the system. For extracellular signal-regulated kinase (ERK) activation assay, protein samples were obtained 0.25, 0.5, 1.0, 1.5, 3.0, 6.0, and 12.0 h after 12% single stimulation.

Protein production inhibition assay for ERK activation

Cycloheximide (10 μg/ml, Sigma, MO, USA), which is an inhibitor for protein synthesis, was added to the culture 60 min prior to the stretch stimulation. After 60 min of stimulation, cultures were harvested.

Alcian blue staining

After the culture period, cells were fixed in 2% acetic acid in 95% ethanol for 10 min and stained with 0.5% alcian blue solution overnight at 4°C. Then the cultures were rinsed thoroughly in PBS.

Reverse transcription-polymerase chain reaction (RT-PCR)

After the culture period, total RNA was isolated from cultures using RNeasy mini kit (Qiagen, Hilden, Germany) by following manufacturer's instruction. They were reverse transcribed and subjected to semi-quantitative RT-PCR of ERK-1/2 and type II collagen gene. The PCR condition used in the present study was described in our previous study.

Western blotting analysis

Protein samples were isolated from the cultures 30, 60, and 90 min, and 2, 3, 6, and 12 h after the beginning of mechanical stimulation. After the concentration of the samples were determined by using BCA protein assay kit (Pierce Biotechnology, IL, USA), they were separated on 12% SDS-PAGE, transblotted onto nitrocellulose membrane, and subjected to western blot analysis of ERK-1/2, JNK-1/2/3, and p38 MAPK. Primary antibodies used were for both phosphorylated and non-phosphorylated MAPKs (Cell Signaling Technology, MA, USA). All primary antibodies were diluted at 1:1,000. HRP-conjugated anti-rabbit IgG secondary antibodies were used for chemiluminescent detection by using chemiluminescence detection kit (Pierce Biotechnology, IL, USA). Optical density of the bands was quantified by using ImageJ software (NIH, MD, USA).

3 Results and discussion

Intracellular signaling activated by mechanical stretch (Figs. 1, 2, 3)

Chondrogenic differentiation of the cells was observed in the culture when assayed by alcian blue staining. Chondrogenic differentiation was inhibited at day 4 after stretch stimulation started as shown in Fig. 1. Alcian blue positive nodules and their area were fewer in the stretched culture than that in the non-stretched culture. Thus, chondrogenic differentiation of rat limb bud cells was inhibited by stepwise stretch stimulation.

Phosphorylation of ERKs was transiently and directly up-regulated (five-fold in ERK-1 and two-fold in ERK-2), 1 or 1.5 h after the stretch stimulation (Fig. 2),

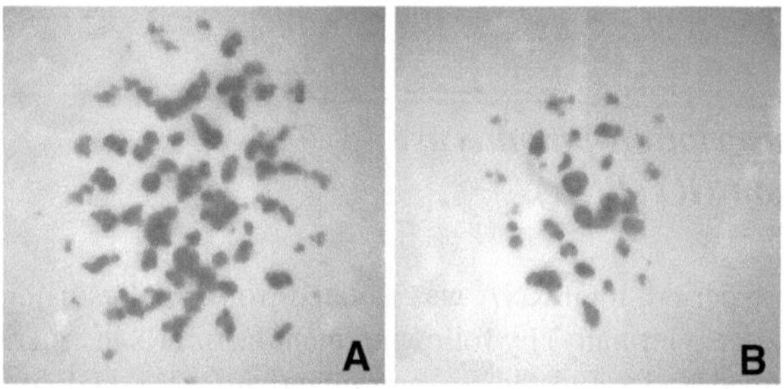

Fig. 1. Alcian blue staining of control micromass culture (**A**) and stretched culture (**B**). Alcian blue positive nodules were fewer in the stretched culture (**B**) than that in the control culture (**A**)

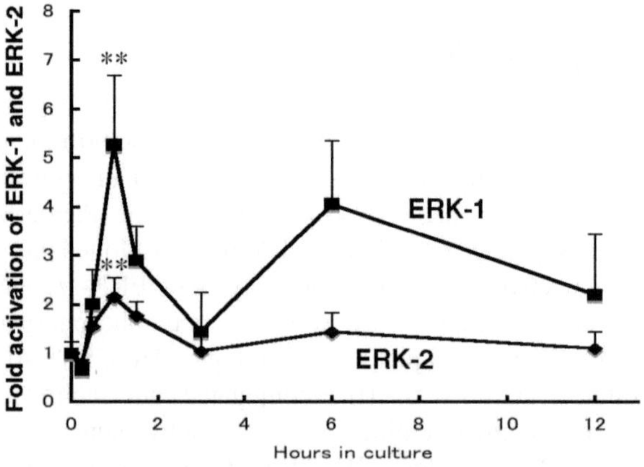

Fig. 2. Time course change in phosphorylation of ERK-1 and ERK-2 after stretch stimulation is indicated. Phosphorylation of the ERKs peaked after 1 h of stimulation. ** $P < 0.01$

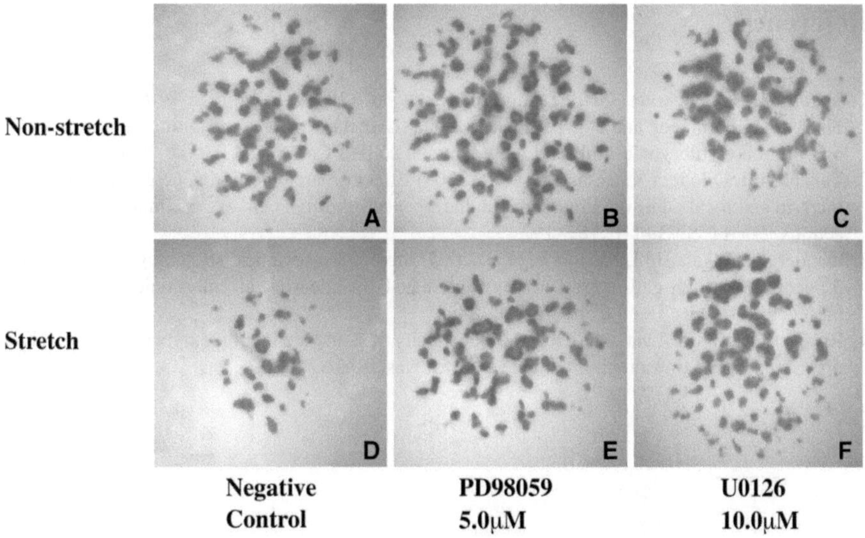

Fig. 3. Inhibitors of ERK-1 and ERK-2 rescued the stretch mediated inhibition of chondrogenesis. Alcian blue staining shows the rescued chondrogenesis under the stretch stimulation, when cultures were supplemented with *U0126* or *PD98059*, MEK-1/2 or MEK-1 inhibitors, respectively

whereas other MAPKs were not (data not shown). In addition, gene expression level of ERK-1/2 did not change throughout the experimental period by semi-quantitative RT-PCR analysis (data not shown). Taken together, mechanical stretch directly activated ERK-1/2, but not JNKs and p38 MAPK. Western blot analysis revealed that phosphorylation of ERK-1/2 increased and peaked at 1.0h after stretch loaded.

MEK-1/2 or MEK-1 inhibitors inhibited the mechanical stress induced inhibition of chondrogenesis as shown in Fig. 3. Alcian blue staining revealed that inhibited chondrogenesis by mechanical stretch stimulation was rescued by inhibiting MEK-1 and MEK-2 activities. U0126 and PD98059 completely rescued inhibited chondrogenesis at 10.0 and 5.0μM, respectively. Consequently, gene expression of type II collagen was also rescued by MEK inhibitors from the inhibition by stretch stimulation (data not shown). Thus, it could be considered that the ERK pathway is involved in mechano-response of chondrocytic differentiation in limb bud cells.

4 Conclusion

In summary, chondrogenic differentiation was inhibited by mechanical stretch stimulation through ERK signaling. It was concluded that signaling through ERK-1/2 was activated by stretch stimulation in micromass cultures and was involved in the inhibition of chondrogenesis by stretch stimulation.

References

1. Takahashi I, Nuckolls GH, Takahashi K, et al (1998) Compressive force promotes Sox9, type II collagen and aggrecan and inhibits IL-1beta expression resulting in chondrogenesis in mouse embryonic limb bud mesenchymal cells. J Cell Sci 111:2067–2076
2. Takahashi I, Onodera K, Sasano Y, et al (2003) Effect of stretching on gene expression of b1 integrin and focal adhesion kinase and chondrogenesis through cell–extracellular matrix interactions. Eur J Cell Biol 82:182–192
3. Onodera K, Takahashi I, Sasano Y, et al (2005) Stepwise mechanical stretching inhibits chondrogenesis through cell-matrix adhesion mediated by integrins in embryonic rat limb bud mesenchymal cells. Eur J Cell Biol 84:45–58
4. Ruwhof C, van der Laarse A (2000) Mechanical stress-induced cardiac hypertrophy: mechanisms and signal transduction pathways. Cardiovasc Res 47:23–37

Interface Oral Health Science 2007: 167–172

Development of mechanical strain cell culture system for mechanobiological analysis

Taisuke Masuda[1], Ichiro Takahashi[2], Aritsune Matsui[3], Takahisa Anada[1], Fumihito Arai[4], Teruko Takano-Yamamoto[2], and Osamu Suzuki[1]*
[1]*Division of Craniofacial Function Engineering (CFE);* [2]*Division of Orthodontics and Dentofacial Orthopedics;* [3]*Division of Oral Surgery, Tohoku University Graduate School of Dentistry, Sendai 980-8575;* [4]*Departments of Engineering and Robotics, Tohoku University Graduate School of Engineering, Sendai 980-8578; Japan*
*suzuki-o@mail.tains.tohoku.ac.jp

Abstract. Cytoskeleton is thought to undergo remodeling in response to mechanical stimuli such as compression (hydrostatic pressure or direct platen contact), tensile strain, and fluid shear stress. Despite the wealth of information as to effect of these mechanical stimuli on the structure and function of their cells, there is a paucity of information as to whether mechanical stimuli may facilitate cell differentiation regarding the quality and quantity of the strain and the timing applied. The present study was designed to investigate whether the deformation of scaffold affects the differentiation of chondrogenic cells. We developed a culture system that induces the mechanical stimuli to the inoculated cells with quantitative gradient strain. The culture systems previously developed were reviewed and compared to the present system in designing the mode to load the deformation.

Key words. mechanical stimulation, mechanical strain, chondrogenic cells, polydimethylsiloxane (PDMS), finite element modeling (FEM)

1 Introduction

Mechanical stimulation is considered to be one of the major epigenetic factors regulating the metabolism, proliferation, survival, and differentiation of cells in the skeletal tissues [1]. It is generally accepted that the cytoskeleton can undergo remodeling in response to mechanical stimuli such as compression (hydrostatic pressure or direct platen contact), tensile strain, and fluid shear stress. Mechanical strain has a variety of effects on the structure and function of their cells in the skeletal tissues, such as chondrocytes [2, 3], osteoblasts [4], fibroblasts [5, 6], and myoblasts [7]. Various dynamic mechanical loading methods have been proposed to simulate physiological mechanical environment in the skeletal tissues. The methods include cell culture systems that control compressive stress, tensile strain, and shear stress. Although a large number of methods have been examined in the application of these stimuli, the culture system that gradient strain can be applied, thereby the load mechanical stress could be controlled, has not been proposed. The present study was designed to investigate whether the deformation of scaffold affects the differentiation of chondrogenic cells. We developed a culture system that induces the mechanical stimuli to the cell seeded on the scaffold having quantitative gradient strain in a culture schedule. The culture systems previously developed were reviewed and compared to the present system in the design, functionality, and advantages.

2 Development of a cell culture system

2.1 Design of loading cyclic mechanical strain for culture

Since osteoblasts and chondrocytes play pivotal roles in both endochondral and intramembranous pathways of new bone formation, analysis of the mechanical stimuli should be significant under controls by artificially regulated load on these cells. The first section investigated the efficacy of a new developed culture system, with the quantitative mechanical strain, to the chondrogenic cell differentiation. Figure 1 shows the concept of a mechanical strain loading culture system. In order to apply quantitative mechanical strain to cells, we developed a polydimethylsiloxane (PDMS)-made cell culture system. PDMS is widely used for the fabrication of various micro-devices because of its transparency, biocompatibility, and remarkable advantageous mechanical properties. This system composed of a PDMS chamber and a PDMS membrane. The PDMS membrane adhered to the substrate for cell culture. This substrate was applied to induce the deformation by vacuuming the holes of chamber. Thus, the PDMS membrane can be dented depending on the amount of

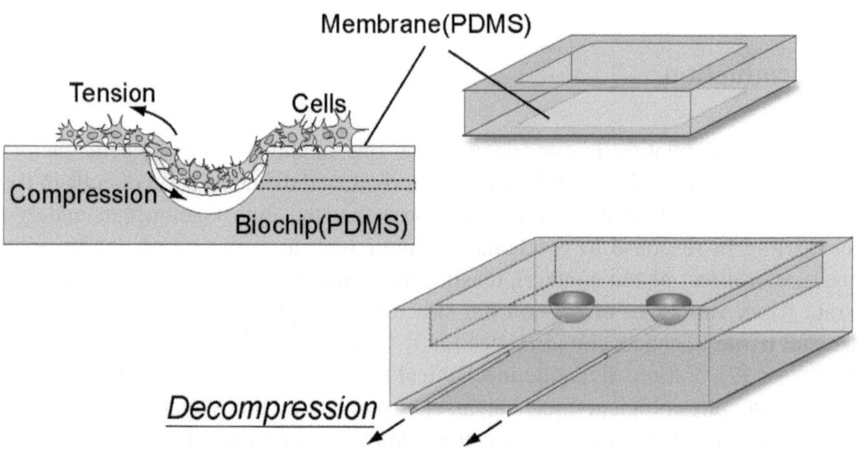

Fig. 1. The concept of the mechanical stress loading culture devices. Effects of various growth factors in combination with mechanical stress on differentiation and proliferation of the cells can be compared when the separate chamber is used. Reproduced from Masuda et al. [8], with permission from IEEE publication

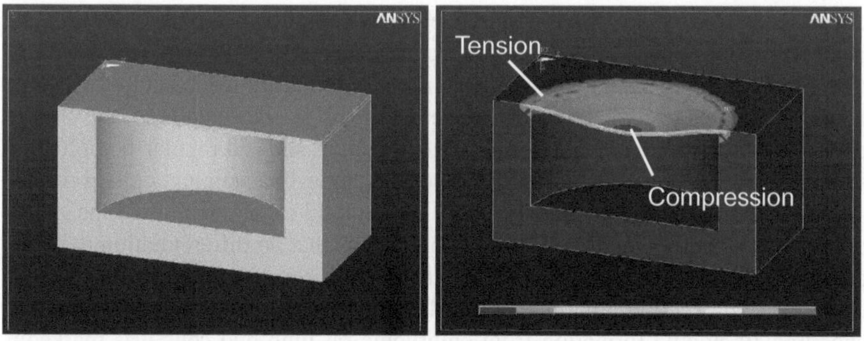

Fig. 2. Equivalent strain distributions in culture devices. This model has a hemisphere hole (ϕ 5 mm) and a membrane of 100 μm thick. Boundary conditions: Young's modulus = 1.1 MPa, Poisson's ratio = 0.4, Gauge pressure = –50 kPa

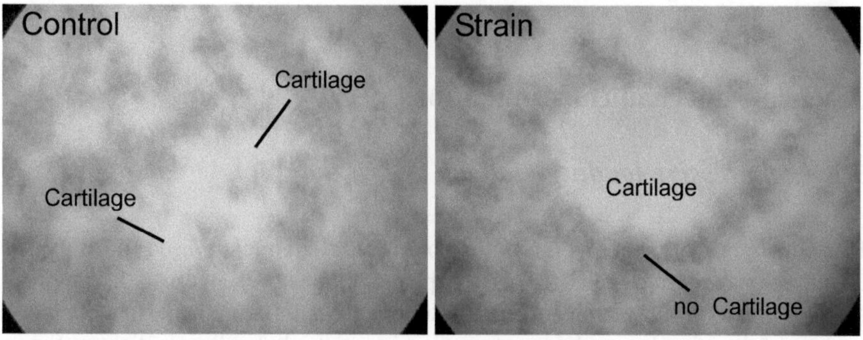

Fig. 3. Microphotographs of ATDC5 cells in which mechanical strain was loaded in the culture system for 5 days. Tensile strain inhibited chondrogenic differentiation of ATDC5 cells. On the other hand, compressive strain enhanced chondrogenic differentiation, when assayed by alcian blue staining

decompression and the thickness of the membrane, due to the intrinsic elastic modulus. The amount of the strain became highest in the center and the outside of the stress of the membranes by the analysis of finite element modeling (FEM, Fig. 2). The center showed the compression, while the outside showed tension. We can see the stress gradation on one membrane from the center to the outside.

2.2 Cell culture

Figure 3 shows the microphotograph where the chondrogenic cells (ATDC5 cell) were stained by the alcian blue after the mechanical strain culture for 5 days. The control experiments were performed under static conditions on a flat PDMS membrane. It was confirmed that the cell proliferation pattern was distinct between strain of compressive mode and non-strain condition. In the strain-loaded cultures, alcian blue positive area occupied the central area of the circle-like staining of the culture device as shown in Fig. 2, while the control culture showed regular homologous staining with alcian blue staining. Inside of the circle-like staining, cells were not stained with alcian blue positively. Thus, the circle-like staining around the center of the mass of cells should be identical to the cells that no differentiation to chondrocytes was promoted [8]. Previous study, using a commercially available Flexercell plate (Flexcell International Inc., Mckeesport, PA, USA), demonstrated that chondrocytic nodule formation from embryonic rat limb bud cells was markedly inhibited by stretch stimulation [9]. It seems likely that the mechanical stimulations, produced by the compressive and the tensile strains, regulate the chondrogenic differentiation. This mechanical strain culture also inhibited chondrogenic differentiation of ATDC5 cells where tensile strain was loaded. Compressive strain enhanced chondrogenic differentiation.

3 Various mechanical stimulations

3.1 Compressive loading systems

Hydrostatic pressurization (Fig. 4a) has been a very frequently used modality for compression of cultured cells, tissue, or explants cultures [10]. Hydrostatic compression holds several substantial attractions: simplicity of the equipment, spatial homogeneity of the stimulus, and ease of configuring multiple loading replicates. Since there is no direct contact, there is no concern regarding local specimen compaction. In addition, an approach for achieving compressive stress has been realized through direct platen (Fig. 4b). This loading regime has been proven to be applicable for tissue preparation of cartilage [11, 12]. Confined compression more closely approximates the physiologic situation at the platen surfaces, but carries the disadvantage of potentially impeding the free flow of growth factor or metabolites transport processes between the specimen's surface and the nutrient medium.

3.2 Substrate strain systems

Loading systems utilizing controlled uniaxial distention of deformable substrates (Fig. 4c) hold many advantages, and have been widespread in the use [13]. The system utilizes the transformation of the substrate to load the mechanical strain in the cultured cell. The system achieves nominal strains up to 10–30% by smooth muscle cells and endothelial cells seeded on the elastic substrates. Another broad class of mechanical stimuli cell culture systems are those in which strains can be

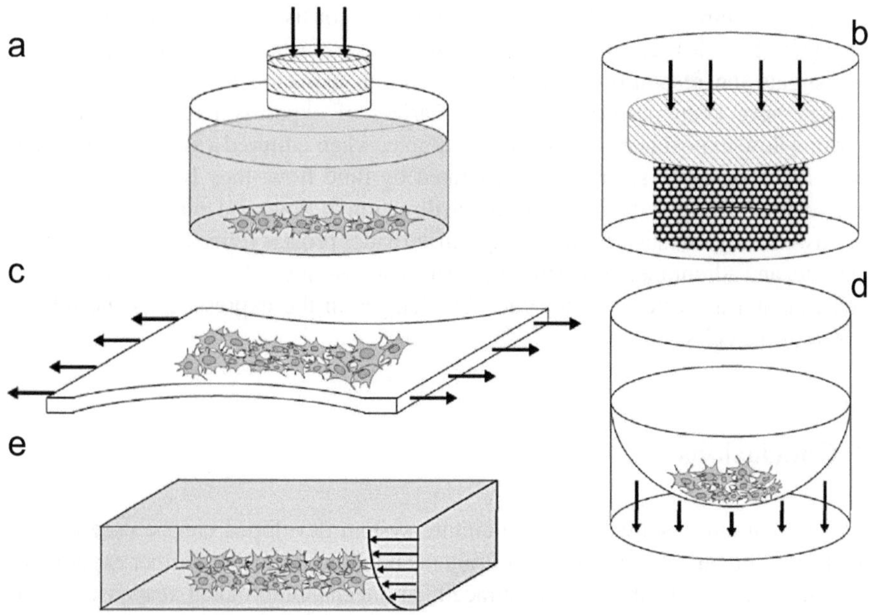

Fig. 4. Various concepts to induce mechanical stimuli to cells. **a** Hydrostatic pressure, the pressure is brought about primarily by the fluid phase. **b** The surface of the skeletal explants is compressed between the platens and the substrate. **c** Uniaxial stretch of substrain. **d** Substrate displacement by means of an applied vacuum. **e** Laminar flow chamber widely utilized in configurations for fluid shear stress input. Cells are exposed to laminar shear stress by using flow loading device

loaded by motions of deformable circular substrates [8]. In 1985, Banes et al. [14] introduced the first flexible-bottomed circular cell culture plates designed expressly for mechanical stimulus work. These plates interfaced with a vacuum manifold system, controlled by a PC-based programmable loading device. When vacuum was applied to the culture well undersurface, the substrate was stretched downward, imparting strain to the culture layer (Fig. 4d). After several units were fabricated on request for the use by investigators affiliated with the Benes' group, the system was commercialized under the name Flexercell.

3.3 Fluid shear systems

The vascular endothelial cells are exposed to fluid shear stress, a mechanical force generated by blood flow or interstitial fluid flow. Another approach to cell culture mechanical stimulus has been attained by applying fluid shear stress [15]. Cellular function is known to be influenced by fluid shear, including both mechanoreception (e.g., ion channels, integrins adhesions) and response (e.g., intracellular calcium, cytoskeletal remodeling). The main configuration consists of the parallel plate flow chamber (Fig. 4e), causing uniform laminar flow across the culture surface. To investigate the response of endothelial cells on exposure to shear stress, many studies have been performed under a variety of experimental conditions with various flow-exposing apparatus. For example, when cultured vascular endothelial cells are exposed to shear stress generated by fluid flow, they became elongated and oriented along to the long axes in the direction of flow, and enhance production of vasodilating substances, including nitric oxide, prostacyclin, C-type natriuretic peptide, and adrenomedulin [16]. These shear stress-induced changes in endothelial cell function are often accompanied by changes in the expression of the related genes of the cells.

4 Conclusions

The present study showed that the culture system developed can be used for the analysis of quantitative mechanical strain in chondrogenic cells. Further experiment is under way to determine the intracellular mechanobiological response of the chondrogenic cells and other calcified tissue-forming cells, such as osteoblasts, using the present culture system.

Acknowledgments. This study was supported by grant-in-aids (17076001, 18800007) from the Ministry of Education, Culture, Sports, Science and Technology of Japan.

References

1. Bassett CAL, Herrmann I (1961) Nature 190:460–461
2. Deschner J, Hofman CR, Piesco NP, et al (2003) Curr Opin Clin Nutr Metab Care 6:289–293
3. Pingguan-Murphy B, El Azzeh M, et al (2006) J Cell Physiol 209:389–397
4. Turner CH, Pavalko FM (1998) J Orthop Sci 3:346–355
5. Ruwhof C, van der Laarse A (2000) Cardiovasc Res 47:23–37
6. Tondreau T, et al (2004) Differentiation 72:319–326
7. Grossi A, Yadav K, Lawson MA (2007) J Biomech
8. Masuda T, et al (2006) Proc IEEE MHS2006 & Micro-Nano COE 418–423
9. Onodera K, et al (2005) Eur J Cell Biol 84:45–58
10. Wong M, Siegrist M, Goodwin K (2003) Bone 33:685–693
11. Parkkinen JJ, et al (1993) Arch Biochem Biophys 300:458–465
12. Kawanishi M, et al (2007) Tissue Eng 13:957–964
13. Naruse K, Yamada T, Sokabe M (1998) Am J Physiol 274:H1532–H1538
14. Banes AJ, et al (1985) J Cell Sci 75:35–42
15. Blackman BR, Barbee KA, Thibault LE (2000) Ann Biomed Eng 28:363–372
16. Yamamoto K, et al (2003) J Appl Physiol 95:2081–2088

Interface Oral Health Science 2007: 173–178

Regulation of osteoprotegerin and RANKL gene expression by Wnt/β-catenin and bone morphogenetic protein-2 in C2C12 cells

Mari Sato[1,2]*, Aiko Nakashima[1], Masayuki Nashimoto[3], Yasutaka Yawaka[2], and Masato Tamura[1]

[1]*Department of Biochemistry and Molecular Biology;* [2]*Dentistry for Children and Disabled Person, Graduate School of Dental Medicine, Hokkaido University, Sapporo 060-8586;* [3]*Department of Applied Life Sciences, Niigata University of Pharmacy and Applied Life Sciences, Niigata 956-8603; Japan*

*satomari@den.hokudai.ac.jp

Abstract. Wnt/β-catenin signaling plays an important role in developing the skeletal system. However, the exact mechanisms by which Wnt/β-catenin signaling regulates bone remodeling remain to be elucidated. Our previous studies demonstrated that canonical Wnt signaling inhibited the ability of bone morphogenetic protein (BMP)-2 to induce myotubes in the C2C12 cell line, and that this inhibition was mediated by Id-1. We also showed that BMP-2 induced β-catenin-mediated lymphoid enhancer factor 1/T cell factor (Lef1/Tcf)-dependent transcription in this cell line. Here, we examined the role of intracellular signaling by Wnt/β-catenin and BMP-2 in the regulation of expression of osteoprotegerin (*OPG*) and receptor activator of NFκB ligand (*RANKL*) in osteoblasts. *OPG* expression was induced by overexpression of activated β-catenin in C2C12 cells and caused an increased OPG concentration in the culture supernatant. Silencing of glycogen synthase kinase-3β (GSK-3β) by sgRNA (tRNaseZL-utilizing gene silencing) also increased OPG concentration in the culture supernatant. BMP-2 acted synergistically with Wnt3a to enhance OPG expression. In contrast, Wnt3a suppressed RANKL expression. Transient transfection analyses using murine OPG gene promoter constructs revealed that activated β-catenin induced transcription activity, and that this induction might be mediated by the Lef1/Tcf response element in the OPG gene promoter. These results show that both Wnt/β-catenin and BMP-2 signaling regulate *OPG* and *RANKL* expression.

Key words. Wnt, β-catenin, osteoblasts, bone morphogenetic protein, osteoprotegerin

1 Introduction

Bone mass levels depend on a balance between bone formation by osteoblasts and bone resorption by osteoclasts. The coupling of osteoblast and osteoclast functions is critical for skeletal modeling, remodeling, and repair. Elucidation of the molecular mechanisms that control the differentiation and function of osteoblasts is one of the major goals of bone cell biology research. Extracellular signals, such as hormones, growth factors, cytokines, and extracellular matrix components, and also their intracellular mediators, regulate cell differentiation or expression of phenotypes by modulating the activities of transcription factors involved in the expression of various target genes [1].

Recently, Wnt/β-catenin signaling has been suggested to be involved in the regulations of bone mass and bone formation [2]. Wnt-secreted proteins belong to a protein family that regulates embryonic development, cell differentiation, proliferation, and migration. Signaling is initiated by the binding of the Wnt ligand to receptor molecules of the Frizzled family and to lipoprotein receptor-related protein (LRP) 5 and 6 [3]. Two types of Wnt protein have been identified: β-catenin-dependent canonical Wnts, such as Wnt1 and Wnt3a; and the so-called noncanonical Wnts that are independent of or inhibit β-catenin signaling. According to the current model of canonical Wnt (Wnt/β-catenin) action, glycogen synthase kinase-3β (GSK-3β) phosphorylates β-catenin and thereby induces rapid degradation of β-catenin in cells that lack Wnt signaling. Stabilized β-catenin interacts in the cytosol with several molecules, including lymphoid enhancer factor 1/T cell factor (Lef1/Tcf). A complex involving the transcription factor Lef1/Tcf and β-catenin regulates expression of several target genes. The Wnt pathway is inhibited by Dickkopf (Dkk), a secreted protein that acts by binding to and antagonizing LRP5/6 [2, 3].

Here, we summarize our recent progress in understanding the function of Wnt/β-catenin signaling in osteoblastic cells. We previously established pluripotent C2C12 cell lines that expressed either Wnt3a (Wnt3a-C2C12 cells), which stimulates canonical Wnt signaling, or Wnt5a (Wnt5a-C2C12 cells), which inhibits canonical Wnt signaling. These cell lines were used to determine whether bone morphogenetic protein (BMP)-2 modulates canonical Wnt signaling [4]. BMP-2 has been reported to induce ectopic bone formation and osteoblast differentiation [5]. In our previous reports, we showed that BMP-2 induced expression of matrix extracellular phosphoglycoprotein (*MEPE*) in Wnt3a-C2C12 cells but not in Wnt5a-C2C12 or C2C12 cells [6]. MEPE is one of the bone matrix proteins present in osteoblasts. Wnt3a was found to downregulate expression of inhibitor of DNA binding/differentiation 1 (*Id1*), a BMP-2-responsive gene that has been shown to inhibit myogenesis and that is a typical early-response gene following BMP treatment of various cell types [4]. A GC-rich region in the *Id1* promoter mediates suppression of gene expression. Using the TOPflash reporter assay, which makes use of six tandem repeats of the Wnt/β-catenin response element, we found that BMP-2 did not alter promoter activity in C2C12 cells but, surprisingly, upregulated activity in cells overexpressing Wnt3a or activated β-catenin. Inducible activity was also observed after transfection with Smad1/4, instead of addition of BMP-2, in cells transfected with either Wnt3a or activated β-catenin [4]. These findings identify the molecular mechanism through which the BMP-2 and Wnt/β-catenin signaling pathways cooperate to regulate gene expression in osteoblasts.

2 Identification of osteoprotegerin as a target gene for Wnt/β-catenin signaling

2.1 Regulation of OPG expression by Wnt/β-catenin signaling

Osteoclasts are multinucleated cells derived from hematopoietic precursor cells of the monocyte/macrophage lineage; osteoclast precursor cells interact with osteoblasts to differentiate into mature osteoclasts. This interaction is mediated by receptor activator of NFκB ligand (RANKL), which is expressed on the osteoblast cell surface, and by receptor activator of NFκB (RANK), a cognate receptor expressed on hematopoietic precursor cells. RANKL and RANK are central regulators of osteoclast formation and function. Osteoprotegerin (OPG), a secreted receptor of the tumor necrosis factor receptor family that lacks a transmembrane domain, is a decoy receptor that blocks the interaction between RANKL and RANK. OPG is a key osteoclastogenesis inhibitory factor that inhibits the formation and activity of osteoclasts in vitro and of bone resorption in vivo [7]. Until recently, little was known of the effects of Wnt/β-catenin signaling on the regulation of OPG/RANKL in osteoblastic cells and their progenitors.

To examine the potential role of Wnt/β-catenin signaling in the regulation of OPG and RANKL in osteoblasts, OPG levels in culture supernatants were measured using ELISA. C2C12 and Wnt5a-C2C12 cells had low or undetectable levels of OPG, whereas OPG was detected in the conditioned media from Wnt3a-C2C12 cells (Table 1). We also examined C2C12 cell lines transfected with an expression plasmid carrying an activating form of β-catenin that lacked sites for phosphorylation by GSK-3β (β-catenin ΔGSK) or carrying Wnt3a. Both cell lines had detectable OPG in their conditioned media. Co-transfection of DKK1 significantly reduced the OPG levels (Table 1). These results indicate that OPG protein production in C2C12 cells mediated both Wnt/β-catenin signaling and an extracellular interaction between the Wnt receptor, Wnt3a and DKK1. In order to determine whether induction of OPG was associated with increased transcription, RT-PCR

Table 1. Osteoprotegerin (OPG) levels in cell culture supernatants

Cell line	Plasmid	OPG (pg/ml)
C2C12	pcDNA3	ND
C2C12	β-cat ΔGSK	828 ± 15.7
C2C12	Wnt3a	960 ± 29.6
Wnt3a-C2C12	—	811 ± 18.7
Wnt5a-C2C12	—	ND
Wnt3a-C2C12	DKK1	565 ± 8.62
Wnt3a-C2C12	SgGSKL	1388 ± 292

Data are means ± SD
ND, not detected

analyses were performed on total RNAs isolated from these cells. OPG mRNA could not be detected in either C2C12 or Wnt5a-C2C12 cells but was present in Wnt3a-C2C12 cells (data not shown).

2.2 *Knockdown of GSK-3β facilitates OPG production*

Since GSK-3β is a known mediator of Wnt/β-catenin signaling, we examined the effect of knockdown of GSK-3β by sgRNA on OPG induction. We previously developed a unique system for downregulating gene expression in which specific mRNAs are cut using the long form of tRNA 3′ processing endoribonuclease (tRNase ZL) under the direction of an sgRNA [8]. A 14-nt linear RNA, sgGSKL, that targets murine GSK-3β mRNA was designed, and an expression plasmid (pRNA Tin-H1.2/Neo-sgGSKL) was constructed that produced sgRNAs from a human H1 promoter [8]. Transient transfection of C2C12 cells with the sgRNA-expressing plasmid increased OPG levels to approximately 40% of the control level (Table 1), indicating that OPG induction is mediated by GSK-3β, and suggesting that phosphorylation of GSK-3β by Wnt/β-catenin signaling may be effective for OPG induction.

2.3 *BMP-2 modulates Wnt/β-catenin mediated OPG and RANKL expression*

Bone morphogenetic protein (BMP)-2 enhanced the level of OPG protein in Wnt3a–C2C12 cells (Table 2), indicating that BMP-2 plays a role in modulating expression of *OPG*. In other words, it may explain why Wnt/β-catenin signaling may yield a high level of OPG production and also indicates that a combination of BMP-2 and Wnt/β-catenin is required for maximum *OPG* expression in osteoblasts in order to inhibit osteoclast-mediated bone resorption. RANKL expression was reciprocally downregulated in relation to the level of OPG by BMP-2 in Wnt3a-C2C12 cells (Fig. 1). These reciprocal changes in the expression of *RANKL* may also contribute to Wnt-mediated regulation of osteoclastogenesis in bone.

Table 2. Bone morphogenetic protein (BMP)-2 enhances over-expression of Wnt/β-catenin mediated OPG protein levels in Wnt3a-C2C12 cells

BMP-2 (ng/ml)	OPG (pg/ml)
0	81.3 ± 10.3
50	227 ± 17.2
300	223 ± 15.4

Data are means ± SD

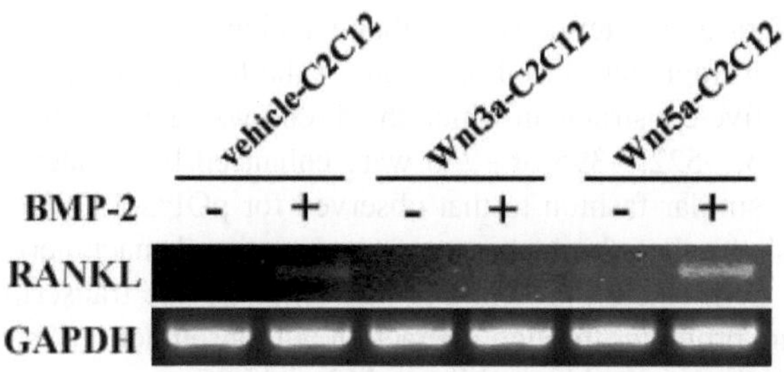

Fig. 1. C2C12, Wnt3a- or Wnt5a-C2C12 cells were cultured. BMP-2 (300 ng/ml) (+) or vehicle (−) was added, after which the cells were cultured for 24 h. Receptor activator of NFκB ligand (*RANKL*) mRNA levels were determined by RT-PCR

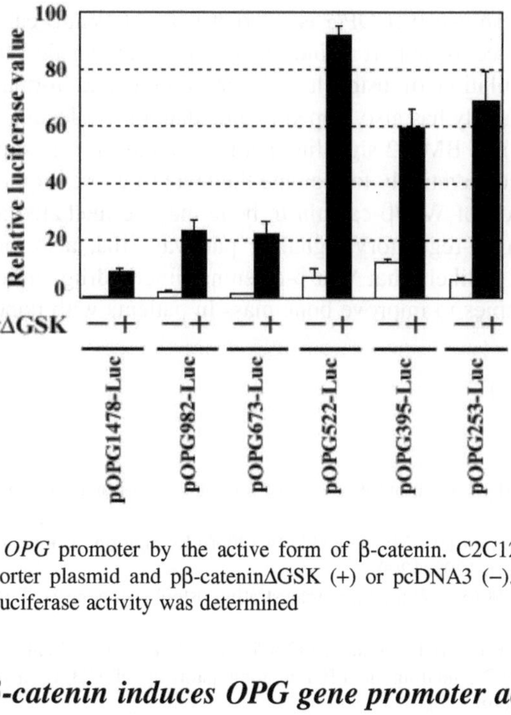

Fig. 2. Activation of *OPG* promoter by the active form of β-catenin. C2C12 cells were co-transfected with a reporter plasmid and pβ-cateninΔGSK (+) or pcDNA3 (−). The cells were cultured for 48 h and luciferase activity was determined

2.4 Activated β-catenin induces OPG gene promoter activity

To investigate the mechanisms by which Wnt/β-catenin signaling activates OPG gene transcription, we cloned a genomic DNA fragment of approximately 1.5 kilobases from the 5′-flanking promoter region of the murine OPG gene (from positions −1478 to +37) The OPG promoter region was ligated into a luciferase reporter expression vector to examine its responsiveness to Wnt/β-catenin stimulation. Transient transfection of this construct (pOPG1478-luc) in C2C12 cells with β-cateninΔGSK resulted in a significant increase in luciferase activity (Fig. 2). To identify the sequences that mediate β-catenin-dependent induction, we constructed a series of luciferase reporter constructs that contained sequential deletion mutants of the 5′-flanking fragment placed upstream of the luciferase gene. The luciferase activities of the five constructs in which the 5′-end was deleted to nucleotide positions –982, –673, –522, –395 or –253 were enhanced by β-catenin ΔGSK overexpression in a similar fashion to that observed for pOPG-1478luc (Fig. 2). This deletion analysis implies the presence of transcriptional machinery that responds to interference by β-catenin protein and that regulates the transcriptional activity of the OPG gene promoter through interaction with a segment of the promoter that resides between the nucleotide positions –253 and +37.

3 Conclusion

Here, we have shown that OPG is a target gene of Wnt/β-catenin. We have also provided evidence to support an indirect role for osteoblastic Wnt/β-catenin signaling in the regulation of osteoclast differentiation and formation via OPG and RANKL. This study has also helped to elucidate the molecular mechanisms of the Wnt/β-catenin and BMP-2 signaling pathways, and the ways in which these pathways interact cooperatively to regulate the level of expression of OPG. The physiological function of Wnt/β-catenin in bone may be mediated at several levels by cross-talk with co-regulatory signaling pathways that also target Wnt/β-catenin target genes. It is likely that Wnt/β-catenin mimetic drugs may be of value in the future for therapies to improve bone mass in patients with bone disorders.

References

1. Aubin JE (2001) Regulation of osteoblast formation and function. Rev Endocr Metab Disord 2:81–94
2. Krishnan V, Bryant HU, Macdougald OA (2006) Regulation of bone mass by Wnt signaling. J Clin Invest 116:1202–1209
3. Hartmann C (2006) A Wnt canon orchestrating osteoblastogenesis. Trends Cell Biol 16:151–158
4. Nakashima A, Katagiri T, Tamura M (2005) Cross-talk between Wnt and bone morphogenetic protein (BMP)-2 signaling in differentiation pathway of C2C12 myoblasts. J Biol Chem 280:37660–37668
5. Katagiri T, Yamaguchi A, Komaki M, et al (1994) Bone morphogenetic protein-2 converts the differentiation pathway of C2C12 myoblasts into the osteoblast lineage. J Cell Biol 127: 1755–1766
6. Nakashima A, Tamura M (2006) Regulation of matrix metalloproteinase-13 and tissue inhibitor of matrix metalloproteinase-1 gene expression by Wnt3a and bone morphogenetic protein-2 in osteoblastic differentiation. Front Biosci 11:1667–1678
7. Theoleyre S, Wittrant Y, Tat SK, et al (2004) The molecular triad OPG/RANK/RANKL: involvement in the orchestration of pathophysiological bone remodeling. Cytokine Growth Factor Rev 15:457–475
8. Nakashima A, Takaku H, Shibata HS, et al (2007) Gene silencing by the tRNA maturase tRNase Z(L) under the direction of small-guide RNA. Gene Ther 14:75–85

Interface Oral Health Science 2007: 179–180

Application of electroporation to mandibular explant culture system for gene transfection

Fumie Terao[1], Ichiro Takahashi[1], Hidetoshi Mitani[2], Naoto Haruyama[2], Osamu Suzuki[3], Yasuyuki Sasano[4], and Teruko Takano-Yamamoto[1]*

[1]*Division of Orthodontics and Dentofacial Orthopedics;* [2]*Division of Oral Dysfunction Science;* [3]*Division of Craniofacial Function Engineering;* [4]*Division of Craniofacial Development and Regeneration, Tohoku University Graduate School of Dentistry, Sendai 980-8575; Japan*

*t-yamamo@mail.tains.tohoku.ac.jp

Abstract. Gene transfer by electroporation in combination with mandibular explant culture is expected to give better understanding of the molecular mechanisms during maxillofacial development. In the present study, we attempted to analyze the gene transfer efficiency of mandibular explant in relation with vector size or the magnitude of electroporation voltage in comparison with gene transfer using lipofection method. Ectopic expression of enhanced green fluorescent protein (EGFP) was accomplished in the mesenchymal area where the gene was transferred in electroporation group, whereas the lipofection group showed minimal transfection efficiency. We established a method for in vitro electroporation, which induced transient gene delivery without tissue damage.

Key words. electroporation, gene transfer, vector size, electric voltage, mandibular organ culture

1 Introduction

Embryonic mandibular organ culture system provides various and important information regarding pattern formation and tissue differentiation during craniofacial development. On the other hand, electroporation is quite a simple and easy strategy of gene transfer among a variety of gene delivery systems [1, 2]. To investigate the gene transfer efficiency, we utilized a mammalian expression vector carrying an enhanced green fluorescent protein (EGFP) gene. In the present study, we have optimized electrical parameters and application designs to establish the method for electroporation to mandibular explant culture system without serious dismorphogenetic effects. We also attempted to analyze the gene transfer efficiency to mandibular explant culture in relation with vector size.

2 Materials and methods

Mandibules of embryonic day 12 rats were microdissected and cultured by modified Trowell method [3] in BGJb medium (Invitrogen, Carlsbad, CA, USA) in 5% CO_2 and at 37°C. We used pCMS-EGFP plasmid vector (5.5 kb) (Clontech, Lo Jolla, CA, USA) and reconstructed EGFP expression vectors (3.8, 4.2, 4.8 Kb) driven by CMV promoter and injected into the mandibular explant. Following microinjection of the vector, the injected site of the explant was placed between a custom-made platinum electrode prepared for the present study, and then four electric pulses were applied under various magnitudes of voltage ranging from 10 to 25 V (EP group). Plasmid was microinjected to the mandibular explant with lipofection reagent with various gene/reagent ratios (LF group). After 7 days culture, we carried out hematoxylin and eosin staining and whole-mount Alcian blue staining to evaluate histological changes.

3 Results

Extreme expression of EGFP was observed in the mesenchymal area in EP group compared with LF group. Transfection efficiency was the highest under 25 V, while histological damage was observed. However, 20 V did not affect the Meckel's cartilage formation. Gene expression of EGFP was statistically significantly higher in 3.8 kb group than in 5.5 kb group. Gene expression level of EGFP showed significant correlation ($r = -0.78$, $P < 0.01$) to the size of vector.

4 Conclusions

We have successfully developed an electroporation-based useful transfer system in combination with embryonic mandibular organ culture system.

References

1. Angeli I, James CT, Morgan PJ, et al (2002) Misexpression of genes in mouse tooth germ using in vitro electroporation. Connect Tissue Res 43:180–185
2. Felgner PL, Gadek TR, Holm M, et al (1987) Lipofection: a highly efficient, lipid-mediated DNA-transfection procedure. Proc Natl Acad Sci USA 84:7413–7417
3. Trowell OA (1959) The culture of mature organs in a synthetic medium. Exp Cell Res 16:118–147

Interface Oral Health Science 2007: 181–182

Effects of initially light and gradually increasing force on orthodontic tooth movement

Ryo Tomizuka[1]*, Hiroyasu Kanetaka[1], Yoshinaka Shimizu[2], Akihiro Suzuki[1], Sachiko Urayama[2], and Teruko Takano-Yamamoto[1]
[1]*Division of Orthodontics and Dentofacial Orthopedics;* [2]*Division of Oral and Craniofacial Anatomy, Tohoku University Graduate School of Dentistry, Sendai 980-8575, Japan*
*ryotommy@mail.tains.tohoku.ac.jp

Abstract. This study investigated the effect of initially light and gradual increases in force on tooth movement using the attractive force of magnets in an experimental rat model. The distance between the magnets incrementally decreased from an initial light force in the experimental group, in contrast to no tooth displacement in the control group. There were significant differences in the number of osteoclasts and in the relative hyalinized area on the pressure side of the periodontal tissue between the control group and the experimental group. The application of gradual incremental increases in force induced effective tooth movement in rats, and recruitment of osteoclasts and inhibition of hyalinization.

Key words. initially light force, gradually increasing force, tooth movement, osteoclasts, hyalinization

Introduction

Light continuous force results in a relatively smooth progression of tooth movement by frontal resorption [1]. However, traditional orthodontic appliances are not suitable for generating light force. The objective of this study was to investigate histologically the effect of initially light and gradual increases in force on tooth movement in an experimental rat model.

Material and methods

Sixty male Wistar rats (18 weeks old) were used in the experiment. Cuboids (1.5 mm × 1.5 mm × 0.7 mm) made of magnet (experimental group) or titanium (control group) were bonded on the lingual surface of maxillary first molars. The initial distance between the materials in both groups was 1.5 mm, exerting only a light force in the experimental group. Rats were killed at 1, 3, 7, 10 or 14 days after treatment. Measurement of tooth movement was determined in 28 rats, and the number of TRAP-positive osteoclasts and the relative hyalinized area on the pressure side of periodontal tissue was determined in 32 rats.

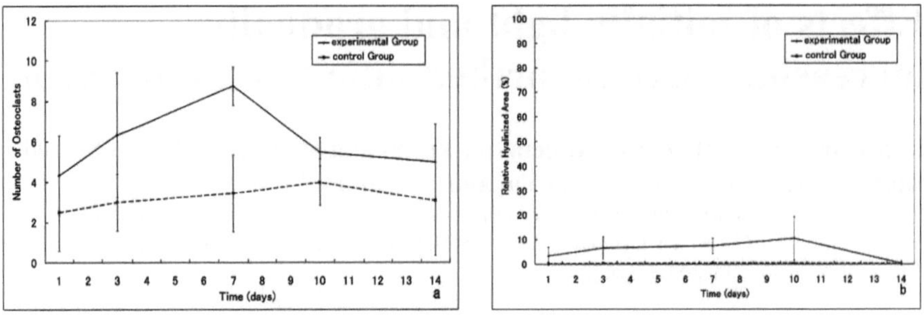

Fig. 1. **a** Time course changes in the number of osteoclasts on the pressure side. **b** Time course changes in the relative hyalinized area on the pressure side

Results and discussion

The distance between the magnets significantly decreased over time in the experimental group, resulting in gradual tooth movement ($P < 0.01$). No tooth displacement occurred in the control group. There was a significant difference in the number of osteoclasts (Fig. 1a) and in the relative hyalinized area between the control group and the experimental group (Fig. 1b) (both $P < 0.01$). We hypothesized that stepwise process may be solved by an application of initially light and gradually increasing force [2]. It is well established that bone resorption by osteoclasts is crucial to orthodontic tooth movement [1, 3]. The formation of resorbed lacunae on the bone surface in the initial stage of force application may be beneficial for the recruitment of osteoclasts and continuous bone resorption, even despite a subsequent increase in force.

Conclusion

The initial application of light force with initially light and gradual increases in force induced effective tooth movement in rats, and recruitment of osteoclasts and inhibition of hyalinization.

References

1. William RP, Henry WF (2000) Contemporary orthodontics, 3rd edn. Mosby, St Louis, pp 296–325
2. Tomizuka R, Kanetaka H, Shimizu Y, et al (2006) Effects of gradually increasing force generated by permanent rare earth magnet for orthodontic tooth movement. Angle Orthod 76:1004–1009
3. Reitan K (1960) Tissue behavior during orthodontic tooth movement. Am J Orthod 46:881–900

Interface Oral Health Science 2007: 183–184

Biomechanical effect of incisors' traction using miniscrew implant

Shota Yoshida[1]*, Koshi Sato[1], Toru Deguchi[2], Kazuhiko Kushima[2], Takashi Yamashiro[2], and Teruko Takano-Yamamoto[1]
[1]*Division of Orthodontics and Dentofacial Orthopedics, Tohoku University Graduate School of Dentistry, Sendai 980-8575;* [2]*Department of Orthodontics and Dentofacial Orthopedics, Okayama University Graduate School of Medicine, Dentistry and Pharmaceutical Sciences, Okayama 700-8525; Japan*
*shota-yo@mail.tains.tohoku.ac.jp

Abstract. The purpose of this study was to compare the biomechanical effects of the incisor movement in three different types of traction by the miniscrew implants as an anchorage using 2D finite element method (FEM). The FEM model was made by tracing the maxilla including tooth, the wire, and the bracket. Three different types of hooks were set up in the model. We suggested that controlled incisor retraction is possible by changing the length and the position of the hooks soldered to the archwire when using miniscrew implants.

Key words. miniscrew, implant, incisors, traction, stress, 2D-FEM

Introduction

The purpose of this study was to compare the biomechanical effects of the incisor movement in three different types of hooks soldered to the archwire during incisor retraction by elastic or coil springs from miniscrew implants as an anchorage using 2D finite element method (FEM).

Materials and method

A cephalometric radiograph of a patient treated by the miniscrew implant was selected. 2D-FEM model was made by tracing the maxilla, upper incisor with bracket, molar, and archwire. Three different types of hooks examined were as follows: Hook1, hook positioned higher than the miniscrew; Hook2, hook positioned at the same height with the miniscrew; Hook3, hook positioned lower than the miniscrew. The experimental results were obtained using supercomputing resources at Information Synergy Center, Tohoku University.

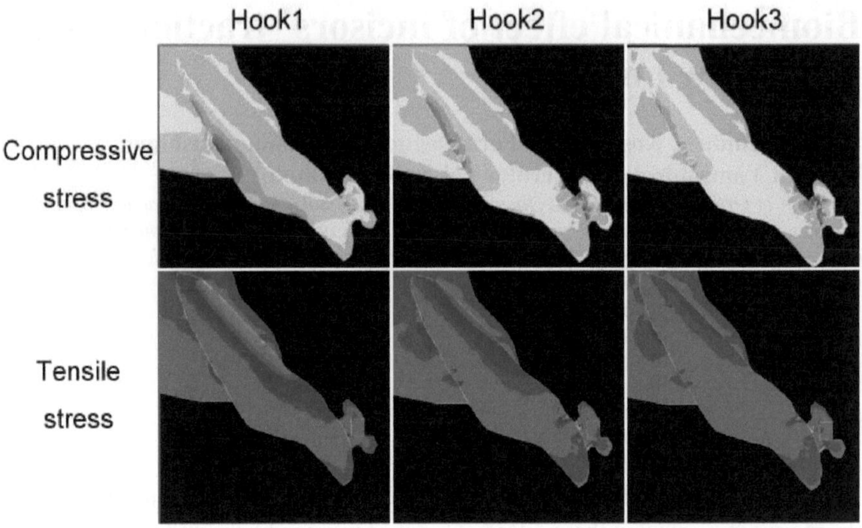

Fig. 1. Distribution of compressive and tensile stresses

Results and discussion

In Hook1, the tensile stress loaded widely at the labial part of the centric incisor root (Fig. 1). In Hook3, the compressive stress loaded at the lingual part of the centric incisor root. These results suggested the centric incisor was tipped. On the other hand, in Hook2, the compressive stress was observed in the distal 1/2, and tensile stress at the medial 1/2 of the centric incisor root. This result suggested that the centric incisor moved bodily.

We suggested that controlled incisor retraction is possible by changing the length and the position of the hooks soldered to the archwire when using miniscrew implants.

Interface Oral Health Science 2007: 185–186

Periodontal tissue activation by resonance vibration

Makoto Nishimura[1], Mirei Chiba[2]*, Toshiro Ohashi[3], Masaaki Sato[3], and Kaoru Igarashi[1]
[1]*Division of Oral Dysfunction Science;* [2]*Division of Orthodontics and Dentofacial Orthopedics, Department of Oral Health and Development Sciences, Graduate School of Dentistry;* [3]*Biomechanics Laboratory, Department of Bioengineering and Robotics, Graduate School of Engineering, Tohoku University, Sendai, Japan*
*mirei@mail.tains.tohoku.ac.jp

Abstract. Accelerating the speed of orthodontic tooth movement should contribute to the shortening of the treatment period, which is one of the unfavorable aspects of orthodontic treatment. This study developed the resonance vibration delivery system and evaluated the effects of resonance vibration on tooth movement, and revealed the cellular and molecular mechanisms of the periodontal ligament responses. The application of resonance vibration might accelerate orthodontic tooth movement via enhanced receptor activator of nuclear factor kappa B ligand expression in the periodontal ligament without additional damage to periodontal tissues, such as root resorption.

Key words. vibration, orthodontic tooth movement, osteoclasts, receptor activator of nuclear factor kappa B ligand (RANKL), root resorption

Introduction

Since orthodontic treatment usually takes place over a long period of time, the problems of caries, periodontal disease, and prolonged treatment period are burdensome for the patient. It has also been reported that total treatment duration proved to be highly correlated with root resorption [1]. The initial response of cells to mechanical stress in vitro appears within 30 min [2–4]. These initial responses were attempted to activate at the cellular level by applying resonance vibrational stimulation to a tooth and its periodontal tissue. Thus, we hypothesized that the application of resonance vibration accelerates tooth movement by increasing the activity of the cells in the periodontal ligament.

The aim of this study was to investigate the effects of resonance vibration on the speed of tooth movement and root resorption, and to elucidate the cellular and molecular mechanisms underlying the acceleration of tooth movement.

Materials and methods

We have developed a vibration-imposed system (resonance vibration delivery system) in conjunction with IMV Corp., Osaka, Japan. The upper first molars of 6-week-old male Wistar rats were moved to the buccal side using an expansive spring for 21 days (control group), and the amount of tooth movement was measured. Vibrational stimulation (60 Hz, 1.0 m/s^2) was applied to the rat's first molars using a loading vibration system on days 0, 7, and 14 during orthodontic tooth movement (experimental group). Sections were used for immunohistochemical analysis of receptor activator of NF kappa B ligand (RANKL) expression. The number of osteoclasts in the alveolar bone was counted, and the amount of root resorption was measured.

Results and conclusion

The average resonance frequency of the upper first molar was 61.02 ± 8.38 Hz. Tooth movement in the experimental group was significantly greater than that of the control group. Enhanced RANKL expression was observed on fibroblasts and osteoclasts in the periodontal ligament of the experimental group. The number of osteoclasts in the experimental group was significantly increased over the control group. Histologically, there were no pathological findings in either group nor significant difference in the amount of root resorption between the two groups.

In conclusion, the application of resonance vibration accelerated the speed of tooth movement, possibly activating the RANK-RANKL signaling, in the absence of collateral damage to periodontal tissues, such as root resorption.

Acknowledgements. This research was supported in part by a Grant-in-Aid for Scientific Research (B) from the Ministry of Education, Culture, Sports, Science and Technology of Japan (Nos. 12557179, 16390602).

References

1. Segal GR, Schiffman PH, Tuncay OC (2004) Meta analysis of the treatment-related factors of external apical root resorption. Orthod Craniofac Res 7:71–78
2. Yamaguchi N, Chiba M, Mitani H (2002) The induction of c-fos mRNA expression by mechanical stress in human periodontal ligament cells. Arch Oral Biol 47:465–471
3. Matsuda N, Morita N, Matsuda K, et al (1998) Proliferation and differentiation of human osteoblastic cells associated with differential activation of MAP kinases in response to epidermal growth factor, hypoxia, and mechanical stress in vitro. Biochem Biophys Res Commun 249:350–354
4. Kikuiri T, Hasegawa T, Yoshimura Y, et al (2000) Cyclic tension force activates nitric oxide production in cultured human periodontal ligament cells. J Periodontol 71:533–539

Interface Oral Health Science 2007: 187–188

Mesenchymal stem cells in human wisdom tooth germs

D. Nishihara[1]*, Y. Iwamatsu-Kobayashi[1], M. Hirata[1], K. Kindaichi[2], J. Kindaichi[3], and M. Komatsu[1]
[1]*Division of Operative Dentistry;* [2]*Division of oral and Craniofacial Anatomy, Tohoku University Graduate School of Dentistry, Sendai 980-8575;* [3]*National Center for Child Medical Health and Development, Tokyo 157-8535; Japan*
*nishihara-th@umin.ac.jp

Abstract. We examined the immunohistochemical localization and isolation of mesenchymal stem cells in human wisdom tooth germs. Pericytes of blood vessels in the dental papilla and ameloblasts were stained strongly, but odontoblasts reacted weakly with STRO-1 and CD146. The STRO-1 positive cells represented approximately 7% of the total cell population. It is suggested that human wisdom tooth germs are useful for the stem cell treatment and the tissue engineering.

Key words. mesenchymal stem cell, wisdom tooth germ

1 Introduction

Mesenchymal stem cells have the possibility for the stem cell treatment and the tissue engineering. They have been isolated from the dental pulp, the deciduous tooth, and the periodontal ligament. If the stem cells of wisdom tooth germs are separated and identified, it is possible to use them for the regenerative medicine. In this research, we examined the immunohistochemical localization and isolation of mesenchymal stem cells in human wisdom tooth germs.

2 Materials and methods

Human wisdom tooth germs were extracted from healthy subjects (aged 8–12 years) after informed consent. The wisdom tooth germs were about 9–10 mm width. For histochemical study, they were fixed with 4% PFA and embedded in paraffin or Tissue Tek®. The specimens were sectioned and stained with H-E, AZAN or antibodies as STRO-1 (MAB1038, R&D Systems, Minneapolis, MN, USA), CD146 (MAB16985X, Chemicon International, Temecula, CA, USA), and Laminin (AB19012, Chemicon International). For isolation of mesenchymal stem cells, tooth germs were dissociated into cells. Using Dynabeads® (Invitrogen, Carlsbad, CA, USA), the mesenchymal stem cells (STRO-1 positive) were isolated and counted.

3 Results

Human wisdom tooth germs were at the early bell stage, which showed the formation of predentin and the polarization of ameloblast, but not the mineralization and formation of enamel. In dental papilla, odontoblasts were arranged but cell-rich zone and cell-free zone were not observed. The predentin was stained with eosin and showed collagen blue-stained with AZAN. Pericytes of blood vessels in the dental papilla and ameloblasts were stained strongly with STRO-1, and some odontoblasts reacted with STRO-1 and CD146. Staining with STRO-1 and CD146 in the dental papilla, both positive reactions appeared in the same area. The STRO-1 positive cells represented $7.35 \pm 4.12\%$ ($n = 8$) of the total cell numbers.

4 Discussion

Seo et al. [1] reported that mesenchymal stem cells which had high differentiation and proliferation potency were separated from periodontal ligament. They were expressed by the mesenchymal stem cell makers STRO-1 and CD146. In this research, mesenchymal stem cells existed in human wisdom tooth germs was confirmed. Moreover, a lot of these cells existed especially around the blood vessels like the stem cells in the adult periodontal ligament. Nagatomo et al. [2] reported that the percentage positivity with STRO-1 was $1.2 \pm 0.1\%$ in the periodontal ligament. In this study, it was $7.35 \pm 4.12\%$ in human wisdom tooth germs. It is suggested that the human wisdom tooth germs could be a good source of stem cells. In this research, ameloblasts and some odontoblasts were also positive with STRO-1 and CD146. There is a possibility that the distribution of STRO-1 positive ameloblasts or odontoblasts is changeable with the differentiation and the growth stage of tooth germs. It is suggested that human wisdom tooth germs are useful for the stem cell treatment and the tissue engineering.

Acknowledgements. This work was supported by Grant-in-Aid for Scientific Research (No. 17591982) from Japan Society for the Promotion of Science, Japan.

References

1. Seo BM, Miura M, Gronthos S, et al (2004) Lancet 364 (9429):149–155
2. Nagatomo K, Komaki M, Sekiya I, et al (2006) J Periodontal Res 41(4):303–310

Interface Oral Health Science 2007: 189–190

Osteoblast apoptosis by compressive force and its signaling pathway

Mirei Chiba[1]*, Yuko Goga[1], Aya Sato[2], and Kaoru Igarashi[2]
[1]*Division of Orthodontics and Dentofacial Orthopedics;* [2]*Division of Oral Dysfunction Science, Department of Oral Health and Development Sciences, Graduate School of Dentistry, Tohoku University, Sendai 980-8575, Japan*
*mirei@mail.tains.tohoku.ac.jp

Abstract. The aim of this study is to examine whether the in vitro application of mechanical stress would induce apoptosis in cultured human osteoblastic cell line MG-63 cells and/or human primary periodontal ligament (PDL) cells. Cells were subjected to continuous compressive force directly. As a result, in vitro compressive force induces apoptosis via caspase-8 activation in osteoblasts, while PDL cells did not show a significant change. The cell-type-specific cell-death reaction may regulate periodontal remodeling at sites where excessive compressive force is applied during orthodontic tooth movement.

Key words. mechanical stress, apoptosis, osteoblasts, periodontal ligament cells

Introduction

Orthodontic tooth movement (OTM) occurs during the sequential bone remodeling induced by therapeutic mechanical stress [1]. The application of light orthodontic force causes direct resorption of alveolar bone, whereas the application of excessive orthodontic force results in excessive compressive force, which induces local ischemia, tissue hyalinization, and cell death in the periodontal ligament (PDL) [2]. Recent studies have demonstrated that, during OTM, the application of pressure caused an increase in the number of apoptotic cells at the pressure site [3–5]. However, little is known about how mechanical force induces apoptotic cell death.

This study examined whether the in vitro application of a continuous compressive force induces apoptosis in osteoblast-like cells and/or PDL cells.

Materials and methods

The protocol for this experiment was reviewed and approved by the Tohoku University School of Dentistry Research Ethics Committee, and informed consent was obtained from all volunteers. Human osteoblast-like cell line MG-63 cells and primary human PDL cells were used. Cells were subjected to continuous compressive force directly [6]. The viability of the cells, apoptosis [terminal deoxy-nucleotidyl transference-mediated nick end labeling (TUNEL) assay], caspase-3 activity were examined.

Results and conclusion

After 24 h of the continuous compressive force application, MG-63 human osteoblast-like cells became aligned irregularly and the spaces between them increased. MG-63 cell viability decreased in a time- and force-dependent manner, suggesting that the continuous compressive force induced cell death. In the TUNEL analysis, the number of apoptotic cells was significantly increased in a time- and force-dependent manner. Moreover, caspase-3 activity increased with the continuous compressive force. Furthermore, the caspase-8 inhibitor significantly reduced the compressive force-induced caspase-3 activation, whereas the caspase-9 inhibitor did not. In vitro compressive force induces apoptosis in osteoblasts, while PDL cells did not show a significant change in the markers related to apoptosis.

We conclude that osteoblast-like cells undergo apoptosis by application of a continuous compressive force, while PDL cells can resist such force. The cell-type-specific cell-death reaction to mechanical stress may regulate periodontal remodeling at sites where excessive compressive force is applied during OTM.

Acknowledgements. This work was supported in part by Grants-in-Aid of Scientific Research (B) from the Ministry of Education, Culture, Sports, Science and Technology, Japan (16390602).

References

1. Roberts WE, Goodwin WC, Heiner SR (1981) Cellular response to orthodontic force. Dent Clin North Am 25:317
2. Reitan K, Rygh P (1994) Biomechanical principles and reactions. In: Graber TH, Vanarsdall RL (eds) Orthodontics, current principles and techniques. C.V. Mosby, St. Louis, pp 96–192
3. Hatai T, Yokozeki M, Funato N, et al (2001) Apoptosis of periodontal ligament cells induced by mechanical stress during tooth movement. Oral Dis 7:287–290
4. Rana MW, Pothisiri V, Killiany DM, et al (2001) Detection of apoptosis during orthodontic tooth movement in rats. Am J Orthod Dentofacial Orthop 119:516–521
5. Mabuchi R, Matsuzaka K, Shimono M (2002) Cell proliferation and cell death in periodontal ligaments during orthodontic tooth movement. J Periodontal Res 37:118–124
6. Goga Y, Chiba M, Shimizu Y, et al (2006) Compressive force induces osteoblast apoptosis via caspase-8. J Dent Res 85:240–244

Interface Oral Health Science 2007: 191–192

Effects of a selective cyclooxygenase-2 inhibitor, celecoxib, on osteopenia and increased bone turnover in ovariectomized rats

Hitoshi Yamazaki and Kaoru Igarashi*
Division of Oral Dysfunction Science, Department of Oral Health and Development Sciences, Tohoku University Graduate School of Dentistry, Sendai 980-8575, Japan
*igarashi@mail.tains.tohoku.ac.jp

Abstract. The purpose of this study was to determine whether a selective cyclooxygenase-2 inhibitor, celecoxib, could control osteopenia and increased bone turnover associated with ovariectomy in rats. Female Sprague-Dawley rats were subjected to either bilateral ovariectomy (OVX) or sham surgery (Sham). Sham and OVX rats were treated daily with vehicle alone or celecoxib (20 mg/kg) for 5 weeks. When determined by peripheral quantitative computed tomography (pQCT), celecoxib had no apparent effects on osteopenia induced by OVX. However, it inhibited osteoclast differentiation and bone resorption at a serum marker level. These results suggest that celecoxib could be used as a therapeutic agent for bone metabolic disorders such as osteoporosis.

Key words. osteoporosis, cyclooxygenase-2 inhibitor, osteoclast number, type I collagen C-terminal telopeptide, rats

1 Introduction

Postmenopausal osteoporosis is a disease that results from estrogen deficiency. Clarification of underlying pathological mechanisms and development of new therapies are important issues even in the dental field. The purpose of this study was to determine the effects of a selective cyclooxygenase-2 inhibitor, celecoxib, on osteopenia and increased bone turnover in ovariectomized rats.

2 Materials and methods

Twenty-six virgin female Sprague-Dawley rats were subjected to either bilateral ovariectomy (OVX) or sham surgery (Sham). Sham and OVX rats were treated daily with vehicle alone or celecoxib (20 mg/kg). The food consumption of OVX rats was restricted to limit OVX-induced weight gain. After 5 weeks, all animals were sacrificed and limb bones were removed for peripheral quantitative computed tomography (pQCT) and histological evaluations. The blood samples were also collected for assays for serum levels of markers of bone turnover.

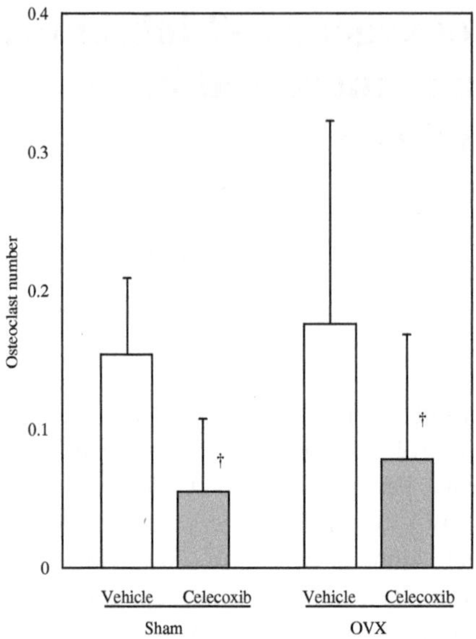

Fig. 1. Osteoclast number in cancellous bone. Celecoxib decreased the number of osteoclasts regardless of ovariectomy (*OVX*) (†$P < 0.05$ vs. Vehicle)

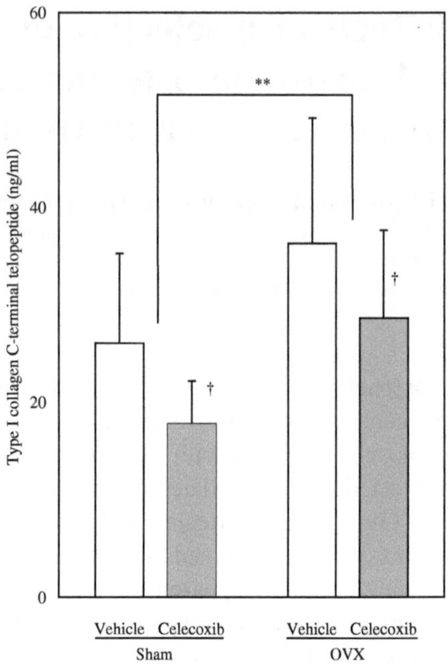

Fig. 2. Serum level of type I collagen C-terminal telopeptides (CTx). OVX increased CTx (**$P < 0.01$ vs. Sham) and celecoxib decreased CTx (†$P < 0.05$ vs. Vehicle). These effects were independent of each other

3 Results

In OVX rats, trabecular bones and their connectivity in the femoral metaphysis decreased as compared with Sham rats, indicating relatively severe osteopenia. pQCT also revealed decreased bone mineral density in the cancellous bone and polar stress strain index of the femoral distal metaphysis. However, there were no significant effects of celecoxib on these parameters. On the other hand, celecoxib significantly decreased the number of osteoclasts that appeared in the cancellous bone (Fig. 1) and the serum level of type I collagen C-terminal telopeptides, a bone resorptive marker (Fig. 2).

4 Conclusion

Although celecoxib had no apparent effects on osteopenia in early stage of estrogen deficiency in rats, it inhibited osteoclast differentiation and bone resorption at a serum marker level, suggesting that celecoxib could be used as a therapeutic agent for bone metabolic disorders such as osteoporosis.

Interface Oral Health Science 2007: 193–194

Amelogenin splicing variant promotes chondrogenesis

Junko Hatakeyama[1]*, Yuji Hatakeyama[2], Naoto Haruyama[3], Ichiro Takahashi[4], Ashok B. Kulkarni[5], and Yasuyuki Sasano[1]

[1]*Division of Craniofacial Development and Regeneration;* [3]*Division of Oral Dysfunction Science;* [4]*Division Orthodontics and Dentofacial Orthopedics, Tohoku University Graduate School of Dentistry, Sendai 980-8575, Japan;* [2]*Center for Craniofacial and Molecular Biology, University of Southern California, Los Angeles, CA;* [5]*Craniofacial Developmental Biology and Regeneration Branch, National Institute of Dental and Craniofacial Research, NIH, DHHS, Bethesda, MD; USA*

*junko-h@mail.tains.tohoku.ac.jp

Abstract. Isoforms of amelogenin, a highly conserved protein in mammalians that constitutes 90% of the enamel organic matrix, are mainly produced by ameloblasts. One of the amelogenin splicing variants, leucine rich amelogenin peptide (LRAP) has been reported to promote maturation of osteoblasts and periodontal ligament (PDL) cell proliferation. However, it is still unclear whether the amelogenin splicing variants are involved in chondrogenesis. The aim of the present study was to characterize the function of LRAP in chondrogenesis. Micromass culture of embryonic limb bud mesenchymal cells was performed as in vitro model for chondrogenesis. The recombinant LRAP was added to micromass culture system to analyze the effect on proliferation of mesenchymal cells and chondrogenic differentiation. LRAP increased the number of 5-bromo-2-deoxyuridine (BrdU) positive cells and nodules stained with alcian blue. In addition, LRAP increased mRNA expression of Sox9 and type II collagen. These findings suggest that the amelogenin splicing variant, LRAP, promotes chondrogenesis.

Key words. amelogenin, splicing variant, LRAP, chondrogenesis, proliferation

1 Introduction

Isoforms of amelogenin, a highly conserved protein in mammalians that constitutes 90% of the enamel organic matrix, are mainly produced by ameloblasts. Leucine rich amelogenin peptide (LRAP) is one of the amelogenin splicing variants, and has been suggested as a signaling peptide that promotes maturation of osteoblasts and suppresses maturation of osteoclasts and periodontal ligament (PDL) cell migration and proliferation [1–3]. Recently, enamel matrix protein derivatives were reported to induce chondrogenesis [4]. However, it is unclear whether the amelogenin splicing variants are involved in chondrogenesis. The aim of the present study was to characterize the function of LRAP in chondrogenesis.

2 Materials and methods

Timed pregnant C57BL/6 mice were obtained, and embryos were collected at embryonic day 10. Micromass culture of embryonic limb bud mesenchymal cells was performed as described previously [5]. Thirty microliter drops of cell suspension was plated and cultured with the recombinant porcine LRAP (85% homology to mouse LRAP) or GDF5 as used for a positive control of chondrogenesis. Alcian blue staining and quantification of staining intensity in cartilaginous nodules were performed as described previously [5]. Cell proliferation was determined by measuring the incorporation of 5-bromo-2-deoxyuridine (BrdU) into DNA using a cell proliferation kit. The gene expression levels of *Sox9*, collagen 2 alpha 1(*Col2a1*), and aggrecan were examined by semi-quantitative reverse transcription-polymerase chain reaction (RT-PCR).

3 Results and discussion

Leucine rich amelogenin peptide (LRAP) promoted proliferation of mesenchymal cells isolated from limb buds in micromass culture dose-dependently. LRAP increased mRNA expression of both *Sox9* and *Col2a1*, which are chondrogenic differentiation-related marker genes. In addition, the intensity of alcian blue stainings for detection of cartilaginous matrix nodules was increased. These findings suggest that the amelogenin splicing variant, LRAP, promotes proliferation of chondroprogenitor cells and their differentiation.

References

1. Hatakeyama J, Sreenath T, Hatakeyama Y, et al (2003) The receptor activator of nuclear factor-kappa B ligand-mediated osteoclastogenic pathway is elevated in amelogenin-null mice. J Biol Chem. 278:35743–35748
2. Hatakeyama J, Philp D, Hatakeyama Y, et al (2006) Amelogenin-mediated regulation of osteoclastogenesis, and periodontal cell proliferation and migration. J Dent Res 85:144–149
3. Viswanathan HL, Berry JE, Foster B, et al (2003) Amelogenin: a potential regulator of cementum-associated genes. J Periodontol 74:1423–1431
4. Kim NH, Tominaga K, Tanaka A (2005) Analysis of eosinophilic round bodies formed after injection of enamel matrix derivative into the backs of rats. J Periodontol 76:1934–1941
5. Hatakeyama Y, Tuan RS, Shum L (2004) Distinct functions of BMP4 and GDF5 in the regulation of chondrogenesis. J Cell Biochem 91:1204–1217

Interface Oral Health Science 2007: 195–196

The relationship between the laser Doppler blood-flow signals and the light intensity in the root canals in human extracted teeth

Motohide Ikawa* and Hidetoshi Shimauchi
Division of Periodontology and Endodontology, Department of Oral Biology, Tohoku University Graduate School of Dentistry, Sendai 980-8575, Japan
*moto-i@mail.tains.tohoku.ac.jp

Abstract. The relationship between the laser Doppler blood-flow signals and the laser light intensity in the root canal was examined. The root canals of extracted teeth were enlarged from the apical end to the 2 mm incisal to the level of the cement-enamel junction (CEJ). Human peripheral blood was pumped through the apical foramen of the teeth, and the blood flow signal was monitored. Both the light intensities in the root canal and pulpal blood-flow (PBF) signals were small when the root canal was enlarged 3 or 4 mm apical to the CEJ. Both intensities increased with the progress of root canal enlargement to the tooth. There was a significant relationship between the laser Doppler blood-flow signals and the light intensity in the root canals. The results indicated that the laser light intensity in the root canal effected on the PBF in human teeth using Laser Doppler flowmetry (LDF).

Key words. human, tooth, pulp, blood-flow, laser-Doppler

1 Introduction

Laser Doppler flowmetry (LDF) has been used as one of the measurement techniques in monitoring pulpal blood-flow (PBF) in humans [1, 2]. The blood-flow signal is considered to relate with the amount and the location of PBF. In this study, the artificial blood circulation in the root canals of extracted human upper central incisors was prepared, and the relationship between the blood-flow signal and the laser light intensity in the root canal was examined.

2 Materials and methods

Resin cap to cover both labial and palatal sides of the examined teeth (n = 6) was prepared. A hole for the insertion of the LDF probe was drilled into each side of the tooth crown. The apical portions of examined teeth were severed and discarded. The root canals were enlarged from the apical end to the 2 mm incisal to the level of the cement-enamel junction (CEJ).

The intensity of the laser light in the root canal was measured with tap water filled in the canal. Heparinized human peripheral blood collected in advance and diluted three times with physiological saline was pumped through the apical foramen of the teeth, and the blood flow signal was monitored (FLO-HP1, Omegawave Inc., Tokyo, laser output 5 mW).

Statcel 2 was used as the statistical analysis. Repeated measures ANOVA was used to examine the effect of the probe location and the degree of root canal enlargement on blood flow signal. Scheffe's F test was used as Post hoc test. Relationship between the blood flow signal and the light intensity in the root canal was examined. P value less than 0.05 was considered as significant.

3 Results and discussion

Both the blood flow signal and the light intensities in the canal significantly increased with the progress of the root canal enlargement (n = 6). Both the blood flow signal and the light intensities in the canal were significantly higher when the probe was placed in the palatal surface of the tooth (n = 6). There was a significant relationship between the the blood flow signal and the light intensities in the canal (n = 84).

The LDF is considered to detect a limited area of pulp: the pulp chamber and the coronal part of the root canal. The blood flow signals were affected by the location of the pulp horn. This is considered to relate to the optical path of LDF. The incident light in LDF reaches pulp tissues (blood vessels), via mainly enamel prisms and dentinal tubules [3], and returns to the probe by the reverse route.

As the results in this study showed, more light could reach the pulp with much larger blood flow signal; suggesting the advantage of the palatal location of the recording probe. The optimal probe location would produce more light intensities in the root canal (pulp) and more blood flow signals.

References

1. Amess TR, Andrew D, Son H, et al (1993) The contribution of periodontal and gingival tissues to the laser Doppler blood flow signal recorded from human teeth. J Physiol (Lond) 473: 142P
2. Hartman A, Azerad J, Boucher Y (1996) Environmental effects on laser Doppler pulpal blood-flow measurements in man. Arch oral Biol 41:333–339
3. Odor TM, Watson TF, Pitt Ford TR, et al (1996) Pattern of transmission of laser light in teeth. Int Endod J 29:228–234

Interface Oral Health Science 2007: 197–198

Pulpal blood flow in human primary teeth with different root resorption

Hideji Komatsu[1]*, Motohide Ikawa[2], and Hideaki Mayanagi[1]
[1]*Division of Pediatric Dentistry, Department of Oral Health and Development Sciences;*
[2]*Division of Periodontology and Endodontology, Department of Oral Biology, Tohoku University Graduate School of Dentistry, Sendai 980-8575, Japan*
*komatsu@pedo.dent.tohoku.ac.jp

Abstract. The purpose of this study was to examine the relation between the root resorption and the pulpal blood flow (PBF) in human primary teeth using laser Doppler flowmeter (LDF). Recordings were made on 15 clinically healthy upper primary central incisors in nine healthy participants (age: 3 years 11 months–7 years 3 months). The state of roots of the teeth examined were confirmed by radiographs. The mean PBF signals tended to decrease with the progress of the root resorption. Results indicated that PBF could indicate the status of the root resorption in the human primary teeth.

Key words. pulp, blood flow, laser Doppler, primary teeth, root resorption

1 Introduction

Komatsu et al. [1] reported that the pulpal blood flow (PBF) in the human primary tooth (subjects' age: 3 years 11 months–7 years 3 months, n = 12) reduces with age. To our knowledge, no evidence on the relation between the root resorption and the PBF in primary tooth pulp has been obtained in previous studies. Therefore, the purpose of this study was to examine the relation between the root resorption and the PBF in human primary teeth using laser Doppler flowmeter (LDF).

2 Materials and methods

Recordings were made on 15 clinically healthy upper primary central incisors in nine healthy participants (age: 3 years 11 months–7 years 3 months). Six examined teeth of three subjects were measured three times and one examined teeth of one subject was measured twice on different root resorption. The interval between the repeated measurements in each individual ranged between 6 and 18 months (mean 10.7 months).

Prior to the measurement, individual resin caps with a hole at approximately 1.5 mm incisal to the gingival margin of the examined tooth were prepared, and the probe of the LDF (MBF3D, Moor instruments, UK) was fitted in the hole during the recording. Measurements were made with and without opaque black rubber dam application to the examined tooth. Recordings were stored in a computer (Power Macintosh G3, Apple Computer Inc.) via a laboratory interface (MacLab/8s, ADInstruments Pty Ltd., Australia) and signal processing software (Chart, ADInstruments Pty Ltd., Australia).

The state of roots of the teeth examined were confirmed by radiographs. According to Yamashita's method [2], the state of roots of the teeth were classified into four groups; Group1: roots of the primary teeth were slightly resorptive, Group2: roots of the primary teeth were resorptive to 1/4 from apex, Group3: roots of the primary teeth were resorptive to 3/4 from apex, Group4: roots of the primary teeth were almost resorptive.

3 Results

Results obtained were as follows: the mean PBF signals tended to decrease with the progress of the root resorption.

4 Discussion

The authors [1, 3] reported the age-related reduction of PBF in human primary teeth, and the reduction is considered to be due to the decrease in blood supply caused by the root resorption. In the human primary teeth, the size and volume of the pulp is not reduced with age [4]. Therefore the age-related reduction of PBF shown in the present study is considered to be due to the reduction of the blood supply caused by the root resorption.

References

1. Komatsu H, Ikawa M, Mayanagi H (2007) Age-related changes of pulpal blood flow in primary teeth measured by laser Doppler blood flowmetery. Pediatr dent J 17:27–31
2. Yamashita H (1977) Pediatric dentistry (Syounishikagaku: souron). Ishiyaku, Tokyo, pp 265–269 (in Japanese)
3. Komatsu H, Ikawa M, Mayanagi H (2003) Age-related changes of pulpal blood flow in human primary tooth. J Dent Res 82:C-402 (Abst #58)
4. Yamashita H (1977) Pediatric dentistry (Syounishikagaku: souron). Ishiyaku, Tokyo, pp 240–242 (in Japanese)

Interface Oral Health Science 2007: 199–200

Sympathetic nerve fibers in rat normal and inflamed dental pulp: absence from dentinal tubules

Y. Shimeno*, Y. Sugawara, M. Iikubo, N. Shoji, and T. Sasano
Division of Oral Diagnosis, Tohoku University Graduate School of Dentistry, Sendai 980-8575, Japan
*shime-no@umin.ac.jp

Abstract. This study was designed to determine if sympathetic nerve fibers exist in dentinal tubules in rat normal dental pulp, and if they sprout into the tubules in the inflamed condition. Sympathetic nerve fibers in rat molar dental pulp were labeled using an anterograde axonal transport technique involving injection of wheat germ agglutinin-horseradish peroxidase (WGA-HRP) into the superior cervical ganglion. In the normal dental pulp, scattered WGA-HRP reaction products were observed in unmyelinated nerve endings in odontoblast layer and subodontoblastic region, mostly close to odontoblast cell bodies. More reaction products were detected in the above areas in the inflamed dental pulp than that in the normal pulp. However, no labeled nerve fibers were observed in dentinal tubules in either normal or inflamed dental pulp. These results indicate that although sympathetic nerve fibers do indeed sprout in inflamed dental pulp, they do not penetrate into the dentinal tubules.

Key words. sympathetic nerve fiber, anterograde axonal transport, wheat germ agglutinin-horseradish peroxidase (WGA-HRP), dentinal tubule, dental pulp

1 Introduction

The autonomic nerve fibers found in the dental pulp are mainly sympathetic, and principally routed from the superior cervical ganglion. The present experiment was designed to study whether cavity preparation caused the sprout of sympathetic nerve fibers into the dentinal tubules, or not.

2 Materials and methods

A total of 14 adult male Wistar rats were used. Cavity preparation was made on left maxillary first molar in the experimental group (n = 7), while examined teeth were left intact in the control group (n = 7). Sympathetic nerve fibers in the rat molar dental pulp were labeled using an anterograde axonal transport technique

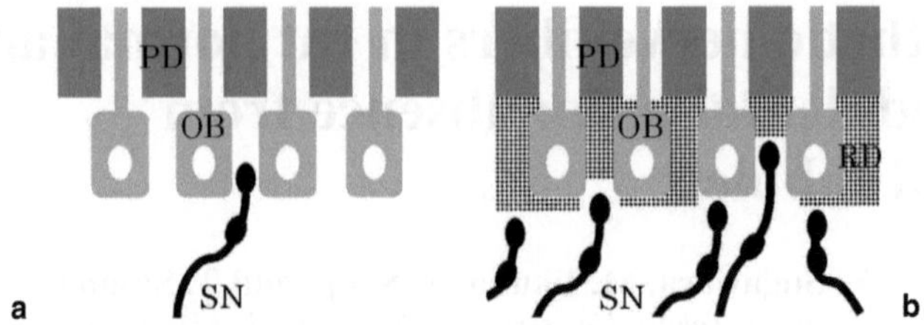

Fig. 1. Illustrations show relationship between odontoblasts and sympathetic nerves. In the controls, one sympathetic nerve (*SN*) (or none) is close to each odontoblast cell body (*OB*) [**a**; *PD* predentin]. In contrast, in the experimental (cavity preparation) group sympathetic nerves are sprouting toward the odontoblast cell bodies [**b**; *RD* reparative dentin]. Neither original nor sprouted sympathetic nerve fibers are present in the dentinal tubules

involving the injection of wheat germ agglutinin-horseradish peroxidase (WGA-HRP) into the left superior cervical ganglion (SCG) in the control group and 18 days after cavity preparation in the experimental group. All rats were perfused through 4% paraformaldehyde in 0.1 mol/l phosphate buffer 3 days after WGA-HRP injection. The pulps of the left maxillary first molars in both groups were then observed using electron microscopes.

3 Result and conclusion

In the controls, most of the WGA-HRP reaction products were situated very close to the odontoblast cell bodies in the odontoblastic layer. In the dentinal tubules, odontoblast processes and unmyelinated nerve endings were clearly observed; however, no WGA-HRP reaction products were found (Fig. 1a).

In the experimental group, WGA-HRP reaction products were observed more densely in the odontoblast layer than that in the control pulp. As in the controls no WGA-HRP reaction products were found in the dentinal tubules (Fig. 1b).

These results indicated that the cavity preparation did not cause any sprout of sympathetic nerve fibers into the dentinal tubules.

Interface Oral Health Science 2007: 201–202

Difference of brain function between normal occlusions and malocclusions using NIRS

Koshi Sato[1]*, **Maiko Hayashi**[1], **Teruko Takano-Yamamoto**[1], **Masaki Nakamura**[2], **and Hiroo Matsuoka**[2]
[1]*Division of Orthodontics and Dentofacial Orthopedics, Tohoku University Graduate School of Dentistry, Sendai 980-8575;* [2]*Division of Psychiatry, Tohoku University Graduate School of Medical Sciences, Sendai 980-8574; Japan*
*koshi@mail.tains.tohoku.ac.jp

Abstract. The mastication is not only related with the peripheral function, but also the central control system. The purpose of this study was to clarify the interface between the brain function and the occlusion by investigating the stimulation of the frontal association cortex, which relates to the function of the human nature. Subjects were healthy right-handed young adult volunteers with acceptable normal occlusion and class III malocclusion. Near-infrared spectroscopic topography (NIRS) was used to investigate the stimulation of the frontal association cortex during chewing gum without flavors. It was proved that the frontal association cortex was also stimulated by chewing gum. However, the behavior of the stimulation depended on the individual occlusion type. This study suggests that orthodontic treatment would improve the quality of life (QOL) of patients from the point of healthy brain function.

Key words. brain, NIRS, frontal association cortex, malocclusion, chewing gum

There are many reasons why orthodontists should treat malocclusion. They might be esthetic concerns of the patient and prevention of dental caries, periodontal diseases, and temporomandibular disorders (TMD) problems. Furthermore, the mastication is not only related with the peripheral function, but also the central control system. As a technique to measure a brain function without any invasion like radiation and pains, functional MRI (fMRI), and magnetoencephalography (MEG) are well known. We have already clarified the difference in brain activation between normal occlusions and malocclusions using positron emission tomography (PET) [1] and fMRI [2]. However, these devices are not suitable for performing the task during movement of the head such as the chewing. Near-infrared spectroscopic topography (NIRS) which was developed recently is an optical method that allows noninvasive measurement of concentration in oxygenated and deoxygenated hemoglobin in cortical areas; and NIRS is a method to know the activation state of a brain. The purpose of this study was to compare the difference in the brain function during chewing between normal occlusions and malocclusions by investigating the stimulation of the prefrontal cortex, which relates the function of the human nature and working memory.

Subjects were ten healthy right-handed young adult volunteers with acceptable normal occlusion (n = 9, 3 male, 6 female, mean 26.5 years) and Class III

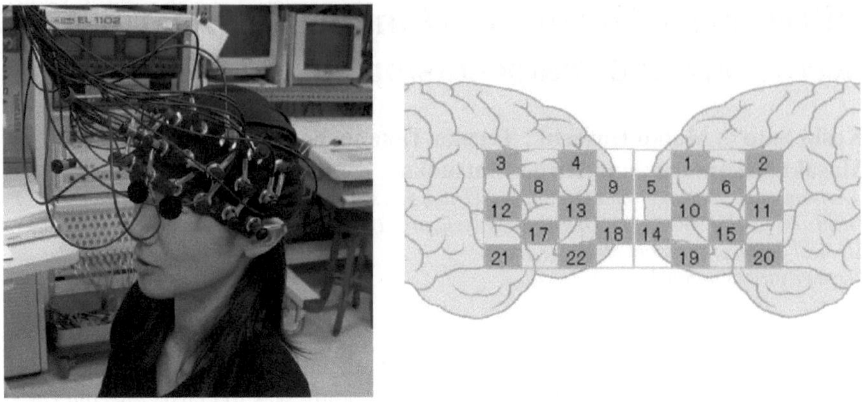

Fig. 1. Channel number of the probe for detecting prefrontal cortex

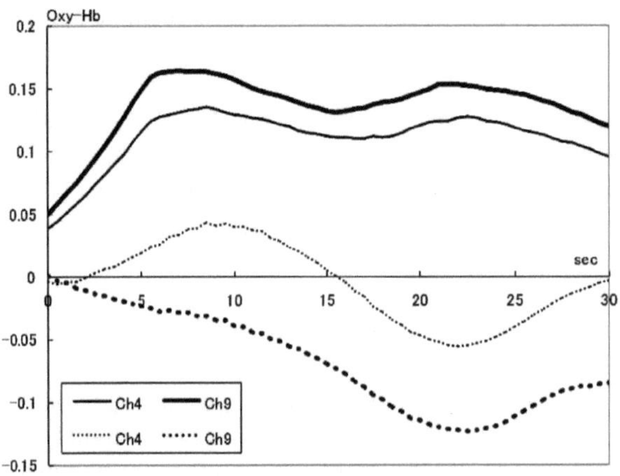

Fig. 2. Comparison of free chewing between normal occlusions (*solid lines*) and reversed occlusion (*broken lines*); prefrontal areas (*Ch4* and *9*) were not stimulated on the reversed occlusion

malocclusion (n = 1, male, 30.9 years). Subjects were given the left side, the right side, and free chewing tasks. The oxygen-hemoglobin was measured with optical topography system ETG-7000 (Hitachi Medical Corp, Tokyo, Japan) during the task procedures (Fig. 1).

Brain activation was measured with an interval of 0.1 s units during chewing gum. In particular, activation was recognized in not only primary motor and sensory area but also prefrontal cortex. However, the behavior of the stimulation depended on the individual occlusion type (Fig. 2).

References

1. Sato K, Mitani H, Mejia MA, et al (1996) Relationships between regional cerebral blood flow during mastication and morphological features of occlusion. J Jpn Orthod Soc 55:300–310
2. Nakajo T, Sato K, Mitani H, et al (2006) Effects of artificial premature contact on dynamics of cerebral blood flow–Examination using functional MRI–. Orthod Waves-Jpn Ed 65:101–111

Interface Oral Health Science 2007: 203–204

Measurement of human cerebral function caused by oral pain

Shin Kasahara[1]*, Toshinori Kato[2], and Kohei Kimura[1]
[1]*Division of Fixed Prosthodontics, Department of Restorative Dentistry, Tohoku University Graduate School of Dentistry, Sendai 980-8575;* [2]*Department of Brain Environmental Research, Katobrain Co., Ltd., Tokyo 108-007; Japan*
*kasahara@mail.tains.tohoku.ac.jp

Abstract. Pain is one of the most important factors in dentistry. Objective measurement of pain may be useful for dental diagnosis and therapy. However, until now, we have had no modality to measure the location and degree of pain objectively. The purpose of this study is to measure reactions to pain in cerebral cortex during dental treatment using Cerebral functional mapping of Oxygen Exchange (COE), and to assess the application of COE in the field of dentistry. We obtained the following results: (1) reactions in cerebral cortex apparently related to dental pain were measured at Brodmann's area 10, (2) COE measurement responses were sensitive and reactions could be shown within a few seconds during an event, and (3) the reactions also disappeared quickly after the pain disappeared.

Key words. oral pain, cerebral function, oxygen exchange, near-infrared light

1 Introduction

Most dental diseases can cause oral pain, and most dental treatments also cause pain. Since pain provides important information about dental disease and must be considered in planning treatment, the objective and quantitative measurement of oral pain is considered to be useful in dental diagnosis and treatment. However, there has been no way to measure location, kind, and degree of pain objectively.

Cerebral functional mapping of Oxygen Exchange (COE) is a system for measuring changes in brain functions as changes of active oxygen exchange in the capillaries at the cerebral cortex, using the diffusion and scattering of near-infrared light through the skin and the bone of the head [1, 2].

In this study, changes in cerebral function caused by oral pain were measured using NISR-imaging and displayed using COE, and changes in cerebral function related to the degree of pain and its disappearance were measured.

2 Materials and methods

The subject was a 72-year-old female patient who uses complete dentures. Oxygen exchange (oxygenated Hb, deoxygenated Hb, and total Hb) was measured (Shimadzu OMM 3000 series) to indicate brain function, with and without painful

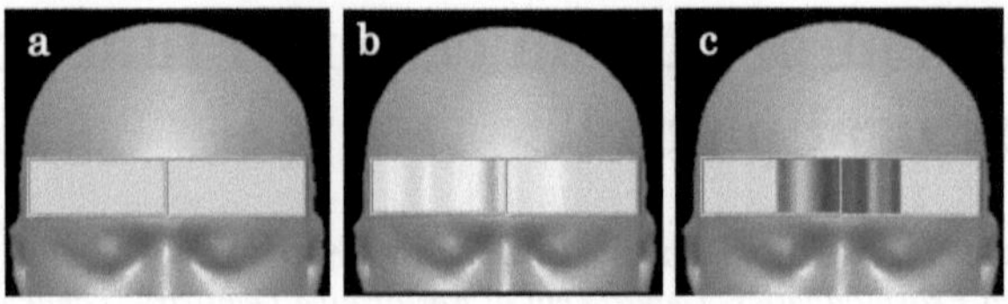

Fig. 1. Cerebral functional mapping of oxygen exchange (COE) results with pain

stimuli due to denture. Measurements were performed five times in each condition, and average measurement data were analyzed by the COE system.

3 Results and discussion

3.1 Changes in active oxygen exchange at the cerebral cortex

Changes in active oxygen exchange at the cerebral cortex when pain was caused by the dentures were as follows. While pain was felt, total Hb increased markedly, and it decreased rapidly after the pain disappeared. These measurements were very sensitive, and changes in active oxygen exchange were measured within 2–3 s during the event.

3.2 Results of COE analysis

Cerebral functional mapping of oxygen exchange results are showed in Fig. 1. Figure 1a is under control conditions. When pain was felt (Fig. 1b), increases in oxygenated Hb and deoxygenated Hb were observed in the frontal region. Then, after the pain disappeared (Fig. 1c), a decrease in deoxygenated Hb was observed.

The results of this research are the first findings relating brain function to oral pain. As a clinical application of measurements of this kind, a system for monitoring changes in pain during dental treatment can be envisioned that would have the potential to reduce dental anxiety and pain, and would also be useful in assessing the effects of a dental therapy. Further research is necessary to analyze these results quantitatively and improve the measuring device.

References

1. Kato T (2004) Int Congr Ser 1270:85–90
2. Kato T, Kamei A, Takashima S, et al (1993) J Cereb Blood Flow Metab 13:516–520

Interface Oral Health Science 2007: 205–206

Physiological characteristics of temporomandibular joint mechanosensitive neurons in the trigeminal ganglion of the rabbit

Yasuo Takafuji[1]*, Akito Tsuboi[1], Shintaro Itoh[1], Kazuki Nagata[1], Takayoshi Tabata[2], Haruhide Hayashi[2], and Makoto Watanabe[1]
[1]*Division of Aging and Geriatric Dentistry;* [2]*Division of Oral Physiology, Tohoku University Graduate School of Dentistry, Sendai 980-8575, Japan*
*takafuji@mandible.dent.tohoku.ac.jp

Abstract. The response properties of mechanosensitive neurons innervating the temporomandibular joint (TMJ neurons) in the rabbit trigeminal ganglion were investigated. Three types of discharge patterns were found: a slowly adapting type (39%), a slowly adapting type with spontaneous discharges (59%), and a rapidly adapting type (2%).

Key words. trigeminal ganglion, mechanosensitive neurons, temporomandibular joint, condyle, rabbit

Introduction

Mechanoreceptors in the temporomandibular joint (TMJ) play an important role in controlling chewing, based on sensory information about the position, velocity, direction of mandibular movement, and load applied to the TMJ. The aim of this study was to investigate the response properties of TMJ neurons in the trigeminal ganglion of rabbit.

Materials and methods

Twenty adult female rabbits (Japanese white, 1.9–2.4 kg) were used in the study. Animals were initially anesthetized with a urethane–chloralose mixture (α-chloralose 40 mg/kg, urethane 500 mg/kg) via the right auricular vein. The ramus of the mandible was cut off approximately 1.5 cm below the condyle. A metal rod was fixed to the upper part of the ramus of the mandible with a metal screw and acrylic resin, such that the condyle could be moved easily using the rod. The heads of the animals were fixed in a stereotaxic apparatus. A part of the left skull and cerebrum over the trigeminal ganglion was removed to expose the left trigeminal ganglion. The discharges of sensory neurons in response to mechanical stimulation of the left TMJ were recorded from the left trigeminal ganglion with a metal

Table 1. Response properties of temporomandibular (TMJ) units

	Spontaneous (+), n (%)			
Adaptation	Non-stopped	Stopped	Spontaneous (–), n (%)	Total, n (%)
RA	0 (0)	0 (0)	1 (2)	1 (2)
SA	25 (42)	10 (17)	23 (39)	58 (98)
Total	25 (42)	10 (17)	24 (41)	59 (100)

microelectrode. The response properties of the TMJ neurons were investigated when the left condyle was moved passively in the rostral, caudal, ventral, dorsal, medial, and lateral directions.

Results and discussion

Fifty-nine TMJ units that responded to mechanical stimulation of the TMJ were recorded from the left trigeminal ganglions of 20 rabbits. The TMJ neurons were divided into slowly adapting (SA) units, spontaneously active units, and rapidly adapting (RA) units (Table 1). Most of the TMJ units (98%) were of the SA type, and 60% of the SA units discharged spontaneously. The spontaneous discharges of 10 SA units disappeared when the condyle was moved in the dorsal and caudal directions (firing stopped type: Stopped), whereas the spontaneous discharges of 25 SA units did not disappear (firing non-stopped type: Non-stopped). Only one TMJ neuron responded transiently at the onset and offset of condyle movement in the ventral direction, whereas almost all SA units responded to condyle movement in the rostral and/or ventral directions. The relationship of the displacement of the condyle and the frequency were expressed with Stevens' power law [power component: 0.47 ± 0.43 (mean $\pm$ SD, $n = 10$)].

The majority of TMJ units (98%) showed slowly adapting responses, in reasonable agreement with previous reports [1, 2]. The slowly adapting units are assumed to carry sensory information about the jaw position and/or the amount of condyle movement to the central nervous system. Sixty percent of SA units were spontaneously active, and two explanations may account for the spontaneous discharges: first, neurons may always be firing without a load on the condyle, and second, the neurons are responsive to a shift in the condyle after separation from the mandible. To resolve this issue, an investigation on the animal with an intact condyle is required.

References

1. Kawamura Y, Abe K (1974) Role of sensory information from temporomandibular joint. Bull Tokyo Med Dent Univ 21:78–92
2. Lund JP, Matthews B (1981) Responses of temporomandibular joint afferents recorded in the Gasserian ganglion of the rabbit to passive movements of the mandible. In: Kawamura Y, Dubner R (eds) Oral-facial sensory and motor functions. Quintessence, Tokyo, pp 153–160

Interface Oral Health Science 2007: 207–208

Pressure measurement of human gingiva by tonometer

Kyoko Ikawa[1]*, Motohide Ikawa[2], and Takeyoshi Koseki[1]
[1]*Division of Preventive Dentistry, Department of Oral Health and Development Sciences;*
[2]*Division of Periodontology and Endodontology, Department of Oral Biology, Tohoku University Graduate School of Dentistry, Sendai 980-8575, Japan*
*kkikw@mail.tains.tohoku.ac.jp

Abstract. In the present study, we examined the applicability of tonometer to evaluate the hardness of human gingival tissue. For this purpose, induction-based impact method for measuring intraocular pressure was applied. The measurement was made on the young male subjects (n = 7, aver. 24.8 years) who had clinically healthy gingival tissue. The labial surfaces of attached gingival site between teeth were selected as examination. The gingival pressure ranged between 18 and 77 (n = 70, mean ± SD; 44.4 ± 8.1 mmHg). The results indicated that tonometric measurement of the gingival tissue could give quantitative information of the gingival inflammation.

Key words. human, gingiva, pressure, tonometer

1 Introduction

It is commonly experienced that the gingival tissue tends to soften with the progress of gingival inflammation. This is considered to be due to the loosened gingival cells' connection. In the present study, we examined the applicability of tonometer to evaluate the hardness of human gingival tissue.

2 Materials and methods

Induction-based impact method for measuring intraocular pressure [1] was applied; a very light probe was used to make momentary contact with the gingiva. The probe hits the gingival surface and bounces back, and the solenoid detected the movement and impact of the probe. The probe was a 40-mm-long and 1.0 mm-diameter stainless steel tube. Inside one end of the tube there was a 20-mm-long permanent magnet solenoid.

The measurement was made on the young male subjects (n = 7, aver. 24.8 years) who had clinically healthy gingival tissue. Prior to the measurement, the purpose and the method of the study were explained, and informed consent was obtained from all of them.

The labial surfaces of proximal area of attached gingival sites of teeth (13 to 23, 33 to 43) were selected as examination sites (out of total 10 sites). The depths of the periodontal pockets of the recording sites were measured by one of the authors who was qualified as a periodontal specialist. The probe hit the gingival surface and bounced back, and the solenoid detected the movement and impact of the probe. The device displays the digital value of the gingival pressure (mmHg).

Spearmans rank order correlation was performed between the pressure of gingiva and the periodontal pocket depth. The comparison of the gingival pressure between the upper gingival sites and the lower was made by Mann–Whitney *U* test. *P*-value less than 0.05 was considered as significant.

3 Results and discussion

No subjects claimed any pain or discomfort to the measurement. The gingival pressure ranged between 12 and 77 mmHg (n = 70, mean ± SD; 41.0 ± 16.2 mmHg). The gingival pressure was higher with upper gingival sites than the lower gingival sites ($P < 0.05$). The pocket depths ranged between 1.0 and 3.0 mm (n = 70, mean ± SD; 1.67 ± 0.44 mm). There was a significant relationship between the pocket depths and the gingival pressures ($P < 0.05$).

In the present study, we measured the hardness of gingiva by a tonometer. This measurement technique has the advantage that elicits no pain in the measurement [2]. The tonometric measurement gives us quantitative information of the physical property of human gingiva. Further study is required for elucidating the relationship between the gingival pressure and the status of the gingival health.

References

1. Kontiola AI (2000) New induction-based impact method for measuring intraocular pressure. Acta Ophthalmol Scand 78:142–145
2. Sakalliolu EE, Ayas B, Sakalliolu U, et al (2006) Osmotic pressure and vasculature of gingiva in periodontal disease: An experimental study in rats. Arch Oral Biol 51:505–511

Interface Oral Health Science 2007: 209–210

Retrospective study on factors that affect removable partial denture usage

Shigeto Koyama[1]*, Tomohiro Atsumi[2], Kouki Hatori[2], Toru Ogawa[2], Tomohumi Sasaki[2], Masayoshi Yokoyama[2], Kei Kubo[2], Soushi Hanawa[2], Mika Inoue[2], Kenji Kadowaki[2], Shintaro Gorai[2], Tetsuo Kawata[2], Kohei Kimura[3], Makoto Watanabe[4], and Keiichi Sasaki[2]
[1]*Maxillofacial Prosthetics Clinic, Tohoku University Hospital, Sendai 980-8575;* [2]*Division of Advanced Prosthetic Dentistry;* [3]*Division of Fixed Prosthodontics;* [4]*Division of Aging and Geriatric Dentistry, Tohoku University Graduate School of Dentistry, Sendai 980-8575; Japan*
*koyama@mail.tains.tohoku.ac.jp

Abstract. The purpose of this retrospective study was to investigate the factors that affected the usage of removable partial dentures (RPDs) and the patients' satisfaction with their RPDs 5 years after insertion. Sixty-seven patients treated with 90 RPDs that were inserted at the Tohoku University Hospital between 1996 and 2001 participated in this study. The usage rate and 12 factors that might affect usage were examined, and 15 factors regarding satisfaction were evaluated. Statistically significant associations were found between denture usage and patient's age, location of edentulous area, pain, color of artificial teeth, and arrangement. These findings suggest that the acceptance of RPDs among patients was related to factors such as patient's age, oral status, and satisfaction.

Key words. RPD, retrospective study, RPD usage, patients' satisfaction

1 Introduction

Prosthetic intervention using a removable partial denture (RPD) has been applied for oral rehabilitation in partial edentulous patients. However, factors that affect the treatment outcome have not been identified sufficiently. The aim of this retrospective study was to investigate the factors that affected RPD usage and patient satisfaction 5 years after denture insertion.

2 Materials and methods

Sixty-seven patients treated with 90 RPDs that were inserted at the Tohoku University Hospital between 1996 and 2001 participated in this study. Those patients were re-examined 5 years after insertion. Data were collected from patients' clinical records and a questionnaire concerning their assessment and use of the dentures. The status of wearing the RPD was categorized as successful (wearing their original

RPDs constantly for 5 years), replaced (wearing RPDs re-fabricated within 5 years), and failed (not wearing the RPD or interrupted wearing during 5 years). The usage rate and 9 factors that might affect usage were examined, and 15 factors regarding satisfaction were evaluated. The analyzed variables were as follows: (1) Gender, (2) Age, (3) Location of RPD, (4) Location of edentulous area, (5) Number of occlusal supports, (6) Number of missing teeth, (7) RPD on opposite jaw, (8) Material of RPD, and (9) Number of abutment teeth. The patient satisfaction factors regarding function were as follows: (1) Ability of speech, (2) Facial expression, (3) Stability of worn RPDs, (4) Fitness of worn RPDs, (5) Ease of putting on and taking off, (6) Pain of worn RPDs, (7) Mastication ability at anterior teeth, (8) Mastication ability at posterior teeth, (9) Ease of mastication and swallowing, and (10) Sense of taste. Those regarding structure were as follows: (1) Color of artificial teeth, (2) Shape of artificial teeth, (3) Size of RPD, (4) Tooth arrangement, and (5) Weight of RPD. Stepwise logistic regression analysis was used to assess statistical significance.

3 Results

There were 52 RPDs regarded as successful, 17 as replaced, and 14 as failed. The usage percentage was 61% at 5 years after insertion. Statistically significant correlations between denture usage and respective age ($P = 0.007$, RR: 5.579, 95% CI: 1.689–21.519), location of edentulous area ($P = 0.014$, RR: 12.821, 95% CI: 1.916–125), pain ($P = 0.006$, RR: 1.965, 95% CI: 1.215–3.175), color of artificial tooth ($P = 0.019$, RR: 2.827, 95% CI: 1.185–6.744), and tooth arrangement ($P = 0.003$, RR: 4.917, 95% CI: 1.724–12.987) of a patient regarding satisfaction were found.

4 Discussion

Older patients tend to pay less attention to their appearance. As a result the need for an RPD and the rate of denture usage would decrease with age, and there would be a correlation between the age of patients and their regard for aesthetics. Aesthetics related to Kennedy Class IV; on the other hand Class I and II RPDs are likely to create pain. This can be explained by the progression of the resorption in the edentulous parts. This resorption probably led to the pain caused by denture usage in older patients. Age, oral status, pain, and aesthetics are critical factors affecting the satisfaction of patients using RPDs and thus play an important role in determining RPD usage.

Section II:
Host-parasite interface

Interface Oral Health Science 2007: 213–218

Profiling of subgingival plaque biofilm microflora of healthy and periodontitis subjects by real-time PCR

Yuki Abiko[1,2]*, Takuichi Sato[1]*, Gen Mayanagi[3], and Nobuhiro Takahashi[1]
[1]*Division of Oral Ecology and Biochemistry, Department of Oral Biology;* [3]*Division of Periodontology and Endodontology, Department of Oral Biology, Tohoku University Graduate School of Dentistry, Sendai 980-8575;* [2]*Department of Oral Disease Research, National Center for Geriatrics and Gerontology, Morioka, Ohbu, Aichi 474-8522; Japan*
*tak@mail.tains.tohoku.ac.jp; abiko@nils.go.jp

Abstract. *Porphyromonas gingivalis* and *Mogibacterium timidum* were more frequently detected in subgingival microflora of periodontitis than that of healthy subjects. However, few have been investigated to quantify these bacteria in subgingival microflora. This study aimed to quantify *P. gingivalis* and *M. timidum*, as well as other 11 bacteria, e.g., *Tannerella forsythia* and *Eubacterium saphenum*, in subgingival plaque, and to evaluate their relationship with periodontitis. Subgingival plaque was obtained from 12 periodontally healthy (mean 26.4 years) and 28 periodontitis (mean 62.4 years) subjects. Total and target bacteria were quantified by real-time polymerase chain reaction (PCR) using universal and species-specific primers, respectively. The proportion of obligate anaerobes, including *P. gingivalis*, *T. forsythia*, and *E. saphenum*, was higher in periodontitis subjects than in healthy subjects. These results suggest an environmental shift in subgingival area toward more suitable for obligate anaerobes in periodontitis subjects.

Key words. 16S ribosomal RNA, microflora, periodontitis, polymerase chain reaction

1 Introduction

Human oral cavity is in healthy status when oral microflora is composed of indigenous bacteria. Some environmental shifts in oral microflora lead to accumulation of periodontitis-associated bacteria in subgingival sulcus, and resulting in the initiation of human periodontitis. Culturing and molecular biological methods have enabled us to isolate and detect various periodontitis-associated bacteria from subgingival plaque biofilm [1–4]. It has been estimated that more than 600 species of bacteria inhabit the human subgingival plaque biofilm [2–4]. In particular, *Porphyromonas gingivalis* and *Mogibacterium timidum* were more frequently detected in subgingival microflora of periodontitis than that of healthy subjects [1].

However, few have been investigated to quantify these periodontitis-associated bacteria in subgingival plaque biofilm. This study aimed to quantify nine periodontitis-associated bacteria (*Porphyromonas gingivalis*, *Mogibacterium timidum*, *Aggregatibacter actinomycetemcomitans*, *Campylobacter rectus*, *Eubacterium saphenum*, *Prevotella tannerae*, *Prevotella intermedia*, *Slackia exigua*, and *Tannerella forsythia*) and four *Streptococcus* species (*S. gordonii*, *S. oralis*, *S. sanguinis*, and *S. salivarius*) in subgingival plaque of periodontitis and periodontally healthy subjects by real-time polymerase chain reaction (PCR), and to evaluate the relationship between periodontitis and these bacteria.

2 Periodontitis and healthy subjects

Twenty-eight patients with periodontitis (mean age 62 ± 9.9 years; range 41–80 years) and 12 periodontally healthy subjects (mean age 26 ± 1.8 years; range 22–29 years) were included in the present study. They had not received periodontal treatment or antimicrobial therapy for at least 6 months and were free from systemic diseases. Informed consent was obtained from each subject. Probing depths were measured in all teeth at six sites per tooth in each subject, and the teeth with the deepest probing depths were chosen as the target sites of sampling. The deepest probing depths were <4 mm (range 2.0–3.0 mm; mean 2.4 ± 0.5 mm) in periodontally healthy subjects (n = 12) and ≥4 mm (range 4.0–10.0 mm; mean 6.7 ± 1.9 mm) in subjects with periodontitis (n = 28). The samples were then collected using sterile periodontal pocket probes.

3 Real-time PCR analysis

Real-time PCR is currently thought to be a suitable method for quantitative detection of bacteria of subgingival plaque biofilm [5]. In the present study, total and target bacteria were quantified by real-time PCR using universal [6, 7] and species-specific primers [1, 5, 7, 8], respectively, and the proportion of each bacterium was calculated. Quantitative real-time PCR amplification was performed with iQ SYBR Green Supermix (Bio-Rad Laboratories, Richmond, CA, USA) in an iCycler (Bio-Rad) programmed for 3 min at 95°C for initial heat activation, followed by 40 cycles of 15 s at 95°C for denaturation, 30 s at 55°C for annealing and 30 s at 72°C for extension. During the extension step, fluorescence emissions were monitored, and data were analyzed using iCycler iQ Software (Bio-Rad). The genomic DNA of *P. gingivalis* W83 was used as a standard for quantitative analysis.

4 Proportion of bacteria in subgingival plaque biofilm

In healthy subjects, the proportion of streptococci, in particular *S. oralis* (5.5 ± 9.8%) and *S. sanguinis* (0.33 ± 0.53%), was higher (Figs. 1, 2), suggesting that these *Streptococcus* species are constituents of healthy subgingival biofilm. While

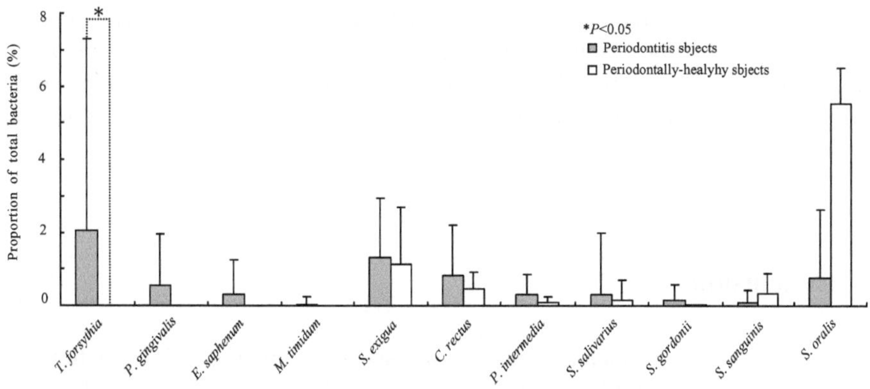

Fig. 1. Proportion of obligate anaerobes and streptococci in healthy and periodontitis subgingival plaque biofilm

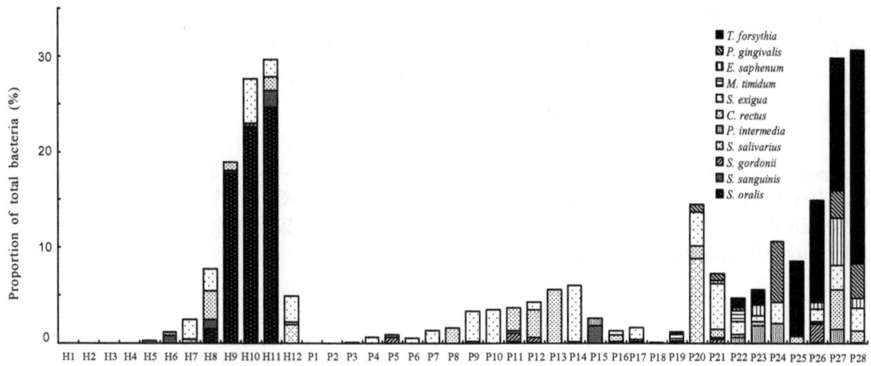

Fig. 2. Proportion of bacteria in healthy and periodontitis subgingival plaque biofilm of each subject

the proportion of obligate anaerobes, including *T. forsythia* (2.0 ± 5.3%), *P. gingivalis* (0.54 ± 1.4%), and *E. saphenum* (0.30 ± 0.96%), was higher in periodontitis subjects than that in healthy subjects (Figs. 1, 2). *T. forsythia*, *P. gingivalis*, and *E. saphenum* are known to be frequently isolated and detected from human periodontal pockets [1, 2, 4, 9]. *M. timidum* was detected frequently (7.1%) in periodontitis subjects, in agreement with the previous study [1], although its proportion was low (0.04 ± 0.21%) (Fig. 1).

It seems that total streptococcal proportions decreased, as *T. forsythia*, *P. gingivalis*, and *E. saphenum* increased in periodontitis subjects (Fig. 2). These suggest that environmental conditions, e.g., oxygen concentration, nutritional supply, and pH, in subgingival area of periodontitis subjects shift suitably toward not for *Streptococcus* species but for obligate anaerobes.

In periodontitis subjects, *S. gordonii* was detected when *P. gingivalis* existed at a low proportion (Fig. 2, subjects P19, P21, and P26). *S. gordonii* is known to have an activity to aggregate with human platelets [10], and to co-adhere with *P. gingivalis* [11]. These suggest that *S. gordonii* can introduce *P. gingivalis* to subgingival area.

5 Conclusion

Facultative anaerobes, such as *Streptococcus* species, were predominant in healthy subgingival plaque biofilm microflora, whereas obligate anaerobes, such as *T. forsythia*, *P. gingivalis*, and *E. saphenum*, were predominant in periodontitis subgingival plaque biofilm microflora, suggesting an environmental shift in subgingival area toward more suitable for obligate anaerobes. Predominance of obligate anaerobes such as *T. forsythia*, *P. gingivalis* and *E. saphenum* in periodontitis subgingival plaque biofilm microflora seems to be associated with the disease status of human periodontitis. These results support that bacterial quantification is informative to understand the oral microbial complexity and thus applicable to the diagnosis/prognosis of periodontitis.

Acknowledgments. This study was supported in part by Grants-in-Aid for Scientific Research (B) (16390601, 18390605 & 19390539), for Scientific Research (C) (17591985), for Exploratory Research (17659659) and for Young Scientists (18926014 & 19890031) from the Ministry of Education, Culture, Sports, Science and Technology, Japan.

References

1. Mayanagi G, Sato T, Shimauchi H, et al (2004) Detection frequency of periodontitis- associated bacteria by polymerase chain reaction in subgingival and supragingival plaque of periodontitis and healthy subjects. Oral Microbiol Immunol 19:379–385
2. Moore WEC, Moore LVH (1994) The bacteria of periodontal diseases. Periodontol 2000 5:66–77
3. Paster BJ, Boches SK, Galvin JL, et al (2001) Bacterial diversity in human subgingival plaque. J Bacteriol 183:3770–3783
4. Uematsu H, Hoshino E (1992) Predominant obligate anaerobes in human periodontal pockets. J Periodontal Res 27:15–19
5. Kuboniwa M, Amano A, Kimura RK, et al (2004) Quantitative detection of periodontal pathogens using real-time polymerase chain reaction with TaqMan probes. Oral Microbiol Immunol 19:168–176
6. Lane DJ (1991) 16S/23S rRNA sequencing. In: Stackebrandt E, Goodfellow M (eds) Nucleic acid techniques in bacterial systematics. Wiley, Chichester, pp 115–175
7. Yamaura M, Sato T, Echigo S, et al (2005) Quantification and detection of bacteria from postoperative maxillary cyst by polymerase chain reaction. Oral Microbiol Immunol 20:333–338

8. Hoshino T, Kawaguchi M, Shimizu N, et al (2004) PCR detection and identification of oral streptococci in saliva samples using *gtf* genes. Diagn Microbiol Infect Dis 48:195–199
9. Teles RP, Haffajee AD, Socransky SS (2006) Microbiological goals of periodontal therapy. Periodontology 2000 42:180–218
10. Takahashi Y, Yajima A, Cisar JO, et al (2004) Functional analysis of the *Streptococcus gordonii* DL1 sialic acid-binding adhesin and its essential role in bacterial binding to platelets. Infect Immun 72:3876–3882
11. Lamont RJ, El-Sabaeny A, Park Y, et al (2002) Role of the *Streptococcus gordonii* SspB protein in the development of *Porphyromonas gingivalis* biofilms on streptococcal substrates. Microbiology 148:1627–1636

Interface Oral Health Science 2007: 219–224

Hydrogen sulfide production by oral *Veillonella*: effects of substrate and environmental pH

J. Washio*, S. Matoba, T. Seki, N. Yamamoto, M. Yamamoto, and N. Takahashi
Division of Oral Ecology and Biochemistry, Department of Oral Biology, Tohoku University Graduate School of Dentistry, Sendai 980-8575, Japan
*j-washio@mail.tains.tohoku.ac.jp

Abstract. Oral *Veillonella* is one of the predominant hydrogen sulfide (H_2S)-producing bacteria in the tongue coating. Metabolic properties of the H_2S production, however, have not been known well. Therefore, this study was aimed to determine the metabolic substrates of the H_2S production and examine whether environmental pH affects the H_2S production. *Veillonella atypica* ATCC17744, *Veillonella dispar* ATCC17748, and *Veillonella parvula* ATCC10740 grew anaerobically in tryptone-yeast extract medium containing 1.8% sodium lactate at pH 7, and all the *Veillonella* species produced higher amounts of H_2S in the presence of 1 mM cysteine or glutathione, when compared with that in the absence of these sulfur compounds. These sulfur compounds, however, did not stimulate bacterial growth. The cell suspensions of these bacteria were able to produce H_2S from 1 mM cysteine or glutathione. The amounts of H_2S produced from cysteine and glutathione were the highest at pH 6–7 and pH 8, respectively. These results indicate that oral *Veillonella* species utilized cysteine and glutathione as substrates for H_2S production, and that their H_2S production was efficient around neutral pH, suggesting a high oral malodor level between meals when environmental pH in the oral cavity is expected to be around neutral.

Key words. cysteine, hydrogen sulfide, oral malodor, pH, *Veillonella*

Introduction

Oral malodor is due to metabolic products such as sulfur compounds by bacteria in the oral cavity, particularly those living on the dorsum of tongue [1, 2]. It is known that some cases of oral malodor are related with periodontitis [3, 4]. Various periodontitis-related bacteria, such as *Fusobacterium nucleatum*, *Treponema denticola*, *Peptostreptococcus* species, and *Porphyromonus gingivalis*, have been detected in the tongue coating [5, 6], suggesting that the tongue coating plays a role as a reservoir of these bacteria [6]. Most of these bacteria have been found to have the ability to produce sulfur compounds such as hydrogen sulfide (H_2S) [7, 8].

On the other hand, we have focused on the oral malodor of the patients who do not have oral disease such as periodontitis and dental caries. We detected the H_2S-producing bacteria in the tongue dorsum, the major source of oral malodor, and examined the relationship between H_2S-producing bacteria in the tongue coating and oral malodor [9]. Consequently, the numbers of H_2S-producing bacteria in the tongue coating were higher in the odorous group. Furthermore, the predominant H_2S-producing bacteria were not periodontitis-related bacteria, but mainly indigenous bacteria of the oral cavity such as *Veillonella* and *Actinomyces* species. Among them, *Veillonella* species including *V. atypica*, *V. dispar*, and *V. parvula* were dominant species.

Veillonella species are a Gram-negative and anaerobic micrococcus, and detected frequently in the tongue coating [6, 9]. Several studies have reported that *Veillonella* species produce H_2S [2, 7, 10, 11]. However, the metabolic properties regarding substrates for producing H_2S have not been investigated. Furthermore, in the tongue coating, the environmental pH may change continuously, suggesting that the pH change affects the H_2S production by *Veillonella* speceis. Therefore, we will review the newest information on substrate for H_2S production by *Veillonella* species and effects of pH on the H_2S production based on our recent study.

H_2S production by oral **Veillonella** *from various sulfur compounds during bacterial growth*

Veillonella atypica ATCC17744, *V. dispar* ATCC17748, and *V. parvula* ATCC10740 were anaerobically grown in tryptone-yeast extract media containing 1.8% sodium lactate with or without 1 mM cysteine, glutathione (L-γ-glutamyl-L-cysteinylglycine) or methionine at pH 5–8. Bacterial growth was monitored by optical density (OD) at 660 nm, and H_2S production was measured by the methylene-blue method.

All the *Veillonella* species grew at pH 7 and produced H_2S in the presence of cysteine and glutathione at similar levels, although both substrates did not stimulate bacterial growth (Fig. 1). Methionine had no effect on growth and H_2S production.

These results indicate that *Veillonella* species are able to produce H_2S from cysteine and glutathione. Cysteine is detected in both saliva [12] and serum [13]. In addition, various forms of peptides containing cysteine are available in these fluids and supplied from desquamation of epithelium. Particularly keratin, the major protein in desquamation of epithelium, contains a number of cysteine molecules as a form of cystine, and may serve as a cysteine source in the oral cavity. Although most of the H_2S producing bacteria are known to have a cystathionine-γ-lyase for cysteine degradation, the enzyme of *Veillonella* species is still unknown.

The H_2S production from glutathione was comparable to that from cysteine, suggesting that *Veillonella* species have a metabolic system to incorporate glutathione into the cells and/or degrade glutathione efficiently, in which glutathione is

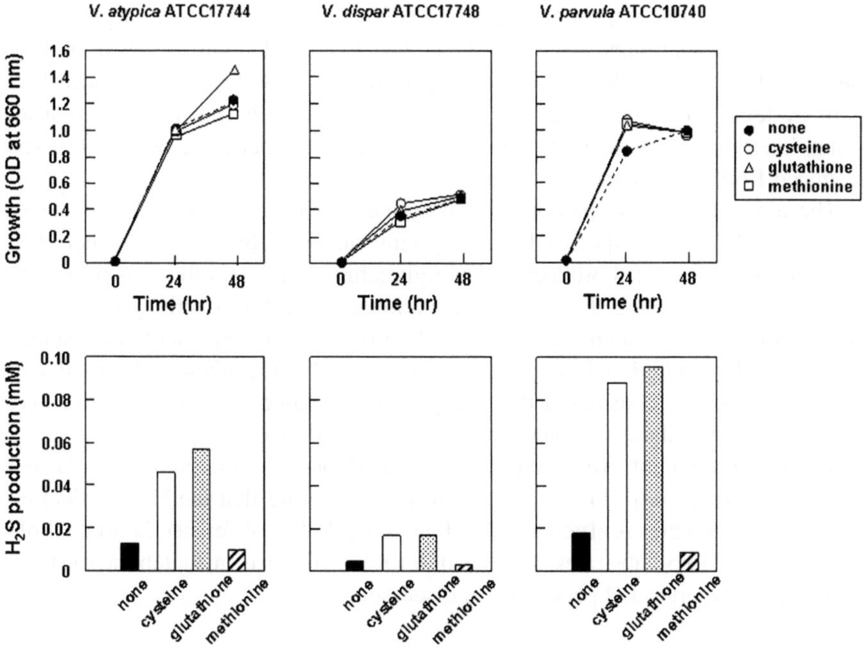

Fig. 1. Growth of *Veillonella* species in the presence of cysteine, glutathione, and methionine and the H_2S production during the growth

metabolized to H_2S in the cooperation with cysteine degrading system. It has been reported that *Fusobacterium nucleatum* has a L-γ-glutamyl peptidase that splits L-cysetinylglycine off from glutathione, and subsequently incorporates L-cysetinylglycine into the cells and degrades it into H_2S [14]. *Peptostreptococcus micros* has been suggested to have an active transport system for glutathione [15]. It is necessary to clarify whether *Veillonella* species have a transport system and/or an extracellular peptidase for glutathione metabolism.

No stimulation on bacterial growth by cysteine and glutathione indicates that the metabolic products (probably carboxylic acids and amino acids) from cysteine and glutathione are not utilized efficiently as energy sources in *Veillonella* species. When enough amounts of lactate are available as an energy source like in this study, *Veillonella* species might not need to utilize these compounds.

H_2S production by cell suspensions of oral Veillonella *from various sulfur compounds and effects of pH on H_2S production*

Veillonella atypica, *V. dispar*, and *V. dispar* were grown anaerobically in tryptone-yeast extract media containing 1.8% sodium lactate at pH 7. The cells were harvested, washed, and suspended in 50 mM potassium phosphate buffer (pH 5–8).

The cell suspensions (OD at 660 nm = 1.33) were incubated with 1 mM cysteine, glutathione or methionine for 3 h at 37°C, and H_2S production was measured by the methylene-blue method. The cell suspensions of these bacteria produced H_2S from cysteine and glutathione at similar levels, although H_2S production from glutathione was much smaller than cysteine (Fig. 2, upper panels). No H_2S was produced from methionine.

These results confirm the observation by the growth experiment (Fig. 1, lower panels) that *Veillonella* species degrade cysteine and glutathione, and produce H_2S. However, the lower H_2S production from glutathione indicates the possibility that the cells grown for the cell suspension experiment had a lower activity to incorporate and/or degrade glutathione. The growth of bacteria in the presence of glutathione may induce additional enzymes necessary for the degradation of glutathione into H_2S. It is also considered that components contained in growth media may function as a stimulator for the degradation of glutathione into H_2S.

The H_2S production from cysteine and glutathione was the highest at pH 6–7 and pH 8, respectively (Fig. 2, lower panels), indicating that the H_2S production by *Veillonella* species is efficient around neutral pH. The difference in optimal pH between these two substrates may be due to a difference in metabolic systems including transporter and peptidase, as discussed above.

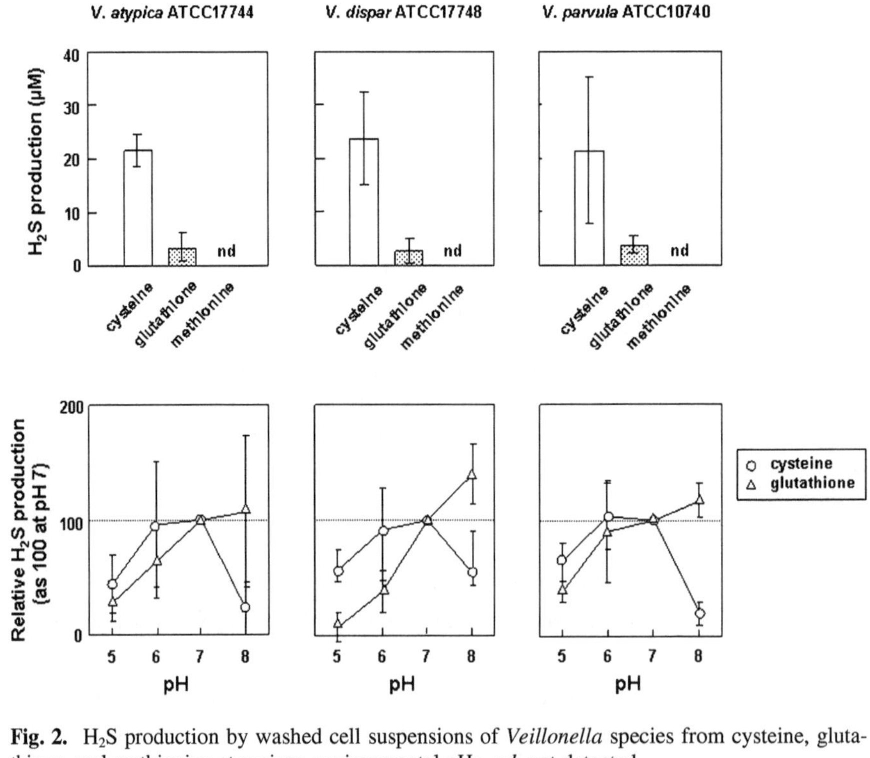

Fig. 2. H_2S production by washed cell suspensions of *Veillonella* species from cysteine, glutathione, and methionine at various environmental pHs, *nd*, not detected

Conclusions

Oral *Veillonella* species utilized cysteine and glutathione as substrates for H_2S production, and their H_2S production was efficient around neutral pH. These results suggest a high oral malodor level between meals when environmental pH in the oral cavity is expected to be around neutral. This pH effect found in our study may lead to a practical method to prevent oral malodor such as wiping tongue surface with acid solution.

This study clarified that oral *Veillonella* species degrade cysteine and cysteine-containing peptide (glutathione) into H_2S, the major components of oral malodor. However, it is still unknown *how* and *why* these bacteria metabolize such sulfur compounds and produce H_2S. We are now conducting several research projects to answer these questions.

Acknowledgments. This study was supported in part by a Grant-in-Aid for Young Scientists (Start-up) (No. 18890031 to JW) and Grants-in-Aid for Scientific Research B (No. 19390539 to NT) and Exploratory Research (No. 17659659 to NT) from the Japan Society for the Promotion of Science.

References

1. Ayers KM, Colquhoun AN (1998) Halitosis: causes, diagnosis, and treatment. N Z Dent J 94:156–160
2. Greenman J (1999) Microbial aetiology of Halitosis. In: Dental plaque revisited: oral biofilms in health and disease (proceedings of a conference held at the Royal College of Physicians, London, 305 Nov., 1999). Newman HN, Wilson M, eds., pp 419–442, Cardiff, Bioline
3. Bosy A, Geller J (1996) Oral malodour-clearing the air. Alpha Omegan 89:25–28
4. Persson S, Edlund MB, Claesson R, Carlsson J (1990) The formation of hydrogen sulfide and methyl mercaptan by oral bacteria. Oral Microbiol Immunol 5:195–201
5. Tyrrell KL, Citron DM, Warren YA, et al (2003) Anaerobic bacteria cultured from the tongue dorsum of subjects with oral malodor. Anaerobe 9:243–246
6. Faveri M, Feres M, Shibli JA, et al (2006) Microbiota of the dorsum of the tongue after plaque accumulation; an experimental study in humans. J Periodontol 77:1539–1546
7. Shibuya K (2001) Constituents and origins of physiological malodor. J Dent Health 51:778–792
8. Fukamachi H, Nakano Y, Yoshimura M, et al (2002) Cloning and characterization of the L-cysteine desulfhydrase gene of *Fusobacterium nucleatum.* FEMS Microbiol Lett 215:75–80
9. Washio J, Sato T, Takahashi N, et al (2005) Hydrogen sulfide-producing bacteria in tongue biofilm and their relationship with oral malodour. J Med Microbiol 54:889–895
10. Rogosa M, Bishop FS (1964) The genus *Veillonella.* III. Hydrogen sulfide production by growing cultures. J Bacteriol 88:37–41
11. Paryavi-Gholami F, Minah GE, Turng BF (1999) Oral malodor in children and volatile sulfur compound-producing bacteria in saliva: preliminary microbiological investigation. Pediatr Dent 21:320–324
12. Battistone GC, Burnett GW (1961) The free amino acid composition of human saliva. Arch Oral Biol 3:161–170

13. Mills BJ, Richie JP, Lang CA (1990) Sample processing alters glutathione and cysteine values in blood. Anal Biochem 184:263–267
14. Carlsson J, Larsen JT, Edlund MB (1994) Utilization of glutathione (L-γ-glutamyl-L-cysteinylglycine) by *Fusobacteirum nucleatum* subspecies *nucleatum*. Oral Microbiol Immunol 9:297–300
15. Carlsson J, Larsen JT, Edlund MB (1993) *Peptostreptococcus micros* has a uniquely high capacity to form hydrogen sulfide from glutathione. Oral Microbiol Immunol 8:42–45

Interface Oral Health Science 2007: 225–230

Expression of various Toll-like receptors, NOD1, and NOD2, in human oral epithelial cells, and their function

Yumiko Sugawara[1], Akiko Uehara[2]*, Yukari Fujimoto[3], Koichi Fukase[3], Takashi Sasano[1], and Haruhiko Takada[2]
[1]*Division of Oral Diagnosis, Department of Oral Medicine and Surgery;* [2]*Department of Microbiology and Immunology, Tohoku University Graduate School of Dentistry, Sendai 980-8575;* [3]*Department of Chemistry, Graduate School of Science, Osaka University, Toyonaka 560-0043; Japan*
*kyoro@mail.tains.tohoku.ac.jp

Abstract. Oral epithelium is endowed with innate immune receptors for bacterial components, which play important roles in host defense against bacterial infection. We examined the expression of various Toll-like receptors (TLRs), NOD1 and NOD2 in oral epithelial cells, and the production of β-defensin 2 and peptidoglycan recognition proteins (PGRPs) upon stimulation with their respective chemically synthesized ligands. We found a clear expression of TLR4 as well as TLR2, and a strong expression of NOD1 and NOD2 in normal oral epithelial tissues by immunohistochemical analysis. In the inflamed oral epithelium, cell-surface localizations of TLR2 and TLR4 were more clearly observed than in the healthy tissue. We also showed that oral epithelial cells in culture constitutively expressed TLR3 and TLR7 in addition to TLR2, TLR4, NOD1, and NOD2, and stimulation with synthetic ligands for these receptors (TLR2 agonistic lipopeptide, TLR3 agonistic poly I:C, TLR4 agonistic lipid A, TLR7 agonistic single-stranded RNA, NOD1 agonistic iE-DAP and NOD2 agonistic muramyldipeptide) markedly up-regulated the expression of antibacterial factors, such as β-defensin 2 and PGRPs, but not the proinflammatory cytokines. These findings indicate that these molecules in oral epithelial cells are functional receptors that induce antibacterial responses without excessive inflammatory responses.

Key words. Toll-like receptors, NODs, human oral epithelial cells, β-defensin 2, peptidoglycan recognition proteins

1 Introduction

In the innate immune system, pattern recognition of microorganisms should initiate host defense against invasive pathogens, where pathogen-associated molecular patterns (PAMPs) are recognized by the molecules of hosts. In bacteria,

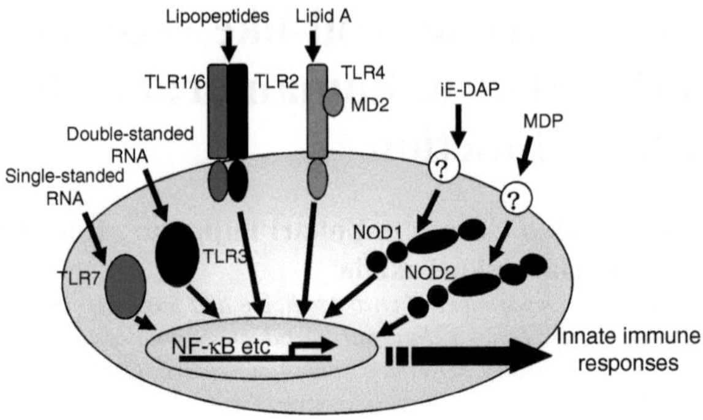

Fig. 1. Recognition of pathogen-associated molecular patterns (PAMPs) by Toll-like receptors (*TLRs*) and NODs molecules

representative PAMPs are distributed mainly on the cell surface such as peptidoglycans (PGNs), lipoproteins and lipopolysaccharides (LPS). Recent studies have demonstrated that, in mammals, these PAMPs are recognized specifically by their respective NODs in addition to Toll-like receptors (TLRs) (Fig. 1) [1, 2]. In this decade, the importance of TLRs in the process of recognition by myeloid cells, such as macrophages or dendritic cells, has been exhaustively studied. Concerning TLR expression on oral epithelial cells, we have reported that primary oral epithelial cells, oral squamous cell carcinoma HSC-2 and HO-1-u-1 cells constitutively expressed TLR2 and TLR4 [3]. In contrast, gingival epithelial cells transfected with HPV-16 constitutively expressed TLR2, but not TLR4 [4]. NOD proteins in human oral epithelium have not been reported so far, except for our report on their function [5]. Here, we will review our findings on the expression of various TLRs and NODs in human oral epithelial cells.

2 Recognition of PAMPs by TLRs and NODs

The innate immune system recognizes microorganisms through a series of pattern recognition receptors that are highly conserved in evolution, specific for common motifs found in microorganisms but not in eukaryotes, and designated as PAMPs [6]. Representative PAMPs are the lipid A moiety of lipopolysaccharides from Gram-negative bacteria, lipopeptides from various bacteria including mycoplasma, and peptidoglycans (PGNs) from either Gram-positive or Gram-negative bacteria. Recent studies have demonstrated that in mammals, these PAMPs are recognized specifically by their respective TLRs (Fig. 1) [1]. More recently, NOD1 and NOD2 were revealed to be an intracellular receptor for a PGN motif containing diaminopimelic acid (DAP) and muramyldipeptide (MDP), respectively (Fig. 1) [2].

3 Oral epithelial cells constitutively expressed functional TLRs and NODs

3.1 Histological analysis showed the expression of TLR2, TLR4, NOD1, and NOD2 in human gingival epithelial tissues

Using immunohistochemistry, we examined whether human oral tissues express TLR2, TLR4, NOD1, and NOD2 molecules. The NOD1 and NOD2 molecules were markedly expressed in the epithelial layer of healthy gingival tissue. Expression of NOD1 and NOD2 was also detected in gingival tissue from adult periodontitis patients, similar to that found in healthy gingival tissue. TLR2 and TLR4 molecules were also detected in healthy and inflamed gingival tissues. It should be noted that cell-surface localizations of TLR2 and TLR4 were more clearly observed in the inflamed gingival tissue than in healthy gingival tissue [7].

3.2 Oral epithelial cells in culture constitutively expressed TLR2, TLR3, TLR4, TLR7, NOD1, and NOD2

We examined whether human epithelial cells in culture expressed TLR2, TLR3, TLR4, TLR7, NOD1, and NOD2 molecules by RT-PCR, flow cytometry, and immunostaining. It was found that human oral HSC-2 and HO-1-u-1 epithelial cells expressed the mRNA of these molecules (Fig. 2a) [7, 8]. Consistent with the results of RT-PCR, tongue, salivary gland epithelial cells, and primary oral epithelial cell lines in addition to these oral epithelial cells constitutively expressed these molecules, as determined by immunostaining assay. In flow cytometry analysis, we could clearly detect cell-surface expression of TL2 and TLR4 and the intracellular expressions of TLR3, TLR7, NOD1, and NOD2 (Fig. 2b) [8].

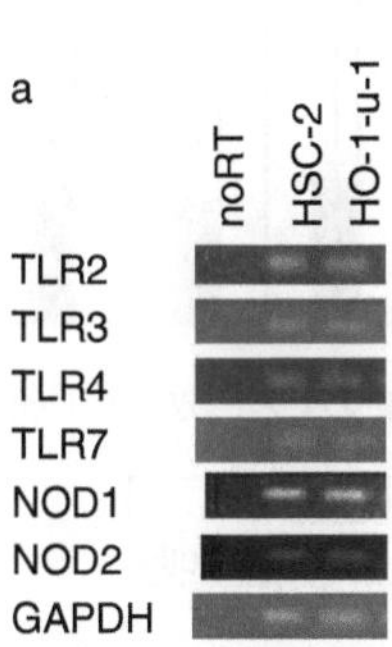

b

Positive cells (%)	HSC-2	HO-1-u-1
TLR2	56.2	48.9
TLR3	32.1	38.3
TLR4	36.4	33.0
TLR7	40.7	41.6
NOD1	90.1	85.4
NOD2	77.4	70.8

Fig. 2. Expression of TLRs, NOD1 and NOD2 in oral epithelial cells. **a** The mRNA expression of TLRs and NODs were analyzed by PCR. **b** The cells were stained with anti-TLRs and NODs Ab to analyze by flow cytometry [8]

3.3 Oral epithelial cells do not secrete proinflammatory cytokines upon stimulation with PAMPs

Although oral epithelia cells constitutively expressed various TLRs and NODs, these cells did not secrete IL-6, IL-8, and MCP-1 upon stimulation with PAMPs [3, 5, 8].

3.4 Induction of β-defensin 2 triggered by PAMPs in human oral epithelial cells

We examined whether the TLRs and NODs expressed in these oral epithelial cells actually functioned as receptors in terms of β-defensin 2 generation upon stimulation with their respective ligands. It was found that Pam_3CSSNA (TLR2/6 agonist), poly I:C (TLR3 agonist), lipid A (TLR4 agonist), ssPoly U (TLR7 agonist), iE-DAP

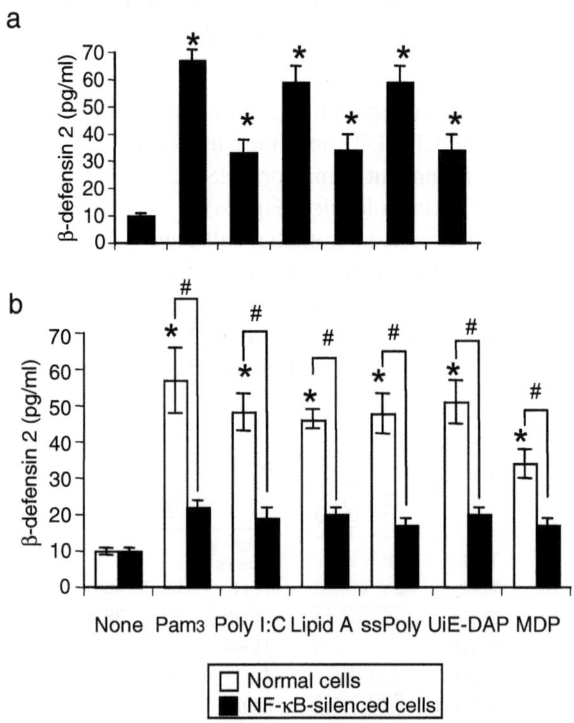

Fig. 3. Induction of β-defensin 2 in oral epithelial cells upon stimulation with synthetic PAMPs via NF-κB. **a** The cells were incubated with PAMPs, and levels of β-defensin 2 were determined by ELISA. **b** HSC-2 cells transfected with siRNA targeting NF-κB were stimulated with PAMPs, and the level of β-defensin 2 in the culture supernatants were determined by ELISA [8]

(NOD1 agonist), and MDP (NOD2 agonist) significantly induced the expression of β-defensin 2 mRNA [8]. In accordance with the results of RT-PCR, β-defensin 2 molecules were significantly up-regulated by stimulation with TLRs and NODs ligands, whereas β-defensin 2 molecules were only slightly expressed on unstimulated cells (Fig. 3a) [7, 8]. These results demonstrated that the TLRs and NODs on oral epithelial cells actively function as pattern recognition receptors and signaling molecules.

3.5 Suppression of β-defensin 2 induction upon stimulation with synthetic PAMPs in oral epithelial cells using siRNA for NF-κB

To elucidate whether NF-κB are responsible for induction of β-defensin 2 with TLR and NOD ligands, we utilized RNA interference assays targeting NF-κB mRNA. Up-regulation of β-defensin 2 induced by synthetic PAMPs was significantly inhibited in NF-κB p65-silenced HSC-2 cells (Fig. 3b) [8]. These results clearly demonstrated that NF-κB are critical molecules for induction of β-defensin 2, triggered by TLR and NOD ligands.

4 Conclusion

Clear expression of TLRs and NODs in oral epithelial cells was found; nevertheless, oral epithelial cells did not secrete proinflammatory cytokines upon stimulation with the PAMPs. Contrary to proinflammatory cytokines, oral epithelial cells secrete β-defensin 2 molecules upon stimulation with respective PAMPs. In addition, up-regulation of β-defensin 2 upon stimulation with PAMPs in oral epithelial cells occurred via NF-κB. These findings indicate that the TLRs and NODs in oral epithelial cells are functional, and epithelial cells might actively participate in bacterial clearance in the mucosa without an accompanying excessive inflammatory response, which might induce tissue destruction.

Acknowledgments. This work was supported in part by a Grant-in-Aid for Science Research from the Japan Society for the Promotion of Science (18390484 to H.T.) (17591959 to Y.S.), from the Ministry of Education, Culture, Sports, Science and Technology, Japan (18689901 to A.U.)

References

1. Akira S, Uematsu S, Takeuchi O (2006) Cell 124:783–801
2. Fritz JH, Ferrero RL, Kahn M, et al (2006) Nat Immunol 7:1250–1257
3. Uehara A, Sugawara S, Tamai R, et al (2001) Med Microbiol Immunol 189:185–192

4. Asai Y, Ohyama Y, Gen K, et al (2001) Infect Immun 69:7387–7395
5. Uehara A, Sugawara Y, Kurata S, et al (2005) Cell Microbiol 7:675–686
6. Janeway Jr CA, Medzhitov R (2002) Annu Rev Immunol 20:197–216
7. Sugawara Y, Uehara A, Fujimoto Y, et al (2006) J Dent Res 85:524–529
8. Uehara A, Fujimoto Y, Fukase K, et al (2007) Mol Immunol 44:3100–3111

Interface Oral Health Science 2007: 231–236

Inflammatory stimuli regulate the binding of gingival fibroblasts to dendritic cells via integrin β2

Maiko Minamibuchi[1]*, Eiji Nemoto[1], Sousuke Kanaya[1], Tomohiko Ogawa[2], and Hidetoshi Simauchi[1]
[1]*Division of Periodontology and Endodontology, Tohoku University Graduate School of Dentistry, Sendai 980-8575;* [2]*Department of Oral Microbiology, Asahi University School of Dentistry, Gifu 501-0296; Japan*
*mai@mail.tains.tohoku.ac.jp

Abstract. We investigated the mechanism and impact of various inflammatory stimuli on dendritic cell (DC)–gingival fibroblast (GF) adhesion. Human immature (im) DCs were generated from monocytes by culturing with interleukin (IL)-4 and granulocyte macrophage-colony stimulating factor (GM-CSF). GFs were outgrown from the human gingival specimen. DCs were co-cultured with GFs with/without pretreatment with various stimuli. Adhered cells were measured by fluorometer. Expression of adhesion molecules was analyzed by flow cytometry. Pretreatment of GFs with tumor necrosis factor (TNF)-α, interferon (IFN)-γ, and *Escherichia coli* (*Ec*) lipopolysaccharides (LPS) significantly increased the adhesion to imDCs and enhanced intercellular adhesion molecule (ICAM)-1 expression. A significantly increased DC–GF adhesion was also observed when imDCs were pretreated with *Ec* LPS, *Porphyromonas gingivalis* (*Pg*) fimbriae, and peptideglycan but not with *Pg* LPS. Expression of LFA-1 and Mac-1 on DCs was not altered by the pretreatment with these stimuli. However, LFA-1 and Mac-1 blockade of imDC significantly reduced the adhesion to TNF-α-stimulated GFs. These results showed that inflammatory stimuli increased the imDC–GF adhesion via lymphocyte function-associated antigen (LFA)-1/Mac-1-ICAM-1 ligation. Adhesion of DC to GFs may be important not only for the localization of DCs in the inflammatory periodontal lesion, but also for the modulation of immune responses.

Key words. dendritic cells, gingival fibroblasts, LFA-1, Mac-1, ICAM-1

1 Introduction

Dendritic cells (DCs) are the only and professional antigen-presenting cells that can initiate the specific immune responses [1, 2]. DCs express a variety of adhesion molecules mediating the cell–cell contact that is important for T cell activation. DCs are reported to be found in periodontal tissues and interact with periodontal pathogens including *Porphyromonas gingivalis* (*Pg*) [3]. In the periodontally diseased gingiva, DCs presumably make contact with fibroblasts under the influence of various inflammatory stimuli. This interaction may modify DC maturation that affects the subsequent immune responses to periodontal pathogens. We investigated the mechanism and impact of various inflammatory stimuli on DC–gingival fibroblast (GF) adhesion.

2 Materials and methods

2.1 Bacterial components

Phenol–water extracted LPS and fimbriae from *Pg* stain 381 were prepared as previously reported. LPS from *Escherichia coli* (*Ec*) O55:B5 and peptideglycan (PGN) from *Staphylococcus aureus* were purchased from Sigma (St. Louis, MO, USA).

2.2 Cell culture

The ImDCs were prepared from human peripheral blood monocytes (PBMC). Briefly, adherent PBMC were cultured for 7 days in the presence of recombinant human (rh) interleukin (IL)-4 (Pepro Tech, London, UK) (500 U/ml) and rh granulocyte macrophage-colony stimulating factor (GM-CSF) (Pepro Tech) (800 U/ml), and then purified by magnetic sorting using the VARIO MACS technique (Milteny Biotec, Bergisch Gladbach, Germany). ImDCs were treated with bacterial components or cytokines in RPMI 1640 containing 5% FCS in 24-well (2.0×10^5 cells/well/ml) or 48-well (8.0×10^4 cells/well/ml) multiplates for indicated times. GFs were outgrown from human gingival specimen. Explants were cultured in α-modified Eagle's medium (α-MEM) (Biomedicals, Eschwege, Germany) supplemented with 10% FCS until confluent cell monolayers were formed. After 4 to 5 subcultures by trypsinization, the cells at subculture levels 5 to 10 were applied to this study. GFs were cultured in α-MEM with 10% FCS in 24- or 96-well multiplates until confluent, and then cells were washed with PBS and added to various stimuli in α-MEM supplemented 5% FCS.

2.3 Binding assay

Confluent GFs were stimulated with 100 ng/ml TNF-α, 100 U/ml IFN-γ (Pepro Tech), 500 ng/ml *Ec* LPS or 500 ng/ml *Pg* LPS for 24 h. DCs treated with 5 mM Calcein, AM (Molecular Probes, Netherlands) for 30 min were added to GFs (5.0×10^4 cells/well). After co-culture for 1 h, unbound cells were washed from the plates with warmed α-MEM. Adhered cells were solubilized with 0.1% (v/v) SDS-PBS and measured by fluorometer (Bio-Rad Laboratories, Hercules, CA, USA).

2.4 Blocking assay

Before co-culture of DCs and GFs, blocking antibody was added to DCs. To be concrete, imDCs were treated with mouse anti-human CD11a antibody (clone; 38) (Ancel, Bayport, MN, USA), mouse anti-human CD11b (ICRF44), mouse anti-human CD18 (TS1/18) (Biolegend, San Diego, CA, USA) or mouse anti-human CD49d antibody (9F10) (eBioscience, San Diego, CA, USA) (10 μg/ml) at room temperature for 30 min.

2.5 Flow cytometry

Cells were stained directly or indirectly using mouse anti-human ICAM-1 antibody (84H10, Immunotech, Marseille, France), mouse anti-human VCAM-1 antibody (1G11, Immunotech), and FITC-conjugated anti-fractalkine antibody (R & D Systems, Inc, Minneapolis, MN, USA). Expression of adhesion molecules on DCs and GFs was analyzed by FACScan (Becton Dickinson, Mountain View, CA, USA). Data were collected for 5,000 events which were sorted in list mode and then analyzed with Lysis II software.

2.6 Statistical analysis

All experiments in this study were performed at least three times. Regarding binding assay and blocking assay, we showed data as each means ± standard deviation (SD). Statistical significance was analyzed using ANOVA. *P*-values of <0.05 were considered statistically significant.

3 Results and discussion

3.1 Changes in cell–cell adhesion by various kinds of stimulation

We first investigated the capability of imDC to bind to GFs and the effect of pretreatment of imDC with bacterial preparations or inflammatory cytokines. When GFs were treated with TNF-α or IFN-γ for 24 h, imDC adhesion was significantly increased in comparison with non-stimulated GFs. On the other hand, *Ec* LPS did not significantly increased the adhesion and *Pg* LPS oppositely decreased the adhesion. Next, imDCs treated with 1 μg/ml of various bacterial components for 24 h were co-cultured with stimulated or non-stimulated GFs. *Pg* LPS-stimulated DC, *Pg* fimbriae-, and PGN-stimulated DC showed a significant increase in adhesion to non-stimulated GFs when compared with imDC. Meanwhile, DC treated with *Pg* LPS or PGN increased adhesion to TNF-α-stimulated GFs.

3.2 Expression of ICAM-1, VCAM-1, and fractalkine on GFs

The GFs express various adhesion molecules [4, 5]. We investigated whether ICAM-1, VCAM-1 [6], and fractalkine [7] are involved in the adhesion to DCs or not. GFs were treated with 100 U/ml IFN-γ, 100 ng/ml TNF-α, 500 ng/ml *Ec* LPS or 500 ng/ml *Pg* LPS for 24 h and then alterations were analyzed in these molecule expressions by flow cytometry. ICAM-1, VCAM-1, and fractalkine were poorly expressed on unstimulated GFs, whereas these expressions were markedly increased in treatment with TNF-α. On the other hand, ICAM-1 and fractalkine, but not VCAM-1 increased in the treatment with IFN-γ. *Ec* or *Pg* LPS increased expression of ICAM-1 slightly.

3.3 Expression of adhesion molecule on DC

Since ICAM-1, VCAM-1, and fractalkine were induced on GFs by TNF-α, we examined expressions of ligands to these molecules on DC. As shown in Fig. 1, imDC expressed LFA-1 (CD11a/CD18) [8, 9] and Mac-1 (CD11b/CD18) [10, 11], both known as ligands of ICAM-1. In addition, imDC weakly expressed VLA-4 (CD49d), a ligand of VCAM-1. However, CX_3CR1 [7], a ligand of fractalkine, was not detected. These data suggested that ICAM-1 and VCAM-1, but not CX_3CR1, seemed to be involved mainly in imDC–GF adhesion. Change in expression patterns of adhesion molecule affects DC functions, and these changes are caused by DC maturation. We added bacterial components to imDC to investigate alterations in the expression of CD11a and CD11b. The results indicated expressions of both CD11a and CD11b were slightly increased by the addition of *Ec* LPS. Contrastingly, *Pg* fimbriae and PGN decreased these expressions. These data were not consistent with Fig. 2, in which PGN-stimulated DC increased adhesion to TNF-α-stimulated GFs. It is possible that stimulated-DCs express other adhesion molecules and affect the adhesion to GFs.

3.4 Blocking assay

We examined whether cell adhesion was inhibited by addition of blocking antibodies specific for LFA-1, Mac-1 or VLA-4. The result showed that anti-LFA-1 and anti-Mac-1 pretreatment significantly inhibited the imDC-adhesion to TNF-α-stimulated GFs. In contrast, anti-VLA-1 antibody did not affect the cell adhesion. These results suggest that imDCs adhere to TNF-α-stimulated GFs via LFA-1/ICAM-1 and Mac-1/ICAM-1 pathway. Our findings strongly suggest that DCs migrate and colonize to periodontal connective tissue mainly via LFA-1/Mac-1-ICAM-1 pathway. Inflammatory environment or infectious agents may modify the functions of both cells, resulting in the alteration of inflammatory/immune responses.

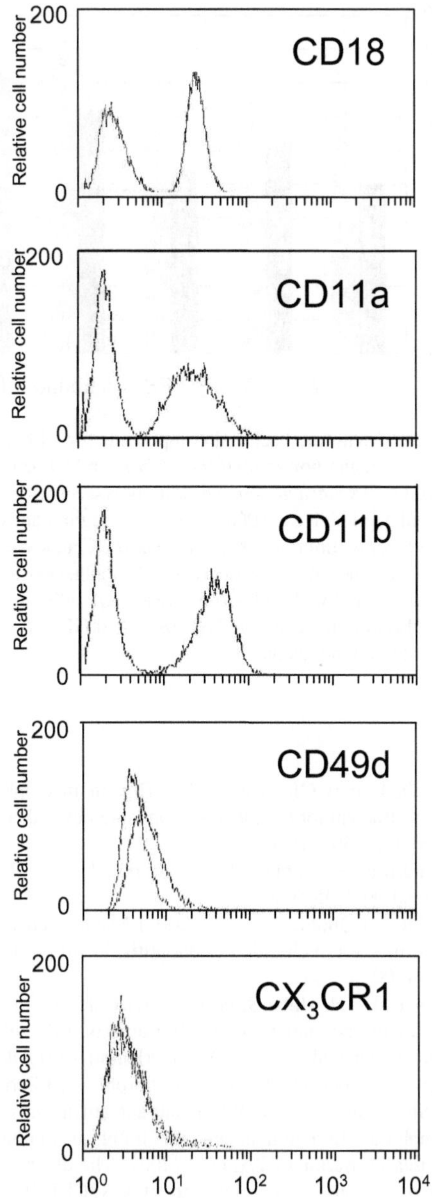

Fig. 1. Expression of adhesion molecules on immature dendritic cells (imDC) were assessed by flow cytometry. Expression of CD18, CD11a, CD11b, and CD49d were observed, but CX_3CR1, a ligand of fractalkine, was not expressed

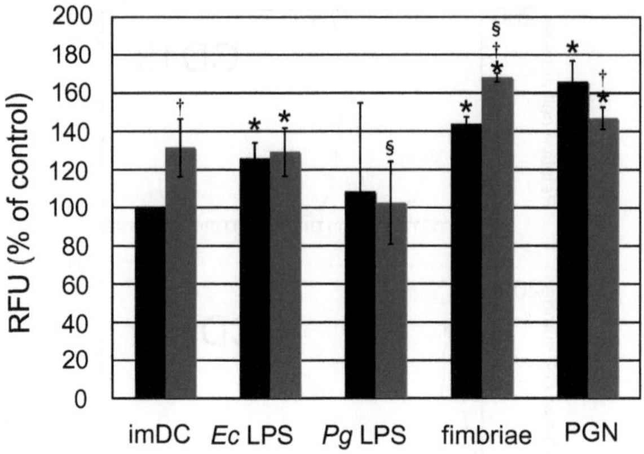

Fig. 2. ImDC were pretreated with the bacterial components for 24 h at the dose of μg/ml, and adhesion assay was performed using non-treated (*black bar*) or TNF-α (100 ng/ml)-treated (*gray bar*) gingival fibroblasts (GFs). *Escherichia coli* (*Ec*) lipopolysaccharides (LPS), *Porphyromonas gingivalis* (*Pg*) fimbriae, and peptideglycan (PGN) treatments significantly increased DC adhesion to non-treated GFs. However, only fimbriae further increased DC adhesion on TNF-α-treated GFs. On the other hand, *Pg* LPS did not affect or decreased DC adhesion to non-treated or TNF-α-treated GFs. $*P < 0.05$ as compared with imDC-non-treated GF adhesion, $^{\S}P < 0.05$ as compared with the adhesion of the adhesion of imDC on TNF-α-treated GFs, $^{\dagger}P < 0.05$ as compared with the stimulated DC-non-treated GF adhesion

References

1. Godiska R, Chantry D, Raport CJ, et al (1997) Human macrophage-derived chemokine (MDC), a novel chemoattractant for monocytes, monocyte-derived dendritic cells, and natural killer cells. J Exp Med 185:1595–1604
2. Hart DNJ (1997) Dendritic cells: unique leukocyte populations which control the primary immune response. Blood 90:3245–3287
3. Jotwani R, Palucka AK, Al-Quotub M, et al (2001) Mature dendritic cells infiltrate the T cell-rich region of oral mucosa in chronic periodontitis: in situ, in vivo, and in vitro studies. J Immunol 167:4693–4700
4. Takahashi K, Takigawa M, Takashiba S, et al (1994) Role of cytokine in the induction of adhesion molecules on cultured human gingival fibroblasts. J Periodontol 65:230–235
5. Joe BH, Borke JL, Keskintepe M, et al (2001) Interleukin-1β regulation of adhesion molecules on human gingival and periodontal ligament fibroblasts. J Periodontol 72:865–870
6. Murakami S, Saho T, Shimabukuro Y (1993) Very late antigen integrins are involved in the adhesive interaction of lymphoid cells to human gingival fibroblasts. Immunology 79:425–433
7. Hundhausen C, Misztela D, Berkhout TA, et al (2003) The disintegrin-like metalloprotease ADAM10 is involved in constitutive cleavage of CX3CL1 (fractalkine) and regulates CX3CL1-mediated cell-cell adhesion. Blood 102:1186–1195
8. Marlin SD, Springer TA (1987) Purified intercellular adhesion molecule-1 (ICAM-1) is a ligand for lymphocyte function-associated antigen 1 (LFA-1). Cell 51:813–819
9. Makgoba MW, Sanders ME, Ginther Luce GE, et al (1988) ICAM-1 a ligand for LFA-1-dependent adhesion of B, T and myeloid cells. Nature 331:86–88
10. Smith CW, Marlin SD, Rothlein R, et al (1989) Cooperative interactions of LFA-1 and Mac-1 with intercellular adhesion molecule-1 in facilitating adherence and transendothelial migration of human neutrophils in vitro. J Clin Invest 83:2008–2017
11. Tonnesen MG (1989) Neutrophil–endothelial cell interactions: mechanisms of neutrophil adherence to vascular endothelium. J Invest Dermatol 2:53s–58s

Interface Oral Health Science 2007: 237–242

Antibodies against proteinase 3 prime human monocytic cells in culture in a protease-activated receptor 2- and NF-κB-dependent manner for various Toll-like receptor-, NOD1-, and NOD2-mediated activation

Akiko Uehara*, Tadasu Sato, Sou Yokota, Atsushi Iwashiro, and Haruhiko Takada
Department of Microbiology and Immunology, Tohoku University Graduate School of Dentistry, Sendai 980-8575, Japan
*kyoro@mail.tains.tohoku.ac.jp

Abstract. Anti-neutrophil cytoplasmic antibodies (ANCA) against proteinase 3 (PR3) have been detected under a wide range of inflammatory conditions, and the interaction of anti-PR3 antibodies (Abs) with leukocytes provoked cell activation. Flow cytometric analysis revealed an increase of CD14, various Toll-like receptors (TLRs), NOD1, and NOD2 expressions during the anti-PR3 priming in human monocytic THP-1 cells. Anti-RP3 Abs resulted in a markedly enhanced interleukin (IL)-8, tumor necrosis factor-α, and monocyte chemoattractant protein-1 liberation on chemically synthesized TLR2-agonistic lipopeptide (FSL-1), TLR3-agonistic Poly I:C, TLR4-agonistic lipid A, TLR7/8-agonistic ssPoly U, TLR9-agonistic bacterial CpG DNA, NOD1-agonistic FK156, and NOD2-agonistic muramyldipeptide in human monocytic THP-1 cells and human peripheral blood mononuclear cells. RNA interference assays revealed that anti-PR3 Abs primed THP-1 cells in a PR3- and protease-activated receptor-2-dependent manner. Furthermore, anti-PR3 Ab-mediated priming was significantly abolished by inhibition of phospholipase C and nuclear factor-κB. These results suggest that anti-PR3 Abs prime human monocytic cells to produce cytokines on stimulation with various microbial components by the up-regulation of the TLR and NOD signaling pathway, and that these mechanisms may actively participate in the inflammatory process in ANCA-related autoimmune diseases.

Key words. proteinase 3, protease-activated receptor-2, Toll-like receptors, NODs, human monocytic cells

1 Introduction

Anti-neutrophil cytoplasmic antibodies (ANCA) were first identified in patients with necrotizing glomerulonephritis [1]. ANCA are autoantibodies directed against the enzymes located in the primary granules of neutrophils and lysosomes of monocyte. An established association does exist between the occurrence of ANCA, especially those targeting proteinase 3 (PR3), and the development of active Wegener's granulomatosis [2]. ANCA have since been detected under a wide range of inflammatory, infectious, and neoplasmtic conditions [3]. In isolated monocytes, anti-PR3 antibodies (Abs) stimulate the release of proinflammatory cytokines [4, 5], and Nowack et al. [6] reported that the expression of CD14 and CD18 was up-regulated on monocytes by ANCA in vitro, as well as by monoclonal Abs against PR3.

Recently, we revealed that proinflammatory cytokines induced the production of PR3 in membrane-bound and secretory forms in human oral epithelial cells, and the addition of anti-PR3 Abs to cytokine-primed oral epithelial cells induced the aggregation of PR3 followed by the activation of protease-activated receptor (PAR)-2, which resulted in remarkable secretion of interleukin (IL)-8 and monocyte chemoattractant protein (MCP)-1 [7].

In this article, we review the priming effects on human monocytic cells by anti-PR3 Abs through PR3 and PAR-2 for Toll-like receptors (TLRs)-, NOD1-, and NOD2, as well as CD-14-dependent activation that we have obtained to date.

2 Recognition of pathogen-associated molecular patterns by TLRs and NODs

The innate immune system recognizes microorganisms through a series of pattern recognition receptors (PRRs) that are highly conserved in evolution, specific for common motifs found in microorganisms but not in eukaryotes, and designated as pathogen-associated molecular patterns (PAMPs) [8]. Representative PAMPs are the lipid A moiety of lipopolysaccharides from Gram-negative bacteria, lipopeptides from various bacteria, including mycoplasma and peptidoglycans (PGNs) from either Gram-positive or Gram-negative bacteria. Recent studies have demonstrated that in mammals, these PAMPs are recognized specifically by their respective TLRs [9]. More recently, NOD1 and NOD2 were revealed to be an intracellular receptor for a PGN motif containing diaminopimelic acid and muramyldipeptide, respectively [10].

3 Activation of human cells through the PAR family

Protease-activated receptor family members are G protein-coupled receptors characterized by a proteolytic cleavage of the N terminus that exposes tethered ligands and autoactivates the receptor function [11]. There are four members of this family. Because PARs are expressed in a wide variety of cell types, they are believed to play important roles in several pathophysiological processes, including growth, development, inflammation, tissue repair, and pain [11]. As oral epithelial cells and gingival fibroblasts also constitutively express PARs, these cells are activated to produce inflammatory cytokines through PARs [12, 13]. In addition, we revealed that the addition of anti-PR3 Abs to cytokine-primed oral epithelial cells induced the aggregation of PR3 followed by the activation of PAR-2, which resulted in remarkable secretion of IL-8 [7].

4 What are ANCA?

Anti-neutrophil cytoplasmic antibodies were first identified in patients with necrotizing glomerulonephritis [1]. It is known that there are two types of ANCA (cANCA and pANCA), and cANCA are autoantibodies against PR3. ANCA have since been detected under a wide range of inflammatory, infectious, and neoplasmtic conditions [3]. Novo et al. [14, 15] described a high rate of the occurrence of ANCA in serum of patients with periodontal disease.

5 Anti-PR3 Abs amplified innate immune responses in human monocytic cells via PAR-2

5.1 Treatment with anti-PR3 Abs up-regulated the expression of CD14, TLRs, NOD1, and NOD2 in human monocytic cells

THP-1 cells constitutively expressed various TLRs and NODs, and the incubation of the cells with anti-PR3 Abs resulted in the up-regulated expression of TLR2 and TLR4, as well as CD14, on the cell surface (Fig. 1). Furthermore, intracellular levels of TLR3, 7, 8, 9, NOD1, and NOD2 also clearly increased (Fig. 1). In addition, THP-1 cells constitutively expressed PAR-1, -2, -3, and PR3 on their surface, and the addition of anti-PR3 Abs also up-regulated the expression of PAR-2 and PR3 (Fig. 1).

5.2 Anti-PR3 Abs enhanced TLR and NOD ligand-induced secretion of proinflammatory cytokines in monocytic cells

We examined whether the up-regulated expression of TLRs and NODs evoked by anti-PR3 Abs induced amplified responses to respective ligands. Stimulation with TLR2-agoinstic FSL-1, TLR3-agonistic poly I:C, and TLR9-agonistic CpG DNA significantly induced the production of MCP-1 and IL-8, and anti-PR3 Abs also

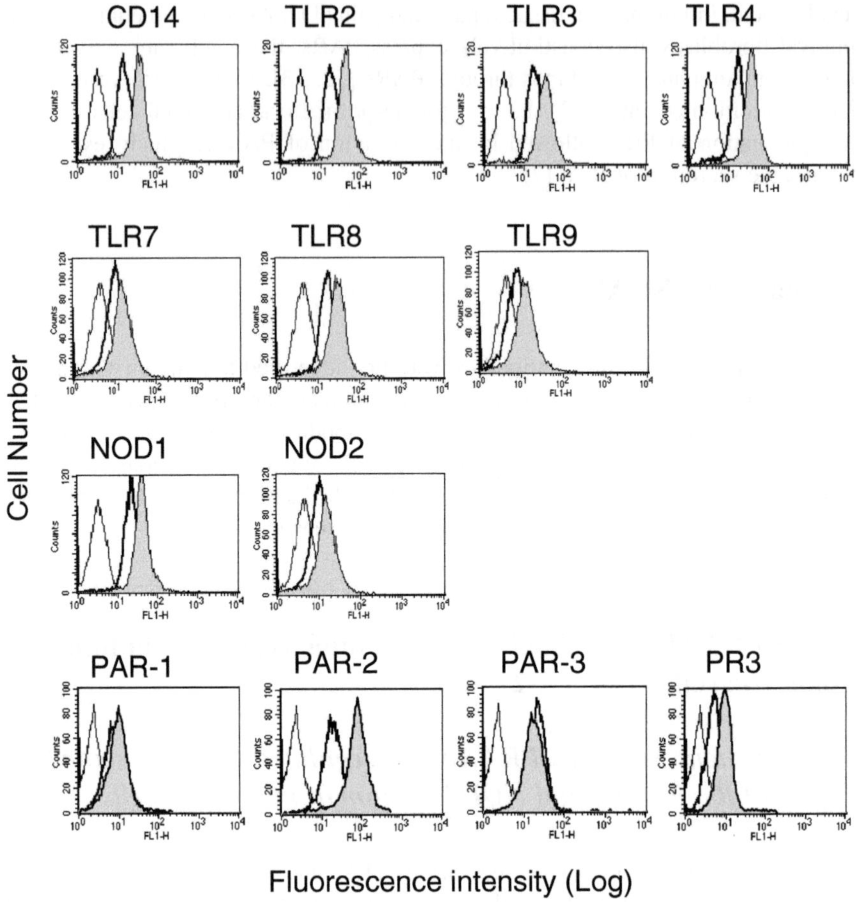

Fig. 1. Up-regulation of the expression of CD14, Toll-like receptors (*TLRs*), NODs, protease-activated receptor (*PAR*)-2, and proteinase 3 (*PR3*) in monocytic cells in response to anti-PR3 antibodies (Abs). THP-1 cells were stimulated with anti-PR3 Abs (*filled graphs*) or isotype-matched control IgG (*bold line*). After 6 h of incubation, the expression of CD14, TLRs, NODs, PARs, and PR3 were assessed by flow cytometry. The *thin lined curve* is the staining with a control Ab [16]

weakly induced the production of MCP-1 and IL-8. When THP-1 cells were pre-incubated with anti-PR3 Abs for 6 h, and subsequently challenged with the various TLR, and NOD ligand-induced production of IL-8 and MCP-1 was observed in THP-1 cells [16]. Next, we examined whether similar priming effects of anti-PR3 Abs were observed in human peripheral blood mononuclear cells (PBMCs). Consistent with the results for THP-1 cells, anti-PR3 Abs promoted the TLR and NOD ligand-induced secretion of IL-8, MCP-1, and tumor necrosis factor-α in PBMCs [16].

5.3 Priming effect of anti-PR3 Abs occurred in a PAR-2-, PR3-, and NF-κB-dependent manner

To examine the signaling molecules of the priming effects of anti-PR3 Abs on CD14, TLR, and NOD expression in THP-1 cells, we used siRNA to diminish the expression of PAR-2, PR3, and nuclear factor (NF)-κB. In transfected cells, the anti-PR3 Ab-mediated priming effect was significantly reduced by PAR-2-, PR3-, and NF-κB-specific siRNA [16]. These results indicated that the secretion of cytokines induced by anti-PR3 Abs occurred through NF-κB, which are located downstream of the PAR-2 and PR3 signaling pathway.

6 Conclusion

Anti-PR3 Abs (ANCA), being weak direct activators of monocytes and neutrophils to release cytokines per se, exert a major priming effect on these leukocytes, enhancing their responsiveness to secondary stimulation with bacterial PAMPs. Up-regulation of various PRRs, including TLRs and NODs, acting as respective PAMPs, was characterized as one mechanism underlying the anti-PR3-elicited priming response. Such cooperation between anti-PR3 Abs and bacterial PAMPs may well trigger exacerbations of disease activity during infections and contribute to the persistence of inflammatory lesions, which might be a novel model for the pathogenesis of autoimmune diseases (Fig. 2).

Acknowledgments. This study was supported in part by Grants-in Aid for Scientific Research from the Ministry of Education, Culture, Sports, Science and Technology, Japan (18689901), and by the Exploratory Research Program for Young Scientists from the President of Tohoku University.

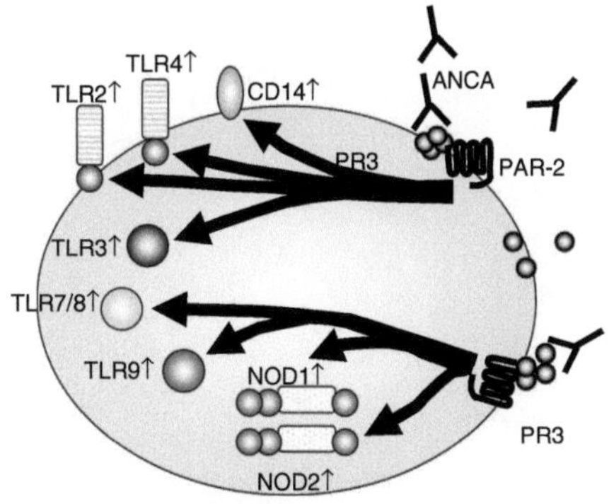

Fig. 2. Anti-PR3 Abs prime human monocytic cells to produce cytokines on stimulation with various bacterial components by up-regulating the TLR and NOD expression. These mechanisms may actively participate in the inflammatory process

References

1. Davies DJ, Moran JE, Niall JF, et al (1982) Br Med J 285:606
2. van der Woude FJ, Rasmussen N, Lobatto S, et al (1985) Lancet 1:425–429
3. Ardiles LG, Valderrama G, Moya P, et al (1997) Clin Nephrol 47:1–5
4. Casselman BL, Kilgore KS, Miller BF, et al (1995) J Lab Clin Med 126:495–502
5. Ralston DR, Marsh CB, Lowe MP, et al (1997) J Clin Invest 100:1416–1424
6. Nowack R, Schwalbe K, Flores-Suárez L-F, et al (2000) J Am Soc Nephrol 11:1639–1646
7. Uehara A, Sugawara Y, Sasano T, et al (2004) J Immunol 173:4179–4189
8. Janeway CA Jr, Medzhitov R (2002) Annu Rev Immunol 20:197–216
9. Akira S, Uematsu S, Takeuchi O (2006) Cell 124:783–801
10. Fritz JH, Ferrero RL, Kahn M, et al (2006) Nat Immunol 7:1250–1257
11. Coughlin SR (2000) Nature 407:258–264
12. Uehara A, Sugawara S, Muramoto K, et al (2002) J Immunol 169:4594–4603
13. Uehara A, Muramoto K, Takada H, et al (2003) J Immunol 170:5690–5696
14. Novo E, Viera N (1996) J Periodont Res 31:365–368
15. Novo E, Gregor EG-M, Nava S, et al (1997) P R Health Sci J 16:369–373
16. Uehara A, Iwashiro A, Sato T, et al (2007) Mol Immunol 44:3552–3562

Interface Oral Health Science 2007: 243–248

Water-insoluble α-glucans from *Streptococcus sobrinus* induce inflammatory immune responses

Shigefumi Okamoto[2]*, Yutaka Terao[1], Hidenori Kaminishi[3], Shigeyuki Hamada[4], and Shigetada Kawabata[1]
[1]*Department of Oral and Molecular Microbiology, Osaka University Graduate School of Dentistry, Osaka 565-0871;* [2]*Laboratory of Virology and Vaccinology, National Institute of Biomedical Innovation, Osaka 567-0085;* [3]*Department of Functional Bioscience, Fukuoka Dental College, Fukuoka 814-0193;* [4]*Department of Life Science, Nihon University Advanced Research Institute for the Sciences and Humanities, Nihon University Kaikan Daini Bekkan, Tokyo 102-8251; Japan*
*sokamoto@nibio.go.jp

Abstract. Recently, several studies have revealed that β-glucans in fungal cell wall components are known to modulate innate immunity. Water-insoluble α-glucans are synthesized from sucrose by glucosyltransferase-I of mutans streptococci and play an important role in the development of dental plaque. However, it remains unknown whether water-insoluble α-glucans also initiate these disease processes because of their innate immune response. In the present study, we showed that water-insoluble α-glucans synthesized by *Streptococcus sobrinus* activated mouse peritoneal exudate macrophages to produce proinflammatory cytokines. Furthermore, human monocytes stimulated by water-insoluble α-glucans produced TNF-α and IL-8, whereas human polymorphonuclear cells were activated by water-insoluble α-glucans, resulting in chemotaxis and hydrogen peroxide production. The results demonstrated that water-soluble α-glucans modulate macrophage- and granulocyte-induced inflammatory immune responses, and suggest that inflammation induced by those α-glucans is associated with the development of periodontal diseases.

Key words. *Streptococcus sobrinus*, water-insoluble α-glucans, inflammatory immune responses

1 It remains unknown whether water-insoluble α-glucans initiate periodontal diseases

Mutans streptococci, including *Streptococcus sobrinus and S. mutans*, have been implicated as primary etiologic agents of dental caries in humans [1, 2]. These bacteria produce water-soluble and water-insoluble α-glucans from sucrose by multiple glucosyltransferases (GTFs). These adhesive glucans, primarily water-insoluble α-glucans, contribute to the development of dental plaque, a type of biofilm composed of microorganisms and their products. Oral bacteria adhere to tooth surfaces in the presence of dental plaque and release acids, which are important factors in their cariogenic potential [1, 3]. On the other hand, bacterial accumulations in dental plaque are necessary to initiate the process leading to periodontal diseases, as well as dental caries [4]. In addition, such accumulation induces vascular changes typical of acute inflammatory reactions, resulting in vascular leakage of fluid, and the active migration of polymorphonuclear cells out of the vessels and into the gingival tissues. Lymphocytes then accumulate adjacent to gingival squamous epithelium, and fibroblasts in the area begin to show morphologic changes. However, it remains unknown whether water-insoluble α-glucans also initiate these disease processes.

Recent studies have revealed that several types of β-glucans in fungal cell wall components modulate innate immunity by activating mouse macrophages *via* complement receptor 3, a scavenger receptor, and/or Toll-like receptors (TLRs) [5]. These β-glucans also initiate inflammatory responses in macrophages *via* NF-κB activation, thus inducing fungi-mediated inflammatory and autoimmune diseases [6–8]. On the other hand, little is known regarding the biological activities of α-glucans. If water-insoluble α-glucans, as well as β-glucans, were shown to activate macrophages and induce inflammatory immune responses, water insoluble α-glucans would be associated with the induction of inflammation of oral squamous tissues. In the present study, we examined whether water-insoluble α-glucans, as well as fungal cell wall β-glucans and water-soluble α-glucans, promote macrophage activation, causing inflammatory immune responses in mice and humans.

2 Water-insoluble α-glucans stimulate monocytes/macrophages resulting in promotion of inflammatory immune responses

We examined the effects of water-insoluble α-glucans on the production of proinflammatory cytokines by mouse peritoneal exudate macrophages. Cytokine ELISA results revealed that peritoneal exudate macrophages produced TNF-α and IL-6 but not IL-1β following stimulation with water-insoluble α-glucans, whereas no such production was seen with sucrose or water-soluble α-glucans from *S. sobrinus* (Fig. 1). The cytokine production of peritoneal exudate macrophages stimulated with water-insoluble α-glucans was significantly greater than that by stimulation with muramyl dipeptide, or with cell wall extracts from *S. sobrinus* which contain lipoteichoic acid and peptidoglycan. Activation of macrophages by water-insoluble α-glucans was revealed in experiments with both the insoluble and soluble forms (Fig. 1). Macrophage activation by stimulation with water-insoluble α-glucans was not suppressed by depletion of TLR2, TLR4, NOD1, or NOD2, suggesting that the stimulatory pathway of water-insoluble α-glucans is different from that of LPS (TLR4 pathway), peptidoglycan (TLR2 pathway), lipoteichoic acid (TLR2 pathway), and muramyl dipeptide (NOD2 pathway) (data not shown).

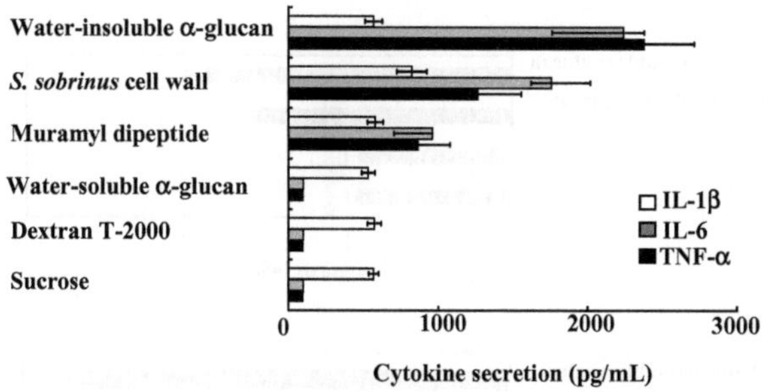

Fig. 1. Cytokine production by mouse immune cells following stimulation with water-insoluble α-glucans. Mouse peritoneal exudate macrophages (5×10^5 cells/ml of complete RPMI 1640 medium) were stimulated for 24 h with 1 mg/ml of the test materials, after which the production of TNF-α (*black bar*), IL-6 (*gray bar*), and IL-1β (*white bar*) was assessed using cytokine ELISA kits

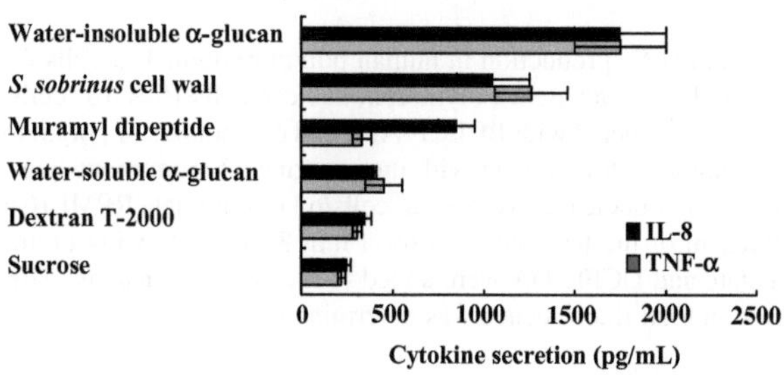

Fig. 2. Cytokine production in human monocytes induced by water-insoluble α-glucans Human monocytes (5×10^5 cells/ml of complete RPMI 1640 medium) were stimulated for 48 h with 1 mg (in case of TNF-α) or 0.1 μg (in case of IL-8) of test materials, after which the production of IL-8 (*black bar*) and TNF-α (*gray bar*) was assessed using cytokine ELISA kits

Next, we attempted to determine whether water-insoluble α-glucans induce cytokine production in human peripheral monocytes. A high level of TNF-α was detected in the supernatants of water-insoluble α-glucan-stimulated monocytes, and IL-8 was produced by monocytes stimulated with 0.1 μg/ml of water-insoluble α-glucans (Fig. 2). IL-8 is known to promote the chemotaxis of human polymorphonuclear cells and memory T cells, and we found that the culture supernatants from human monocytes stimulated with water-insoluble α-glucans, but not sucrose, induced chemotaxis of human polymorphonuclear cells (Fig. 3a). Although the culture supernatants of water-soluble α-glucan-stimulated human monocytes induced chemotaxis in human polymorphonuclear cells, we did not detect IL-8 in the supernatants. Since other factors besides IL-8, including Groα, Groβ, ENA-78, GCP-2, and NAO-2, can induce chemotaxis of polymorphonuclear cells, we speculated that production of chemokines other than IL-8 mediates the ability of water-soluble α-glucans to promote chemotaxis. Since H_2O_2 released by polymorphonuclear

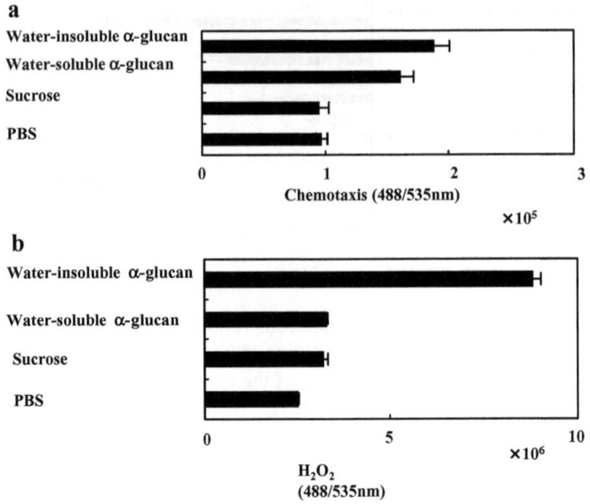

Fig. 3. Chemotaxis and H_2O_2 production in human polymorphonuclear cells following stimulation with water-insoluble α-glucans. **a** Polymorphonuclear cells (5×10^6 cells/ml of complete RPMI 1640 medium) were labeled with BCECF-AM, and chemotaxis of polymorphonuclear cells was measured following a 24-h treatment with the supernatant from monocytes stimulated with the materials. **b** Polymorphonuclear cells (5×10^6 cells/ml of complete RPMI 1640 medium) were pretreated with 10 μg/ml of the test materials for 1 h at 37°C. Following pretreatment, phorbol 12-myristate 13-acetate and DCFH-DA were added to the cell suspensions. After 1-h incubation at 37°C, the increase in total fluorescence was determined

cells is an important mediator of endothelial cell injury and vascular inflammation [9], and we demonstrated that water-insoluble α-glucans but not water-soluble α-glucans induce the production of H_2O_2 in human polymorphonuclear cells. The results suggest that water-insoluble α-glucans induce chemotaxis and inflammatory responses in polymorphonuclear cells (Fig. 3b).

Taken together, water-insoluble α-glucans synthesized by *S. sobrinus* GTF-I, but not sucrose, activated mouse peritoneal exudate macrophages to produce proinflammatory cytokines. Activation of macrophages by water-insoluble α-glucans was revealed in experiments with both the insoluble and soluble forms. Furthermore, the water-insoluble α-glucan preparation activated human monocytes and polymorphonuclear cells, suggesting that α-glucans have an ability to modulate the human immune system. Glucans synthesized by *S. sobrinus* contain α-1,3-glucosyl linkages (mutan) and α-1,6-glucosyl linkages (dextran). Water-soluble α-glucans contain glucose primarily with α-1,6-glucosyl linkages, whereas water-insoluble α-glucans possess a high degree of branching involving α-1,3-glucosyl linkages [4]. We confirmed that much lower levels of TNF-α, IL-8, and H_2O_2 were produced by monocytes/macrophages stimulated with water-soluble α-glucans than with water-insoluble α-glucans. These results suggest that the immunological properties of polysaccharides with α-1,3-glucosyl linkages are different from those with α-1,6 glucosyl linkages.

3 How do water-insoluble α-glucans participate in periodontal diseases?

Periodontal diseases are typical inflammatory diseases that occur in the oral cavity [4] and include inflammation of gingival tissues and the accumulation of dental plaque in the coronal pockets. Several investigators have speculated that a number of different strains of bacteria (e.g., *Porphyromonas gingivalis, Actinobacillus actinomycetemcomitans,* and *Tannerella forsythensis*) in dental plaque invade oral squamous cells and induce inflammation in gingival tissues [4, 10]. Since oral squamous tissues hinder the contact of water-insoluble α-glucans with monocytes/macrophages, polymorphonuclear cells, and lymphocytes in the lamina propria and distinct submucous layer, it is difficult for water-insoluble α-glucans to induce an inflammatory immune response in the gingiva. However, dental plaques contain abundant bacteria that injure oral squamous cells, and the cytotoxicity of bacteria toward squamous cells may allow for water-insoluble α-glucans to reach monocytes/macrophages and polymorphonuclear cells, resulting in inflammatory responses in the gingiva. Therefore, we speculate that water-insoluble α-glucans help in mediating the induction of periodontal diseases. We propose that water-insoluble α-glucans participate in the pathogenicity of dental caries as well as inflammatory diseases in the oral cavity.

References

1. Hamada S, Slade HD (1980) Biology, immunology and cariogenicity of *Streptococcus mutans*. Microbiol Rev 44:331–384
2. Loesche WJ (1986) Role of *Streptococcus mutans* in human dental decay. Microbiol Rev 50:353–380
3. Tamesada M, Kawabata S, Fujiwara T, et al (2004) Synergistic effects of streptococcal glucosyltransferases on adhesive biofilm formation. J Dent Res 83:874–879
4. Schuster GS (1983) Oral microbiology and infectious diseases, 2nd student edn. Williams & Wilkins, Baltimore
5. Brown GD, Herre J, Williams DL, et al (2003) Dectin-1 mediates the biological effects of β-glucans. J Exp Med 197:1119–1124
6. Hahn PY, Evans SE, Kottom TJ, et al (2003) *Pneumocystis carinii* cell wall β-glucan induces release of macrophage inflammatory protein-2 from alveolar epithelial cells via a lactosylceramide-mediated mechanism. J Biol Chem 278:2043–2050
7. Lebron F, Vassallo R, Puri V, et al (2003) *Pneumocystis carinii* cell wall β-glucans initiate macrophage inflammatory responses through NF-κB activation. J Biol Chem 278: 25001–25008
8. Yoshitomi H, Sakaguchi N, Kobayashi K, et al (2005) A role for fungal β-glucans and their receptor Dectin-1 in the induction of autoimmune arthritis in genetically susceptible mice. J Exp Med 201:949–960
9. Lounsbury KM, Hu Q, Ziegelstein RC (2000) Calcium signaling and oxidant stress in the vasculature. Free Radic Biol Med 28:1362–1369
10. Van Dyke TE, Serhan CN (2003) Resolution of inflammation: a new paradigm for the pathogenesis of periodontal diseases. J Dent Res 82:82–90

Interface Oral Health Science 2007: 249–254

Biotin-deficiency up-regulates TNF-α production in vivo and in vitro

Toshinobu Kuroishi*, Yasuo Endo, and Shunji Sugawara
Division of Oral Immunology, Tohoku University Graduate School of Dentistry, Sendai 980-8575, Japan
*kuroishi@mail.tains.tohoku.ac.jp

Abstract. Biotin, a water-soluble vitamin of the B complex, functions as a cofactor of carboxylases that catalyze indispensable cellular metabolism. It was reported that the concentrations of biotin were significantly lower in sera of patients with chronic inflammatory diseases. However, the biological roles of biotin in inflammatory responses are unclear. In this study, we investigated the effects of biotin-deficiency on tumor necrosis factor (TNF)-α production in vivo and in vitro. Mice were fed a basal diet or a biotin-deficient diet for 8 weeks. After intravenous administration of lipopolysaccharide (LPS), serum TNF-α levels in biotin-deficient mice were significantly higher than those in biotin-sufficient mice. A murine macrophage-like cell line, J774.1, was cultured in biotin-sufficient or biotin-deficient medium. Biotin-deficient J774.1 cells produced TNF-α significantly higher than biotin-sufficient J774.1 cells in response to LPS and even without LPS stimulation. Moreover, biotin-supplementation inhibited TNF-α production of biotin-deficient cells. Addition of cyclic guanosine 5′-monophosphate (cGMP) significantly decreased TNF-α production of the biotin-deficient cells, indicating that up-regulation of TNF-α production was regulated by cGMP-dependent signaling pathways. In conclusion, these results suggest that biotin is critically involved in inflammatory diseases via the regulation of TNF-α production in vivo and in vitro.

Key words. biotin, macrophage, TNF-α, cGMP

1 Introduction

Biotin is a water-soluble vitamin of the B complex found in all organisms [1]. Biotin functions as a cofactor of four carboxylases: pyruvate carboxylase, acetyl-CoA carboxylase, propionyl-CoA carboxylase, and 3-methylcrotonyl-CoA carboxylase [1]. These enzymes catalyze the metabolism of glucose, amino acids, and fatty acids. In addition to this classical function as a cofactor of carboxylases, biotin is involved in various cellular events. Biotin regulates the mRNA expression of holocarboxylase synthetase and biotin-dependent carboxylases via cGMP-dependent pathway [2]. Moreover, some transcription factors, such as Sp1, Sp3, and NF-κB, were regulated by biotin, and the biotinylation of histones in human cells was also reported [3, 4]. These reports clearly indicate that biotin regulates the various cellular events at the transcriptional levels.

Biotin-deficiency causes alopecia and scaly erythematous dermatitis [5]. Moreover, it was reported that serum biotin levels are significantly lower in atopic dermatitis patients than in healthy subjects [6]. Biotin has a therapeutic effect on pustulosis palmaris et plantaris, a type of chronic dermatitis which is restricted to the palms and soles [7]. These reports suggest that biotin-deficiency is involved in inflammatory diseases. However, few reports are available on the biological roles of biotin in inflammatory responses.

In this study, we investigate the effects of biotin-deficiency on the production of TNF-α in vivo and in vitro.

2 Experimental procedures

2.1 Mice

Female BALB/c mice (4 weeks old) received a basal diet (AIN-76) or a biotin-deficient AIN-76 diet. The Ethical Board for nonhuman species of the Tohoku University Graduate School of Medicine approved the experimental procedure followed in this study. Concentrations of biotin in serum were measured with ELISA [8].

2.2 Measurement of TNF-α

Concentrations of TNF-α were measured with a commercial ELISA kit.

2.3 Cells and cell culture

Murine macrophage-like J774.1 cells were grown in biotin-sufficient or biotin-deficient medium. The biotin-sufficient medium was RPMI 1640 containing *d*-biotin (0.2 mg/μg) supplemented with 10% FCS. The biotin-deficient medium was biotin free RPMI 1640 supplemented with 10% biotin-deficient FCS. Biotin in FCS was depleted with immobilized avidin-agarose. J774.1 cells were cultured with biotin-deficient medium for 4 weeks, and then further incubated in the medium without biotin (biotin-deficiency) or with biotin (biotin-supplementation) for 2 weeks. J774.1 cells were also cultured in biotin-sufficient medium for 6 weeks (biotin-sufficiency).

2.4 Data analysis

All of the experiments in this study were performed at least three times to confirm the reproducibility of the results. The data shown are representative results. Experimental values are given as the mean ± SD of triplicate assays. Statistical analysis was performed with the unpaired *t*-test or one-way ANOVA using Dunnett's method, and $P < 0.05$ was considered significant.

3 Results

3.1 Augmentation of serum TNF-α levels in biotin-deficient mice injected with LPS

After 8 weeks of feeding with biotin-sufficient or biotin-deficient diets, the serum concentrations of biotin in biotin-deficient group were significantly ($P < 0.01$) lower than those in biotin-sufficient group. No clinical symptoms were detected in the biotin-deficient group, and no significant differences of body weights were detected between biotin-sufficient and biotin-deficient groups. A significant ($P < 0.01$) increase of the serum TNF-α level was induced 90 min after i.v. injection of LPS (2 μg/kg) (Fig. 1). In biotin-deficient group, the concentration of TNF-α was significantly ($P < 0.05$) higher than that in biotin-sufficient group. These results indicated that biotin-deficiency augments TNF-α production in vivo.

3.2 Augmentation of TNF-α production in biotin-deficient J774.1 cells

Next, we analyzed TNF-α production by biotin-sufficient and biotin-deficient J774.1 cells. As shown in Fig. 2, both types of cells were produced TNF-α in a dose-dependent manner with LPS stimulation. The concentration of TNF-α in the culture supernatants of biotin-deficient cells was significantly ($P < 0.01$) higher than that of biotin-sufficient cells even without LPS stimulation. These results clearly indicated that biotin-deficiency induces the augmentation of TNF-α production in vitro.

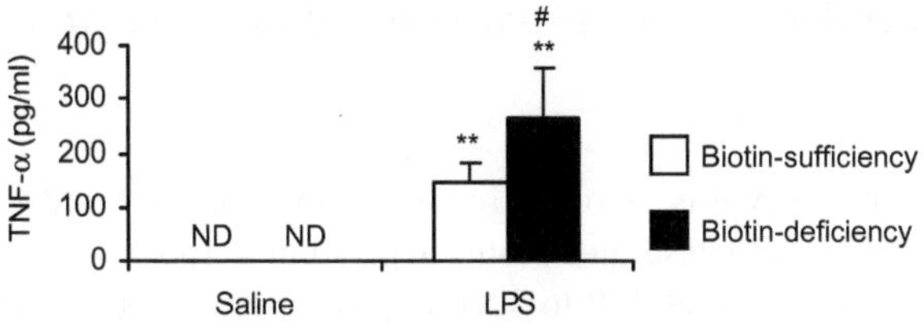

Fig. 1. Serum level of TNF-α in biotin-sufficient and biotin-deficient mice. Biotin-sufficient and biotin-deficient mice were challenged i.v. with LPS (2 μg/kg) or saline alone, and blood was taken at 90 min after injection. The results were expressed as mean ± SD for four mice. *ND*, Not detected. **, $P < 0.01$, compared with saline. #, $P < 0.05$, compared with biotin-sufficiency

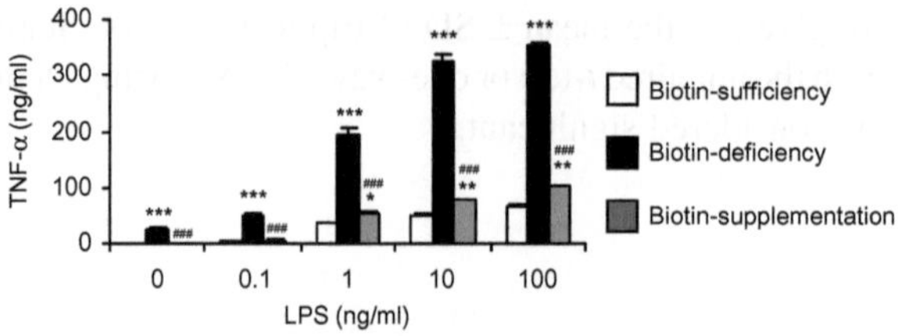

Fig. 2. TNF-α production of biotin-sufficient, -deficient and -supplemented J774.1 cells. Cells (2×10^5 cells/200 μl/well) were stimulated with LPS at 37°C for 24 h. *, $P < 0.05$, **, $P < 0.01$, ***, $P < 0.001$, compared with biotin-sufficiency. ###, $P < 0.001$, compared with biotin-deficiency

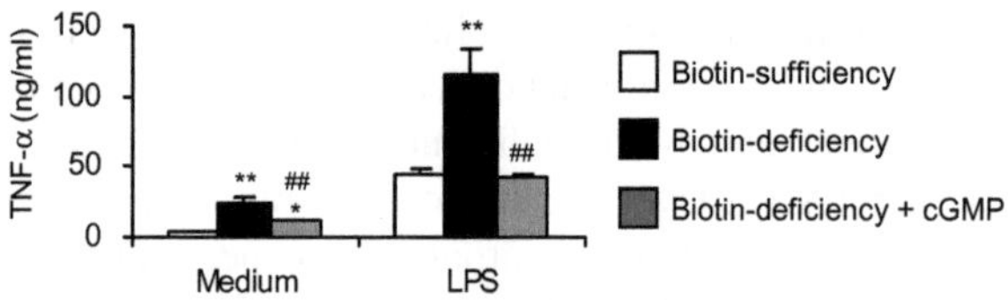

Fig. 3. Effects of cGMP on TNF-α production of biotin-deficient J774.1 cells. Cells (2×10^5cells/200 μl/well) were stimulated with 10 ng/ml of LPS in the presence of 1 mM of 8-Br-cGMP (cGMP) at 37°C for 24 h. * $P < 0.05$, ** $P < 0.01$, compared with biotin-sufficiency. ## $P < 0.01$, compared with biotin deficiency (without cGMP)

To further confirm the effects of biotin deficiency, J774.1 cells were cultured with biotin-deficient medium for 4 weeks and then further incubated with biotin-sufficient medium for 2 weeks (biotin-supplemented cells). The concentrations of TNF-α in the culture supernatants of biotin-supplemented cells with and without LPS stimulation were significantly ($P < 0.01$) reduced to near the levels in the supernatants of biotin-sufficient cells (Fig. 2). These results indicated that biotin-supplementation restored the TNF-α production to the basal level.

3.3 cGMP inhibits TNF-α production of biotin-deficient J774.1 cells

It was reported that cGMP is involved in the biotin-dependent mRNA expressions of holocarboxylase synthetase and biotin-dependent carboxylases [2]. Therefore, we analyzed the effects of cGMP to TNF-α production of biotin-deficient J774.1 cells. TNF-α production of biotin-deficient cells significantly ($P < 0.01$) decreased in the presence of 1 mM of cGMP (Fig. 3). These results indicated that up-regulation of TNF-α production was regulated by cGMP-dependent signaling pathway.

4 Conclusion

It is well known that biotin-deficiency causes cutaneous abnormalities, such as alopecia and scaly erythematous dermatitis [5]. In addition, it was reported that the biotin concentration in serum correlates with inflammatory diseases [5]. Although several studies reported the contribution of abnormalities in lipid metabolism to cutaneous abnormalities [5], the pathological mechanisms of disease conditions caused by biotin-deficiency remain to be clarified. In this study, we clearly demonstrated that biotin-deficiency up-regulated TNF-α production in vivo and in vitro. Moreover, biotin-supplementation inhibits TNF-α production of biotin-deficient cells. Therefore, we considered that the augmentable effect of biotin-deficiency on TNF-α production is reversible.

TNF-α plays important roles in the pathogenesis of atopic dermatitis [9], contact hypersensitivity [10], and pustulosis palmaris et plantaris [11], inflammatory diseases which have been reported to be correlated with biotin. Therefore, it is possible that TNF-α up-regulation caused by biotin-deficiency is involved in the pathological mechanisms of dermatitis and other inflammatory diseases. Our results encourage further investigations on biotin treatment for various inflammatory diseases.

References

1. Wood HG, Barden RE (1977) Biotin enzymes. Annu Rev Biochem 46:385–413
2. Solórzano-Vargas RS, Pacheco-Alvarez D, León-Del-Río A (2002) Holocarboxylase synthetase is an obligate participant in biotin-mediated regulation of its own expression and biotin-dependent carboxylases mRNA levels in human cells. Proc Natl Acad Sci USA 99:5325–5330
3. Griffin JB, Rodriguez-Melendez R, Zempleni J (2003) The nuclear abundance of transcription factors Sp1 and Sp3 depends on biotin in Jurkat cells. J Nutr 133:3409–3415
4. Rodriguez-Melendez R, Schwab LD, Zempleni J (2004) Jurkat cells respond to biotin deficiency with increased nuclear translocation of NF-κB, mediating cell survival. Int J Vitam Nutr Res 74:209–216
5. Mock DM (1991) Skin manifestations of biotin deficiency. Semin Dermatol 10:296–302
6. Makino Y, Osada K, Sone H, et al (1999) Percutaneous absorption of biotin in healthy subjects and in atopic dermatitis patients. J Nutr Sci Vitaminol (Tokyo) 45:347–352
7. Maebashi M, Makino Y, Furukawa Y, et al (1993) Effect of biotin treatment on metabolic abnormalities occurring in patients with strenocostoclavicular hyperostosis. J Clin Biochem Nutr 15:65–76
8. Mock DM (1997) Determination of biotin in biological fluids. Methods Enzymol 279:265–275
9. Villagomez MT, Bae S-J, Ogawa I, et al (2004) Tumor necrosis factor-α but not interferon-γ is the main inducer of inducible protein-10 in skin fibroblasts from patients with atopic dermatitis. Br J Dermatol 150:910–916
10. Nakae S, Komiyama Y, Narumi S, et al (2003) IL-1-induced tumor necrosis factor-α elicits inflammatory cell infiltration in the skin by inducing IFN-γ-inducible protein 10 in the elicitation phase of the contact hypersensitivity response. Int Immunol 15:251–260
11. Murakata H, Harabuchi Y, Kataura A (1999) Increased interleukin-6, interferon-γ and tumor necrosis factor-α production by tonsillar mononuclear cells stimulated with alpha-streptococci in patients with pustulosis palmaris et plantaris. Acta Otolaryngol 119:384–391

Interface Oral Health Science 2007: 255–256

Real-time PCR analyses of genera *Veillonella* and *Streptococcus* in healthy supragingival plaque biofilm microflora of children

Junko Matsuyama[1]*, Takuichi Sato[2], Nobuhiro Takahashi[2], Michiko Sato[3], and Etsuro Hoshino[3]
[1]*Division of Pediatric Dentistry; [3]Division of Oral Ecology in Health and Infection, Department of Oral Health Science, Niigata University Graduate School of Medical and Dental Sciences, Niigata 951-8514; [2]Division of Oral Ecology and Biochemistry, Department of Oral Biology, Tohoku University Graduate School of Dentistry, Sendai 980-8575; Japan*
*junko@dent.niigata-u.ac.jp

Abstract. Since *Veillonella* species obtain the energy for their growth by fermenting organic acids, e.g., lactate, this metabolism has the potential to remove a potent, dental-caries producing acid. Therefore, the presence of *Veillonella* in plaque biofilm may reduce the caries-producing potential of plaque biofilm. Quantification of genera *Veillonella* and *Streptococcus* in healthy supragingival plaque biofilm microflora of children was performed in the present study. Total bacteria and the target genera (*Veillonella* and *Streptococcus*) were quantified by real-time PCR using universal and genus-specific primers, respectively, and the proportion of each genus was calculated. The proportion of genera *Veillonella* and *Streptococcus* was 2.1 ± 4.1% and 19.4 ± 16.7%, respectively. The results of the present study showed that *Veillonella* and *Streptococcus* normally inhabit the human mouths of both deciduous and permanent dentition.

Key words. children, plaque biofilm, real-time polymerase chain reaction, *Streptococcus*, *Veillonella*

1 Introduction

The genus *Veillonella* is small, asaccharolytic, anaerobic, gram-negative cocci. Because *Veillonella* species obtain the energy for their growth by fermenting organic acids, e.g., lactate, this metabolism has the potential to remove a potent, dental-caries producing acid. Therefore, the presence of *Veillonella* in plaque biofilm may reduce the caries-producing potential of plaque biofilm. The aim of this study was to quantify genera *Veillonella*, and *Streptococcus*, one of lactate-producers, in healthy supragingival plaque biofilm microflora of children.

2 Real-time polymerase chain reaction

Real-time PCR is at present a powerful and suitable method for quantification of bacterial DNA, and it has been applied to profiling of microflora of periodontitits [1]. In the present study, after informed consent was obtained from each subject, supragingival plaque was obtained from 44 orally healthy children (1–16 years), and the genomic DNA was extracted from the plaque biofilm with the Instagene matrix kit (Bio-Rad Laboratories, Richmond, CA, USA). Total bacteria and the target genera (*Veillonella* and *Streptococcus*) were quantified by real-time PCR utilizing universal and genus-specific primers [1–5], respectively, and the proportion of each genus was calculated. Real-time PCR amplification was performed in an iCycler (Bio-Rad Laboratories) with the iQ SYBR Green supermix (Bio-Rad Laboratories) according to the manufacturer's instructions.

The proportions of genera *Veillonella* and *Streptococcus* were 2.1 ± 4.1% and 19.4 ± 16.7%, respectively, in the present study. It seemed that the proportion of genus *Streptococcus* was high and varied among individuals, whereas that of genus *Veillonella* was relatively low and stable in supragingival plaque biofilm. In a previous study on subgingival plaque microflora of children, aged 7–8 years, by culturing method, the proportions of *Veillonella* and *Streptococcus* were similar between deciduous (8 and 50%) and permanent teeth (7 and 48%), respectively [6]. In accordance with the previous study, the present study using the molecular biological method confirmed that *Veillonella* and *Streptococcus* species normally inhabit the human mouths of both deciduous and permanent dentition of children.

Acknowledgments. This study was supported in part by Grants-in-Aid for Scientific Research (17592133 to J.M. and 17591985 to T.S.) from the Japan Society for the Promotion of Science, Tokyo, Japan.

References

1. Kuboniwa M, Amano A, Kimura RK, et al (2004) Quantitative detection of periodontal pathogens using real-time polymerase chain reaction with TaqMan probes. Oral Microbiol Immunol 19:168–176
2. Lane DJ (1991) 16S/23S rRNA sequencing. In: Stackebrandt E, Goodfellow M (eds) Nucleic acid techniques in bacterial systematics. Wiley, Chichester, pp 115–175
3. Yamaura M, Sato T, Echigo S, et al (2005) Quantification and detection of bacteria from postoperative maxillary cyst by polymerase chain reaction. Oral Microbiol Immunol 20:333–338
4. Rocas IN, Siqueira JF Jr (2006) Culture-independent detection of *Eikenella corrodens* and *Veillonella parvula* in primary endodontic infections. J Endod 32:509–512
5. Kroes I, Lepp PW, Relman DA (1999) Bacterial diversity within the human subgingival crevice. Proc Natl Acad Sci USA 96:14547–14552
6. Kamma JJ, Diamanti-Kipioti A, Nakou M, et al (2000) Profile of subgingival microbiota in children with mixed dentition. Oral Microbiol Immunol 15:103–111

Interface Oral Health Science 2007: 257–258

Inhibitory effects of maltotriitol on growth and adhesion of mutans streptococci

Harumi Miyasawa-Hori[1]*, Shizuko Aizawa[1,2], Jumpei Washio[1], and Nobuhiro Takahashi[1]
[1]*Division of Oral Ecology and Biochemistry, Department of Oral Biology;* [2]*Division of Pediatric Dentistry, Department of Lifelong Oral Health Sciences, Tohoku University Graduate School of Dentistry, Sendai 980-8575, Japan*
*miyasawa@mail.tains.tohoku.ac.jp

Abstract. Maltotriitol is known to inhibit α-glucosidase and maltose metabolism of *Streptococcus mutans*. In this study, we evaluated inhibitory effects of maltotriitol on the growth and adhesion of mutans streptococci. Bacterial adherence to glass surfaces and bacterial culture pH in the presence or absence of maltotriitol were determined. In the presence of maltotriitol, the growth of *Streptococcus sobrinus* with glucose, maltose, and sucrose was decreased, while the adhesion was not inhibited. On the other hand, maltotriitol did not decrease the growth of *S. mutans* except the cell growth with glucose, while maltotriitol inhibited the bacterial adhesion. The culture pH was less acidic when the growth was inhibited by maltotriitol. These results suggest that inhibitory effects of maltotriitol on mutans streptococci are different among species: maltotriitol inhibits the growth of *S. sobrinus*, while it inhibits the adhesion of *S. mutans*.

Key words. adhesion, bacterial growth, maltotriitol, mutans streptococci

1 Introduction

Maltotriitol (4-O-α-D-glucopyranosyl-4-O-α-D-glucopyranosyl-D-glucitol) is known to inhibit α-glucosidase and the acid production from maltose of *Streptococcus mutans* [1]. In this study, we evaluated inhibitory effects of maltotriitol on the growth and the adhesive property of mutans streptococci, *Streptococcus mutans* and *Streptococcus sobrinus*.

2 Effects of maltotriitol on growth and adherence of *S. sobrinus* and *S. mutans*

We used *S. sobrinus* (strains 6715 and OMZ176) and *S. mutans* (strains ATCC 31989 and NCIB 11723). To assess bacterial adherence to glass surfaces, the strains were grown anaerobically in a complex medium containing 1% glucose, maltose or sucrose in the presence or absence of maltotriitol in glass test-tubes at an angle of 30°. After an 18-h incubation at 35°C, cells were separated by decantation to floating cells and adherent cells, and the latter cells were dispersed by ultrasonic oscillation. The amount of cells was estimated by optical density at 660 nm. The culture pH was measured by a pH-electrode. Data were statistically analyzed by paired *t*-test.

In the presence of maltotriitol, the growth of *S. sobrinus* with glucose, maltose, and sucrose was decreased to 15.0–88.5% ($P < 0.01$–0.05). On the other hand, maltotriitol did not decrease the growth of *S. mutans* except for the cell growth with glucose, which was decreased to 53.0–63.0% ($P < 0.01$). Final culture pH was less acidic when the growth was inhibited.

There was no significant change in the ratio of the amount of adherent cells to those of total cells in *S. sobrinus*. On the other hand, in *S. mutans* ATCC 31989, the ratio of adherent cells to total cells when grown with glucose, maltose, and sucrose was decreased to 26.9–47.5% ($P < 0.01$–0.05). In *S. mutans* NCIB 11723, the ratio when grown with glucose was decreased to 21.8% ($P < 0.05$).

3 Inhibitory effects of maltotriitol and its clinical implication

The culture pH was less acidic when the growth was inhibited by maltotriitol, suggesting that maltotriitol inhibits the acid production through glycolysis, and consequently decreases the growth due to the shortage of energy supply. Maltotriitol inhibited bacterial adhesion in *S. mutans* when grown with glucose and maltose as well as sucrose, suggesting that maltotriitol inhibits not only sucrose/glucosyltransferase-dependent adhesion but also other bacterial adhesive systems. Although the mode of inhibition by maltotriitol on mutans streptococci was different among species: maltotriitol inhibits the growth of *S. sobrinus* while it inhibits the adhesive ability of *S. mutans*; maltotriitol may be useful to control the population of mutans streptococci in the oral cavity.

Acknowledgments. This study was supported in part by Research fellowship (no. 16 · 3025 to HH) and a Grant-in-Aid for Scientific Research B (no. 19390539 to NT) from the Japan Society for the Promotion of Science. Maltotriitol was given from TOWAKASEI Co., Ltd., Japan.

Reference

1. Würsch P, Koellreutter B (1985) Maltotriitol inhibition of maltose metabolism in *Streptococcus mutans* via maltose transport, amylomaltase and phospho-alpha-glucosidase activities. Caries Res 19:439–449

Interface Oral Health Science 2007: 259–260

Influence of yogurt products containing *Lactobacillus reuteri* on distributions of mutans streptococci within dental plaque

Kazuo Kato[1]*, Kiyomi Tamura[1], Takuichi Sato[2], and Haruo Nakagaki[1]
[1]*Department of Preventive Dentistry and Dental Public Health, School of Dentistry, Aichi-Gakuin University, Nagoya 464-8650;* [2]*Department of Oral Biology, Tohoku University Graduate School of Dentistry, Sendai 980-8575; Japan*
*kazkato@dpc.aichi-gakuin.ac.jp

Abstract. Placebo-controlled trial demonstrated that consuming yogurt with *Lactobacillus reuteri* could not significantly reduce the habitat for mutans streptococci within plaque in vivo.

Key words. dental plaque, depth-specific analysis, mutans streptococci, *Lactobacillus reuteri*, yogurt product

The probiotic bacterium *Lactobacillus reuteri* produce compounds that exhibit a broad-spectrum of antimicrobial activity [1]. In Japan, one yogurt product containing *L. reuteri* (Reuteri yogurt, Chichiyasu, Hiroshima) is commercially available. Recently, it was demonstrated that consuming this yogurt reduced the salivary level of *S. mutans* [2]. The present study was carried out to estimate the influence of yogurt products containing *L. reuteri* on the presence of mutans streptococci from the outer to the inner plaque.

Eighteen consenting female subjects (18–39 years) consumed a pack (90 g) of Reuteri yogurt or placebo yogurt containing lactic acid bacteria except *L. reuteri* at lunchtime daily for a period of 2 weeks, respectively. On both occasions, in the second week, they were asked to form plaque by abstaining from tooth-brushing. Two in situ plaque-generating devices were set up in the upper molars on the Monday of the second week in the evening. On the following Friday evening, the devices were collected to obtain the experimental or placebo samples for analysis at least 1 h after any food or drink intake. Samples taken with no yogurt products consumed served as negative control. The experimental design was approved by the Ethics Committee of Aichi-Gakuin University.

The devices were snap frozen in liquid nitrogen and freeze-dried, being embedded in a methacrylate mixture for serial plaque sectioning. The sample was separated into 6–10 layered fractions (100 μm thick) using the method reported by the authors [3]. Genomic DNA was extracted from each fraction using the commercial DNA purification matrix. The 16S rRNA gene sequences were amplified by the PCR with universal primers and then with the species-specific primers. The final products were separated on agarose gels and stained to confirm the existence of cariogenic bacteria.

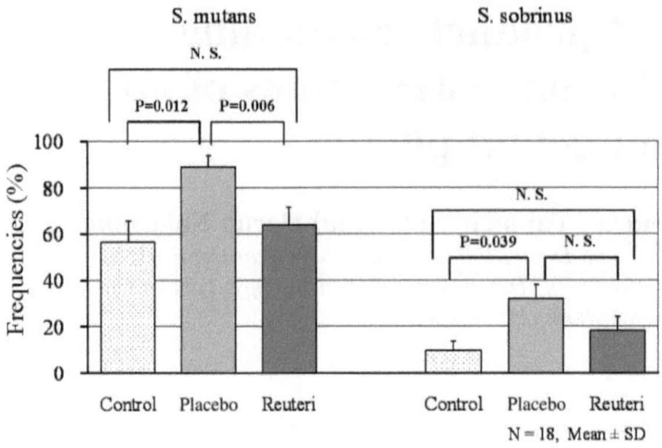

Fig. 1. Frequencies of layers positive for the bacteria per plaque sample among the experimental (reuteri), placebo and control group

Friedman's test and Wilcoxon Matched-Pair Signed-Rank Test were used to evaluate for statistically significant differences in frequencies of layers positive for the bacteria per plaque sample among the groups.

The results showed that the distribution patterns of positive fractions for each strain varied among the subjects. Frequencies of layer containing *S. mutans* and *S. sobrinus* were 64 and 18% in the experimental, 88 and 31% in the placebo and 59 and 11% in the control group, respectively, indicating that the placebo sample was significantly higher than both the experimental and the control samples for *S. mutans* than the control for *S. sobrinus* (Fig. 1). There were no significant differences in the percentage of layer positive for both strains between the experimental and the control group. Therefore, the results demonstrated that consuming yogurt with *L. reuteri* could not significantly reduce the habitat of mutans streptococci within the plaque.

It is known that reuterin, a broad-spectrum antimicrobial substance is produced by *L. reuteri* under anaerobic conditions in the gastrointestinal tract. However, it is doubtful that such antimicrobial compound is produced and exhibits the inhibitory action in the oral cavity. On the other hand, Reuteri yogurt as the commercial product is sweetened with 7% of sucrose, so that the same concentration of sucrose was added to both the experimental and the placebo yogurt. There is a possibility that periodical consuming of yogurt with sucrose makes the habitat of mutans streptococci within the plaque significantly wider, and this might weaken the benefit of consuming yogurt containing *L. reuteri*. Further studies are needed to estimate the influence of probiotic yogurt products on the inhibitory activity against oral mutans streptococci.

Acknowledgment. This work was supported by KAKENHI (C) No. 17592192, Japan.

References

1. Talarico TL, Casas IA, Chung TC, et al (1988) Antimicrob Agents Chemother 32: 1854–1858
2. Nikawa H, Makihara S, Fukushima H, et al (2004) Int J Food Microbiol 95:219–223
3. Kato K, Sato T, Takahashi N, et al (2004) Caries Res 38:448–453

Interface Oral Health Science 2007: 261–262

The effect of amylase and its inhibitors on acid production from starch by *Streptococcus mutans* and *Streptococcus sanguinis*

Shizuko Aizawa[1,2]*, **Harumi Miyasawa-Hori**[2], **Hideaki Mayanagi**[1], **and Nobuhiro Takahashi**[2]
[1]*Divisions of Pediatric Dentistry;* [2]*Oral Ecology and Biochemistry, Tohoku University Graduate School of Dentistry, Sendai 980-8575, Japan*
*shizu-a@mail.tains.tohoku.ac.jp

Abstract. This study aimed to examine effects of α-amylase and its inhibitors on acid production from starch by streptococci. Glucose-grown streptococci were anaerobically incubated with various carbohydrates, including starch in the presence or absence of α-amylase pH fall and acid production rate were measured. α-amylase inhibitors were added to the reaction mixture and the effects were evaluated. In the absence of α-amylase, both pH fall and acid production rate from starch were small. In the presence of α-amylase, pH fall and acid production rate from starch by *Streptococcus mutans* and *Streptococcus sanguinis* were similar to those from maltose. The pH fall from starch by *S. mutans* was similar to that by *S. sanguinis*. The addition of α-amylase inhibitors decreased the acid production rates from starch in the presence of α-amylase. These results suggest that starch is acidogenic similarly to maltose in the presence of α-amylase, and that the acidogenic potential of starch by *S. sanguinis* was comparable to that by *S. mutans*.

Key words. *S. mutans*, *S. sanguinis*, starch, α-amylase, acid production

1 Introduction

Although starch is the major component of human diet, its cariogenicity is still unclear. Starch is degraded into short polymers of glucose by salivary α-amylase and consequently fermented by oral bacteria. This study aimed to examine effects of α-amylase and its inhibitors on acid production from starch by *Streptococcus mutans* and *Streptococcus sanguinis*.

2 Acid production from starch by oral streptococci

Streptococcus mutans NCTC10449 and *S. sanguinis* ATCC10556 were grown anaerobically in complex medium containing 0.3% glucose. The cells were harvested, washed, and suspended. The cell suspensions were anaerobically incubated with glucose, sucrose, maltose or cooked potato starch in the presence or absence of purified human α-amylase. The pH fall from pH 7.0 and the acid production rate at pH 7.0 were measured by pH-electrode and automatic pH-titration, respectively.

In the absence of α-amylase, both pH fall and acid production rate from starch by these streptococci were small. In the presence of α-amylase, however, the lowest pH fall and the highest acid production rate from starch by *S. mutans* were pH 3.9 and 0.24 μmol min^{-1} OD^{-1}, respectively, and these values were similar to those from maltose, but smaller than those from sucrose and glucose. While, those from starch by *S. sanguinis* were pH 4.1 and 0.46 μmol min^{-1} OD^{-1}, respectively, and these values were similar to those from glucose, sucrose, and maltose.

3 Effects of α-amylase inhibitor on acid production from starch

α-amylase inhibitors, acarbose and maltotriitol were added to the reaction mixtures and the inhibitory effects were evaluated. Acarbose is an inhibitor to α-glucosidase in the intestine and often prescribed for diabetes. Maltotriitol is a sugar alcohol and its chemical structure is similar to that of starch, so that maltotriitol may compete with starch for α-amylase.

By the addition of acarbose (1 mM) and maltotriitol (60 mM), the acid production rates from starch by *S. mutans* were decreased to 39 ± 2% and 50 ± 19%, respectively, whereas those by *S. sanguinis* were also decreased to 31 ± 3% and 72 ± 8%, respectively.

4 Conclusion

The present study indicates that starch is acidogenic similarly to maltose in the presence of α-amylase like in the oral cavity, and that the acidogenic potential is enough to demineralize tooth surface. Furthermore, it was suggested that the acidogenic potential of starch by *S. sanguinis* is comparable to that by *S. mutans*. The acidogenic potential of starch-containing snacks might be moderated by α-amylase inhibitors, acarbose and maltotriitol.

Acknowledgments. This study was supported in part by Grants-in-Aid for Scientific Research B (no. 16390601 to NT) and JSPS fellows (no. 16 · 3025 to HH) from the Japan Society for the Promotion of Science. Maltotriitol was kindly provided by Towa Chemical Industry Co. Ltd., Tokyo, Japan.

Interface Oral Health Science 2007: 263–264

Fluoride ion released from glass-ionomer cement is responsible to inhibit the acid production of caries-related oral streptococci

Kazuko Nakajo[1]*, Yusuke Takahashi[2], Wakako Kiba[2], Satoshi Imazato[2], and Nobuhiro Takahashi[1]
[1]*Division of Oral Ecology and Biochemistry, Department of Oral Biology, Tohoku University Graduate School of Dentistry, Sendai 980-8575;* [2]*Department of Restorative Dentistry and Endodontology, Osaka University Graduate School of Dentistry, Osaka 565-0871; Japan*
*nakajo@mail.tains.tohoku.ac.jp

Abstract. This study aimed to evaluate inhibitory effects of glass-ionomer cement (GIC) on the acid production of caries-related oral streptococci, and to identify the components responsible for the inhibition. GIC eluate contained silicon, fluoride, and aluminum, and inhibited the pH fall and the rate of acid production by oral streptococci at acidic pH, with a concomitant decrease in lactic acid production. These inhibitions were comparable to those of a potassium fluoride solution containing the same concentration of fluoride as the eluate. This study indicates that the GIC eluate inhibits the acid production of the caries-related oral streptococci, and suggests that the effect is attributed to the fluoride derived from GIC. Thus, adjacent to GIC fillings, bacterial acid production and the subsequent bacterial growth may decrease, establishing a cariostatic environment.

Key words. acid production, eluate, fluoride, glass-ionomer cement (GIC), *Streptococcus*

1 Introduction

Conventional glass-ionomer cements (GICs) have been known to have cariostatic properties, not only a promotion of tooth remineralization but also an antibacterial property. Previous studies showed that the population of *Streptococcus mutans* on GIC fillings was lower than on composite fillings in vivo, and that pH fall by *S. mutans* cells layered on GIC discs was smaller than that on composite discs in vitro. However, the identity and mechanism of the components of GICs is unclear. Therefore, we evaluated the inhibitory effects of the GIC eluate on the pH fall and the rate of acid production of representative caries-related oral streptococci, *S. mutans* and *Streptococcus sanguinis* under anaerobic conditions like in a deep layer of dental plaque, where tooth demineralization occurs. We also attempted to identify the components of the GIC eluate that were responsible for these effects.

2 Fluoride in GIC eluate inhibits bacterial acid production

The GIC eluate, prepared by immersing set GIC in phosphate-buffered saline, stopped the pH fall by *S. mutans* and *S. sanguinis* completely around pH 4.8–5.0, and markedly decreased the rate of acid production by these bacteria at pH 5.5 with a concomitant decrease in lactic acid production. Although the eluate was found to contain silicon (0.82 ± 0.26 mM), fluoride (0.49 ± 0.02 mM), and aluminum (0.01 ± 0.02 mM), the observation that the GIC eluate inhibited acid production to a degree similar to that of the KF solution with the same fluoride concentration indicates that this effect of the eluate was attributed to the fluoride derived from the GIC. The efficient inhibition of fluoride at acidic pH is due to the fact that hydrogen fluoride (HF) has a relatively high p*K*a value, 3.15. Thus, fluoride ion (F^-) turns to HF easily at acidic pH, and the HF can penetrate into the bacterial cell membrane [1]. The HF can dissociate to F^- and H^+, and F^- inhibits enolase, a glycolytic enzyme, while H^+ promotes intracellular acidification. These phenomena result in the slowdown of glycolytic acid production [2]. The decrease in lactic acid production concomitantly with the inhibition is probably due to the slowdown of entire glycolysis. The slowdown of glycolysis results in the decrease in fructose 1,6-bisphosphate (a glycolytic intermediate which activates lactate dehydrogenase), leading to inactivation of lactate dehydrogenase activity and subsequently the decrease in lactate production [3].

The inhibition of bacterial acid production does not only directly protect against dental caries, but also suppresses the growth of oral streptococci, which obtain most of its energy for growth from sugar fermentation. Thus, adjacent to GIC fillings, the acid production and the subsequent growth are expected to decrease. It appears that the inhibition of bacterial acid production and growth, together with the inhibition of demineralization and the promotion of remineralization at the tooth surface, would establish a cariostatic environment around GIC fillings in vivo.

Acknowledgments. This study was supported in part by Grants-in-Aid for Scientific Research (Nos. 16390601, 17659659, 17791350, 18659562, 19209060) from the Ministry of Education, Culture, Sports, Science and Technology (MEXT), Japan.

References

1. Gutknecht J, Walter A (1981) Hydrofluoric and nitric acid transport through lipid bilayer membranes. Biochim Biophys Acta 644:153–156
2. Hüther FJ, Psarros N, Duschner H (1990) Isolation, characterization, and inhibition kinetics of enolase from *Streptococcus rattus* FA-1. Infect Immun 58:1043–1047
3. Maehara H, Iwami Y, Mayanagi H, et al (2005) Synergistic inhibition by combination of fluoride and xylitol on glycolysis by mutans streptococci and its biochemical mechanism. Caries Res 39:521–528

Interface Oral Health Science 2007: 265–266

Microflora profiling of root canal utilizing real-time PCR and cloning-sequence analyses based on 16S rRNA genes—differences between before and after root canal treatments

Yasuhiro Ito[1,2], Takuichi Sato[2]*, Gen Mayanagi[1], Keiko Yamaki[1], Hidetoshi Shimauchi[1], and Nobuhiro Takahashi[2]
[1]*Division of Periodontology and Endodontology;* [2]*Division of Oral Ecology and Biochemistry, Department of Oral Biology, Tohoku University Graduate School of Dentistry, Sendai, Japan*
*tak@mail.tains.tohoku.ac.jp

Abstract. This study aimed to profile microflora of root canals before and after root canal treatments, using real-time PCR and cloning-sequence analyses based on 16S rRNA genes. Six infected root canals of single-rooted teeth with periapical lesions were included. The quantification of total bacteria was performed by real-time PCR using universal primers based on 16S rRNA genes. PCR products were cloned and partially sequenced, and bacterial identification to the species level was performed by comparative analysis with the GenBank database. The concentrations of bacterial DNA after root canal treatments were less than those before root canal treatments. The cloning-sequence analysis suggested that the root canal microflora in cases of after root canal treatments was clearly distinct from that of before root canal treatments.

Key words. 16S ribosomal RNA, microflora, phylogenetic trees, polymerase chain reaction, root canal

1 Introduction

The periapical periodontitis is an infectious and inflammatory disease surrounding periapical tissue caused by oral bacteria invading the root canal. It is well known that obligate anaerobes, e.g., *Porphyromonas endodontalis*, *Porphyromonas gingivalis*, *Prevotella intermedia*, *Prevotella nigrescens*, *Tannerella forsythia*, and *Treponema denticola*, are the majority among the various invading bacteria, because the root canal is in the anaerobic environment. However, few have been investigated to profile the microflora of the root canal both qualitatively and quantitatively before and after root canal treatments. In the present study, quantification of total bacteria of root canals before and after root canal treatments was performed by real-time PCR, and the phylogenetic analysis of the root canal microflora was performed by the cloning-sequence methods with the GenBank database.

2 Real-time PCR and cloning-sequence analyses

Informed consent was obtained from five patients, and six single-rooted teeth with periapical lesions were investigated in the present study. Upon access opening, dentin sample was collected from the root canal. When the periapical lesion healed clinically through chemo-mechanical cleaning and intracanal medication, the canal dentin was sampled again. The quantification of total bacteria was performed by real-time PCR utilizing universal primers based on 16S rRNA genes [1, 2]. PCR products were cloned with the QIAGEN PCR Cloning plus Kit (QIAGEN GmbH, Hilden, Germany) and partially sequenced, and bacterial identification was performed by comparative analysis with the GenBank database (National Center for Biotechnology Information, Bethesda, MD, USA).

The average of concentrations of bacterial DNA was 52.7 ng/ml upon access opening, and lowered to 10.9 ng/ml after root canal therapy. The cloning-sequence analysis revealed that *Fusobacterium nucleatum* (15%) was initially predominant, whereas *Pseudomonas* (25%), *Bradyrhizobium* (24%), and *Methylobacterium* (18%) prevailed after root canal therapy. *F. nucleatum*, frequently detected only in cases of before root canal treatments in the present study, has been reported to be associated with the formation of periapical periodontal lesions [3–5]. *Pseudomonas*, *Bradyrhizobium*, and *Methylobacterium*, which accounted for the majority of cases of after root canal treatments in the present study, have not been reported to relate to oral diseases including periapical periodontitis. Since the concentrations of bacterial DNA lowered after root canal treatments in the present study, it seems that there is little possibility of the recurrence of periapical periodontitis due to these three bacteria.

The present study suggested that shift in bacterial flora brought by root canal therapy is not only in its quantity but also in its quality, and that the drastic shift in its component might also contribute to the healing process.

Acknowledgments. This study was supported in part by KAKENHI (17591985 to T.S.; 16591905 & 19592193 to K.Y.) from the Japan Society for the Promotion of Science, Tokyo, Japan.

References

1. Lane DJ (1991) 16S/23S rRNA sequencing. In: Stackebrandt E, Goodfellow M (eds) Wiley, Chichester, pp 115–175
2. Yamaura M, Sato T, Echigo S, et al (2005) Oral Microbiol Immunol 20:333–338
3. Lana MA, Ribeiro-Sobrinho AP, Stehling R, et al (2001) Oral Microbiol Immunol 16:100–105
4. Le Goff A, Bunetel L, Mouton C, et al (1997) Oral Microbiol Immunol 12:318–322
5. Siqueira JF, Rocas IN, Moraes SR, et al (2002) Int Endod J 35:345–351

Interface Oral Health Science 2007: 267–268

Detection of periodontopathic bacteria in periodontal pockets by nested polymerase chain reaction

Takuichi Sato[1]*, Yuki Abiko[1,2], Gen Mayanagi[3], Junko Matsuyama[4], and Nobuhiro Takahashi[1]

[1]*Division of Oral Ecology and Biochemistry;* [3]*Division of Periodontology and Endodontology, Department of Oral Biology, Tohoku University Graduate School of Dentistry, Sendai 980-8575;* [2]*Department of Oral Disease Research, National Center for Geriatrics and Gerontology, Morioka, Ohbu, Aichi 474-8522;* [4]*Division of Pediatric Dentistry, Department of Oral Health Science, Niigata University Graduate School of Medical and Dental Sciences, Niigata 951-8514; Japan*

*tak@mail.tains.tohoku.ac.jp

Abstract. A nested polymerase chain reaction (PCR) method was developed for rapid and sensitive detection of periodontopathic bacteria in subgingival plaque samples. Species-specific nested PCR amplification of 16S ribosomal RNA genes was used to detect *Aggregatibacter actinomycetemcomitans, Eubacterium saphenum, Mogibacterium timidum, Porphyromonas gingivalis, Prevotella intermedia, Prevotella nigrescens, Tannerella forsythia, Treponema denticola, Treponema medium, Treponema socranskii, Treponema vincentii*, and *Slackia exigua*. A universal set of PCR primers for bacterial 16S rRNA gene was introduced for the first PCR, and then species-specific primers for the 16S rRNA gene sequences of the respective species were used for the second PCR.

Key words. 16S rRNA gene, anaerobic, detection, polymerase chain reaction, periodontitis

1 Introduction

Oral anaerobic bacteria including Gram negative and positive rods are frequently found in human periodontal pockets and have been reported to be associated with periodontal disease. The detection and identification of periodontopathic bacteria in subgingival plaque samples by conventional culture methods is labor-intensive and time-consuming and sometimes difficult, especially to the species level. Therefore, molecular approaches including polymerase chain reaction (PCR) method have been developed. In particular, a direct PCR method has been reported to detect periodontopathic bacteria directly from subgingival plaque samples using specific PCR primers without the need to culture and isolate bacteria. However, specific bacterial DNA amplification may be influenced by the presence of other bacterial DNA extracted from dental plaque samples, and this may result in reduced sensitivity [1, 2].

2 Nested PCR

Nested PCR amplification has been developed to increase the sensitivity of detecting *A. actinomycetemcomitans*, *S. mutans*, *S. sobrinus*, and oral cultivable treponemes directly from dental plaque samples [3–5]. This method consists of a first-step amplification with universal primers and a second-step PCR with species-specific primers, and it allows amplification of specific DNA regions of the target bacteria with high sensitivity. In the present study, a nested PCR method was used to analyze 14 subgingival plaque samples for the presence of periodontal bacterial species, i.e., *Porphyromonas gingivalis*, *Aggregatibacter actinomycetemcomitans*, *Prevotella intermedia*, *Prevotella nigrescens*, *Tannerella forsythensis*, *Treponema denticola*, *Treponema socranskii*, *Treponema vincentii*, *Treponema medium*, *Eubacterium saphenum*, *Mogibacterium timidum*, and *Slackia exigua*. Fourteen subgingival plaque samples were analyzed, and nested PCR was shown to be more sensitive for detecting the periodontopathic bacteria than direct PCR using only specific primer sets. The 16S rRNA gene-based nested PCR method is a rapid and reliable method for the detection of periodontopathic bacteria in subgingival plaque samples.

Acknowledgments. This study was supported in part by Grants-in-Aid for Scientific Research (17591985 to T.S.; 17592133 to J.M.; 18926014 & 19890031 to G.M.; 18390605 to Reiko Sakashita, T.S. et al) from the Japan Society for the Promotion of Science, Tokyo, Japan.

References

1. Bamford KB, Lutton DA, O'Loughlin B, et al (1998) Nested primers improve sensitivity in the detection of *Helicobacter pylori* by the polymerase chain reaction. J Infect 36:105–110
2. Sugita T, Nakajima M, Ikeda R, et al (2001) A nested PCR assay to detect DNA in sera for the diagnosis of deep-seated trichosporonosis. Microbiol Immunol 45:143–148
3. Leys EJ, Griffen AL, Strong SJ, et al (1994) Detection and strain identification of *Actinobacillus actinomycetemcomitans* by nested PCR. J Clin Microbiol 32:1288–1294
4. Sato T, Matsuyama J, Kumagai T et al (2003) Nested PCR for detection of mutans streptococci in dental plaque. Lett Appl Microbiol 37 :66–69
5. Willis SG, Smith KS, Dunn VL, et al (1999) Identification of seven *Treponema* species in bealth- and disease-associated dental plague by nested PCR. J Clin Microbiol 37:867–869

Interface Oral Health Science 2007: 269–270

Effects of orally administered *Lactobacillus salivarius* WB21 supplement on periodontal clinical parameters and microflora

Gen Mayanagi[1]*, Seigo Nakaya[2], Keiko Yamaki[1], Yasuhiro Ito[1], Maiko Minamibuchi[1], Moto Kimura[2], Haruhisa Hirata[2], and Hidetoshi Shimauchi[1]

[1]*Division of Periodontology and Endodontology, Tohoku University Graduate School of Dentistry, Sendai 980-8575;* [2]*Sagami Research Laboratories, Wakamoto Pharmaceutical Co., Ltd., Kanagawa 258-0018; Japan*

*genm@mail.tains.tohoku.ac.jp

Abstract. The aim of this study was to evaluate whether the oral administration of lactobacilli could change the periodontal condition and microflora compared with placebo. Sixty-six healthy volunteers were randomized into two groups to receive lactobacilli or placebo for 8 weeks. The mean Plaque Index, Gingival Index, and bleeding on probing were significantly improved at 4 and 8 weeks in both groups, but showed the biggest change on smokers in the test group. Occurrence of *Porphyromonas gingivalis* and *Treponema denticola* in the subgingival plaque sample significantly decreased in the test group at 8 weeks. Oral administration of probiotic lactobacilli successfully reduced the prevalence of periodontopathic bacteria from the subgingival plaque, and possibly contributed to the beneficial effects on periodontal conditions.

Key words. *Lactobacillus salivarius*, probiotics, periodontopathic bacteria, plaque, saliva

1 Introduction

Lactobacilli are frequently used as probiotics to induce a beneficial effect on human health. Only a few studies are available regarding the effects of probiotics on oral diseases and bacteria. Several researchers reported that probiotics could successfully reduce dental caries [1, 2] and cariogenic *Streptococcus mutans* colonization [3]; however, little is known about the effects of probiotics on periodontal health and periodontopathic bacteria in the oral cavity.

This study is designed to evaluate probiotic effects of *Lactobacillus salivarius* WB21 (Ls WB21) clinically on periodontal conditions of healthy volunteers and on supragingival and subgingival microflora by detecting selective periodontal pathogens.

2 Materials and methods

Sixty-six healthy volunteers without severe periodontitis were randomized into two groups to receive lactobacilli or placebo for 8 weeks: test group (n = 34) received 2.01×10^9/day of Ls WB21 and xylitol in tablets; control group (n = 32) received placebo with xylitol. Both groups included smokers and non-smokers. Clinical parameters, including probing pocket depth (PPD), Plaque Index (Pl.I), Gingival Index (GI) and bleeding on probing (BOP), were obtained at baseline (BL), 4 weeks (4 W) and 8 weeks (8 W). Plaque and saliva samples were analyzed by a quantitative PCR using 16S rRNA primers specific for *Aggregatibacter actinomycetemcomitans*, *Porphyromonas gingivalis*, *Prevotella intermedia*, *Tannerella forsythia*, *Treponema denticola*, and *Lactobacillus salivarius*.

3 Results and discussion

Oral administration of probiotic Ls WB21 significantly increased the prevalence of lactobacilli in saliva and supragingival plaque in the test group at 8 W, suggesting the colonization of administered *L. salivarius* in the oral cavity.

The mean Pl.I, GI, and BOP values were significantly improved at 4 W and 8 W in test and control groups, but showed the biggest change on smokers in the test group. Furthermore, the differences of Pl.I and PPD from BL at 4 W and 8 W showed the significant change between the smokers in test and control group. After an 8-week intervention, supragingival *A. actinomycetemcomitans* markedly decreased, and the prevalence of *P. gingivalis* and *T. denticola* in the subgingival plaque in the test group also significantly suppressed. Thus, orally administered probiotic lactobacilli successfully reduced the prevalence of periodontopathic bacteria and possibly contributed to the beneficial effects on periodontal conditions.

These data strongly suggest that probiotics targeting microbial etiology in periodontal pocket are useful tools for the improvement of periodontal health.

References

1. Näse L, Hatakka K, Savilahti E, et al (2001) Effect of long-term consumption of a probiotic bacterium, Lactobacillus rhamnosus GG, in milk on dental caries and caries risk in children. Caries Res 35:412–420
2. Nikawa H, Makihira S, Fukushima H, et al (2004) Lactobacillus reuteri in bovine milk fermented decreases the oral carriage of mutans streptococci. Int J Food Microbiol 95:219–223
3. Ahola AJ, Yli-Knuuttila H, Suomalainen T, et al (2002) Short-term consumption of probiotic-containing cheese and its effect on dental caries risk factors. Arch Oral Biol 47:799–804

Interface Oral Health Science (2007) 271–272

Involvement of a tetratricopeptide repeat-containing protein in the virulence of *Porphyromonas gingivalis*

Yoshio Kondo[1,2], Mamiko Yoshimura[1,3], Naoya Ohara[1,4], Mikio Shoji[1], Hideharu Yukitake[1], Mariko Naito[1], Taku Fujiwara[2], and Koji Nakayama[1]*
[1]*Divisions of Microbiology and Oral Infection;* [2]*Pediatric Dentistry, Nagasaki University Graduate School of Biomedical Sciences, Nagasaki 852-8588;* [3]*Department of Host Defense, Osaka City University Graduate School of Medicine, Osaka 545-8585;* [4]*Department of Immunology, National Institute of Infectious Diseases, Tokyo 162-8640; Japan*
*knak@nagasaki-u.ac.jp

Abstract. *Porphyromonas gingivalis* is the most common organism linked to adult forms of periodontal disease. The proteome analysis of *P. gingivalis* cells that were placed in a mouse subcutaneous chamber revealed that ten proteins were upregulated in host tissues, whereas four proteins were downregulated. Among them, three upregulated proteins, PG1089 (DNA-binding response regulator RprY), PG1385 (TPR domain protein), and PG2102 (immunoreactive 61 kDa antigen) were chosen for further analysis. Mouse abscess model experiments revealed that the mutant strains defective in PG1089 and PG1385 were clearly less virulent, whereas the mutant defective in PG2102 was as virulent as the wild type parent strain. These results indicate that PG1089 and PG1385 proteins are involved in virulence of *P. gingivalis*. Using the yeast two hybrid system, PG1385 was associated with 40 proteins including four periplasmic proteins; lack of PG1385 may result in loss of physiological function of these proteins.

Key words. TPR-containing protein, *Porphyromonas gingivalis*, periodontitis

Periodontal diseases are the most common infectious disorders in the oral cavity. Despite identification of more than 500 different bacterial species in the oral cavity, the obligately anaerobic Gram-negative bacterium *Porphyromonas gingivalis* is the most common organism linked to adult forms of periodontal disease. *P. gingivalis* is found to express variety of virulence factors such as fimbriae, lipopolysaccharides (LPS), and various proteases which contribute to the pathogenesis of periodontitis. Expression of these virulence factors is thought to be tightly regulated in response to environmental cues.

Using a mouse subcutaneous chamber model [1], we attempted to identify the molecules which are upregulated or downregulated in the host. *P. gingivalis* cells were grown anaerobically in enriched brain heart infusion medium and concentrated by centrifugation. Then this concentrate was inoculated into the coil-shaped subcutaneous chamber in the mouse. The protein expression in bacteria cells placed in the chamber was analyzed by two-dimensional electrophoresis gels and compared with those of the control-cultured sample. The comparison revealed that protein expression of *P. gingivalis* was significantly altered by infection. The peptide mass analysis revealed that ten *P. gingivalis* proteins were upregulated in mice. These proteins were related to protein synthesis, detoxification, energy metabolism, and degradation of proteins. Four proteins including HSP60 homologue (PG0520) and DnaK (PG1208) were downregulated (published elsewhere).

In the upregulated proteins, three proteins, immunoreactive 61 kDa antigen (PG2102), DNA-binding response regulator RprY (PG1089), and TPR domain protein (PG1385), were selected to construct mutant strains defective in these proteins. Virulence of these mutant strains and the parent strain W83 was determined by inoculating these strains into dorsum of mice subcutaneously [2]. The PG2102 mutant strain, as well as the parent strain W83, induced spreading, ulcerative lesions on the abdomens by 48 h in all mice tested, whereas the percentage of lesion formation and symptom were significantly attenuated in mice inoculated with PG1385 and PG1089 mutants. These results revealed that RprY and TPR domain protein were involved in virulence of *P. gingivalis*.

The TPR motif is a protein–protein interaction module found in multiple copies in a number of functionally different proteins that facilitates specific interactions with a partner protein(s) [reviewed in 3]. *P. gingivalis* strain W83 has at least ten TPR-containing proteins. Recently, it was shown that PG1385 protein was upregulated upon oxidative stress [4]. Using the yeast two hybrid system, we identified 40 proteins including four periplasmic proteins as the PG1385-associated protein. Lack of PG1385 may result in loss of physiological function of these proteins.

Acknowledgement. The initial parts of this study, Dr. Yoshio Nakano in Kyushu University School of Dentistry and Dr. Yoshimitsu Abiko in Nihon University Matsudo Dental School were collaborated.

References

1. Genco CA, Cutler CW, Kapczynski D, et al (1991) Infect Immun 59:1255–1263
2. Neiders ME, Chen PB, Suido H, et al (1989) J Periodontal Res 24:192–198
3. D'Andrea LD, Regan L (2003) Trends Biochem Sci 28:655–662
4. Okano S, Shibata Y, Shiroza T, Abiko Y (2005) Proteomics 6:251–258

Interface Oral Health Science 2007: 273–274

Candida species as members of oral microflora in oral lichen planus

Mika Masaki[1,2], Takuichi Sato[2]*, Yumiko Sugawara[1], Takashi Sasano[1], and Nobuhiro Takahashi[2]
[1]*Division of Oral Diagnosis, Department of Oral Medicine and Surgery;* [2]*Division of Oral Ecology and Biochemistry, Department of Oral Biology, Tohoku University Graduate School of Dentistry, Sendai 980–8575, Japan*
*tak@mail.tains.tohoku.ac.jp

Abstract. Oral lichen planus is a refractory and chronic inflammatory disease with bilateral white reticular lesions on oral mucosa. Identification of *Candida* species including non-*Candida albicans* from buccal mucosa, tongue surfaces, and tooth surfaces of oral lichen planus patients by molecular biological method was performed in the present study. The *Candida* isolates were identified at species level by polymerase chain reaction (PCR) for genes of 18S, 5.8S, and 25/28S ribosomal RNAs and the sequence analyses of these PCR products. The isolation frequency of *Candida* species was higher from oral lichen planus than from healthy subjects. Non-*C. albicans* species (*Candida glabrata*, *Candida fukuyamaensis*, and *Candida parapsilosis*) were isolated only from oral lichen planus patients. In addition, *Candida fukuyamaensis* and *Candida parapsilosis* were genetically close to *C. albicans*, while *C. glabrata* was genetically distinguished from these species. These results support that *Candida* species are related with oral lichen planus, and suggest that *C. glabrata* may have a unique pathogenicity to oral lichen planus.

Key words. *Candida*, detection, oral lichen planus, polymerase chain reaction

1 Introduction

Oral lichen planus is a refractory and chronic inflammatory disease with bilateral white reticular lesions on oral mucosa and often occurs in females more than 40 years. The infection of oral microorganisms, especially *Candida* species has been reported to relate with oral lichen planus [1]. Among *Candida* species, *Candida albicans* was most frequently identified by culture methods. However, recent molecular biological techniques have enabled us to identify *Candida* species other than *C. albicans* (non-*C. albicans*), and subsequently suggested relationship between non-*C. albicans* and various infectious diseases [2–6]. Thus, this study aimed to detect *Candida* species including non-*C. albicans* from oral lichen planus patients by a molecular biological method and reveal their relationship with oral lichen planus.

2 Isolation and identification of *Candida* species from oral lichen planus

Informed consent was obtained from 15 oral lichen planus and 7 healthy subjects. Samples were obtained from buccal mucosa, tongue surfaces, and tooth surfaces of the subjects, and cultured on *Candida*-selective CHROM agar plates (BD, Franklin Lakes, NJ, USA) to determine total colony-forming units (CFUs) of *Candida* species. Statistical comparisons of the CFUs were analyzed by the Fisher's exact test. The *Candida* isolates were identified at species level by polymerase chain reaction (PCR) for genes of 18S, 5.8S, and 25/28S ribosomal RNAs and the sequence analyses of these PCR products, using the GenBank database (National Center for Biotechnology Information, Bethesda, MD, USA). Phylogenetic analysis of 5.8S rRNA genes of *Candida* isolates was performed utilizing the DNASIS Pro V2.6 (Hitachi Software Engineering Co., Ltd., Yokohama, Japan).

The isolation frequency of *Candida* species was higher from oral lichen planus (n = 12; 80%) than from healthy subjects (n = 2; 29%) ($P < 0.05$). *C. albicans* was isolated from both oral lichen planus (n = 10; 67%) and healthy subjects (n = 2; 29%), whereas non-*C. albicans* species (*Candida glabrata*, *Candida fukuyamaensis*, and *Candida parapsilosis*) were isolated only from oral lichen planus patients (n = 4; 33%), and among them, *C. glabrata* was isolated from all the four patients.

In addition, according to the phylogenetic analysis, *C. fukuyamaensis* and *C. parapsilosis* were genetically close to *C. albicans*, while *C. glabrata* was genetically distinguished from these species. Biological and genetic characteristics of *C. glabrata*, e.g., antifungal resistance and phospholipase activity, have been recently reported [4–6], but further investigation is required to elucidate the potential roles in the pathogenesis of *C. glabrata.* The results of the present study support that *Candida* species are related with oral lichen planus, and suggest that *C. glabrata* may have a unique pathogenicity to oral lichen planus.

Acknowledgments. This study was supported in part by Grants-in-Aid for Scientific Research (17591959, 17591985) from the Japan Society for the Promotion of Science, Japan.

References

1. Lundstrom IM, Anneroth GB, Holmberg K (1984) Int J Oral Surg 13:226–238
2. Nguyen MH, Peacock JE Jr, Morris AJ, et al (1996) Am J Med 100:617–623
3. Davies AN, Brailsford SR, Beighton D (2006) Oral Oncol 42:698–702
4. Fidel PL, Jr, Vazquez JA, Sobel JD (1999) Clin Microbiol Rev 12:80–90
5. Kaur R, Domergue R, Zupancic ML, et al (2005) Curr Opin Microbiol 8:378–384
6. Li L, Redding S, Dongari-Bagtzoglou A (2007) J Dent Res 86:204–215

Interface Oral Health Science 2007: 275–276

Meso-diaminopimelic acid and *meso*-lanthionine, amino acids peculiar to bacterial cell-wall peptidoglycans, activate human epithelial cells in culture via NOD1

A. Uehara[1], Y. Fujimoto[2], A. Kawasaki[2], K. Fukase[2], and H. Takada[1]*
[1]*Department of Microbiology and Immunology, Tohoku University Graduate School of Dentistry, Sendai 980–8575;* [2]*Department of Chemistry, Graduate School of Science, Osaka University, Osaka 560–0043; Japan*
*dent-ht@mail.tains.tohoku.ac.jp

Abstract. NOD1 recognizes the diaminopimelic acid (DAP)-containing peptide moiety of bacterial peptidoglycans (PGNs) intracellularly, and a minimum NOD1 ligand has been reported to be γ-D-glutamyl-*meso*-DAP (iE-DAP). In this study, we demonstrated that chemically synthesized *meso*-DAP and *meso*-lanthionine by themselves activated various human epithelial cells through NOD1 to generate anti-bacterial factors and cytokines in specified cases. In human monocytic cells, in the presence of Lipofectamine or cytochalasin D, *meso*-DAP induced production of cytokines. Our findings suggest that *meso*-DAP is a sufficient structure to activate NOD1 when incorporated intracellularly.

Key words. NOD1, peptidoglycans, *meso*-diaminopimelic acid, *meso*-lanthionine, epithelial cells

Peptidoglycans (PGNs) are ubiquitous constituents of the cell walls of Gram-positive and Gram-negative bacteria. The PGNs exhibit various immunobioactivities, most of which have been reproduced by a chemically synthesized muramyldipeptide (MDP; MurNAc-L-Ala-D-isoGln). Another type of PGN fragment, *meso*-DAP-containing peptide moiety also exerted similar bioactivities to MDP, and iE-DAP was reported to be the minimum structure [1]. In 2003, iE-DAP and MDP were demonstrated to be recognized by their respective intracellular proteins carrying a nucleotide-binding oligomerization domain (NOD), NOD1 and NOD2, respectively [2, 3]. It must be noted that the DAP alone was reported to be completely inactive using a commercial (Sigma) DAP, a mixture of *meso*-DAP, LL-DAP and DD-DAP [2], or *meso*-DAP purified from commercial DAP [4].

In the course of a study examining the innate immune system in oral epithelium, we unexpectedly found that a commercial DAP specimen (Sigma) activated human oral epithelial cells. Therefore, we have carried out exhaustive studies using three chemically synthesized stereoisomers of DAP; *meso*-DAP, LL-DAP and DD-DAP, in human epithelial and monocytic cells in culture [5]. We found that the *meso*-DAP itself activated human oral, pharyngeal, esophageal, colonic and cervical epithelial cells through NOD1 to generate anti-bacterial factors, PGN recognition proteins and

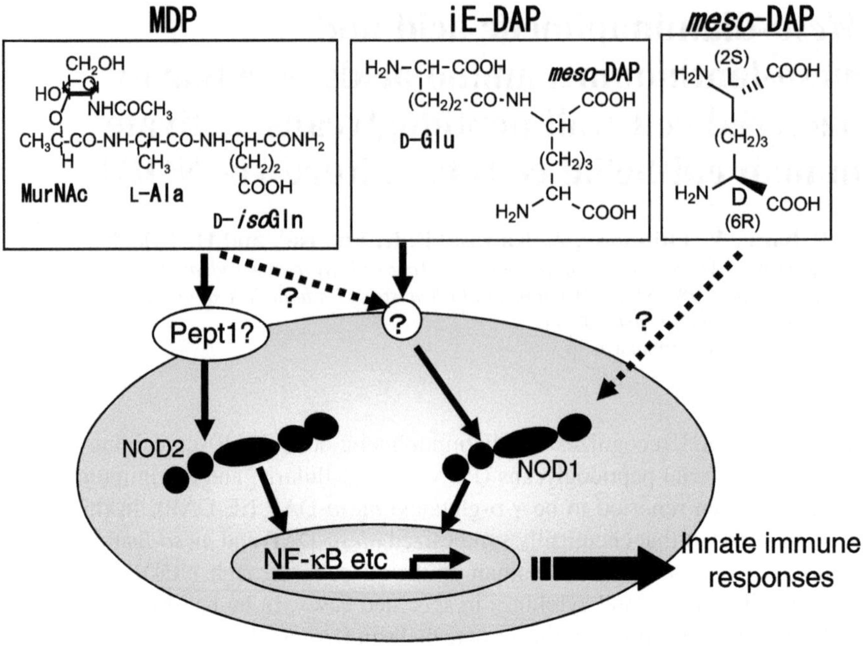

Fig. 1. Which is the minimum structure to activate NOD1? Lipofectamine and cytochalasin D increase the permeability of cells and allow *meso*-diaminopimelic acid (*DAP*) to be internalized into the cytosole. Therefore, even in monocytic cells *meso*-DAP was able to activate NOD1. A plasma membrane transporter PepT1 does not transport NOD1 ligand [6]. The unknown transporter should transport iE-DAP and other NOD1 ligands, but not *meso*-DAP. Therefore, the iE-DAP structure is required to activate NOD1 in most cells such as monocytic cells, whereas epithelial cells might allow *meso*-DAP to permeate the cells. The *meso*-DAP per se is generally capable of activating NOD1 intracellularly

β-defensin 2, and cytokines in specified cases, although the activities of *meso*-DAP were weaker than that of iE-DAP (Fig. 1). Stereoisomers of *meso*-DAP, LL-DAP and LL-DAP were only slightly activated or remained inactive. Synthetic *meso*-lanthionine, which is another PGN component in the specified bacteria such as *Fusobacterium nucleatum*, was also recognized by NOD1. In human monocytic cells, in the presence, but not in the absence, of Lipofectamine or cytochalasin D, *meso*-DAP induced slightly but significantly increased production of cytokines.

References

1. Kitaura Y, Nakaguchi O, Takeno H, et al (1982) J Med Chem 25:335–337
2. Chamaillard M, Hashimoto M, Horie Y, et al (2003) Nat Immunol 4:702–707
3. Girardin SE, Boneca IG, Carneiro LA, et al (2003) Science 300:1584–1587
4. Girardin SE, Travassos LH, Herve M, et al (2003) J Biol Chem 278:41702–41708
5. Uehara A, Fujimoto Y, Kawasaki A, et al (2006) J Immunol 177:1796–1804
6. Ismair MG, Vavricka SR, Kullak-Ublick GA, et al (2006) Can J Physiol Pharmacol 84:1313–1319

Interface Oral Health Science 2007: 277–278

Phagocytic macrophages do not contribute to the induction of serum IL-18 in mice treated with *Propionibacterium acnes* and lipopolysaccharide

T. Nishioka[1,2]*, T. Kuroishi[2], Z. Yu[1,2], Y. Sugawara[1], T. Sasano[1], Y. Endo[2], and S. Sugawara[2]
[1]*Divisions of Oral Diagnosis and* [2]*Oral Immunology, Tohoku University Graduate School of Dentistry, Sendai 980–8575, Japan*
*takashi-n@umin.ac.jp

Abstract. Interleukin (IL)-18 is one of the inflammatory cytokine which is expressed not only in activated macrophages but also in non-immune cells, such as keratinocytes and epithelial cells. It is unclear which type of cell is the major source of serum IL-18. We showed that serum levels of IL-18 were increased in mice treated with *Propionibacterium acnes* and lipopolysaccharide (LPS), whereas, administration of clodronate-liposomes (Clo-lip) to induce depletion of macrophages showed no obvious effect on IL-18 levels. IL-18 levels were marginal in the liver, lung and spleen. Treatment with *P. acnes* alone induced IL-18 in each organ, and *P. acnes* and LPS induced increase in IL-18 levels in the liver and spleen, but decrease in the intestines. The administration of Clo-lip in mice showed only a marginal effect on the IL-18 levels in the organs. These results suggest that IL-18 expressed in keratinocytes/epithelial cells contributes to serum IL-18 levels.

Key words. IL-18, keratinocytes, phagocytic macrophages, *Propionibacterium acnes*, lipopolysaccharide

1 Introduction

Interleukin (IL)-18 was known to be originally identified as an interferon-γ (IFN-γ) inducing factor from a murine liver cell cDNA library, generated from mice primed with heat-killed *Propionibacterium acnes* and subsequently challenged with LPS, and to be intracellularly produced as an inactive 24-kDa precursor form (proIL-18) and secreted as an 18-kDa mature form after cleavage by caspase-1, originally designated IL-1*β* converting enzyme. IL-18 is now recognized as a multifunctional regulator of innate and acquired immune responses through its activation of T helper cell type 1 (Th1) and Th2 responses. A macrophage 'suicide' technique, using liposomes encapsulating dichloromethylene bisphosphonate (clodronate) specifically depletes phagocytic macrophages, but not dendritic cells (DC) or neutrophils, within a day or two of i.v. injection of such liposomes into mice or rats. In addition, Kawase et al. [1] generated Keratin 5/IL-18 transgenic (K5/IL-18 Tg) mice in which mouse mature IL-18 cDNA was fused with the human K5 promoter. Human K5 promoter is active in murine K5-expressing cells in vivo, and consequently IL-18 is overexpressed in K5-expressing cells in the Tg mice. K5 is mainly expressed in stratified squamous epithelia of skin and mucosa and also expressed in a number of epithelial cells, including thymic reticulum, tracheal and glandular epithelia, whereas K5 is not expressed in other tissues such as liver, muscle, spleen and intestine. This technique and the use of the mice have allowed us to investigate whether the major source of serum IL-18 is activated macrophages or not.

2 Materials and Methods

Wild type (female C57BL/6) mice and K5/IL-18 Tg mice were used. The mice were administered with heat-killed *P. acnes* (1 mg dry weight/mouse) i.p. injection, and LPS was injected i.v. 7 days later. Levels of IL-18 in sera and internal organs were collected 2 h later and analyzed using enzyme-linked immunosorbent assay (ELISA) kit and Western Blotting assay.

3 Results and discussion

Serum levels of mature IL-18 with 18 kDa were markedly increased in mice treated with *P. acnes* and LPS, whereas administration of Clo-lip showed no obvious effect on serum IL-18 levels. IL-18 levels were marginal in the liver, lung and spleen, and more pronounced in the intestines, especially in the duodenum. Treatment with *P. acnes* alone induced IL-18 more than two fold in each organ, and *P. acnes* and LPS induced a marked increase in IL-18 levels in the liver and spleen, but decreased in the intestines. Furthermore, serum liver enzyme levels and liver injury induced by *P. acnes* and LPS were moderately reduced by Clo-lip. In untreated K5/IL-18 Tg mice, serum IL-18 levels were already extremely high. Treatment of K5/IL-18 Tg mice with *P. acnes* and LPS induced further increase in serum IL-18 levels comparable to those in WT mice. These results suggest that phagocytic macrophages do not actively contribute to the induction of serum IL-18 and liver injury in mice treated with *P. acnes* and LPS, and the major source of serum IL-18 is non-immune cells, such as keratinocytes and epithelial cells, in mice treated with *P. acnes* and LPS. Part of details was shown in our report [2].

References

1. Kawase Y, Hoshino T, Yokota K, et al (2003) J Invest Dermatol 121:502–509
2. Nishioka T, Kuroishi T, Sugawara Y, et al (2007) J Leukoc Biol 23 2007 [Epub ahead of print]

Interface Oral Health Science 2007: 279–280

Epigenetic regulation of susceptibility to anti-cancer drugs in HSC-3 cells

M. Suzuki[1], F. Shinohara[2], K. Nishimura[2], Y. Sato[2], S. Echigo[2], and H. Rikiishi[1]*

[1]*Department of Microbiology and Immunology;* [2]*Department of Oral Surgery, Tohoku University Graduate School of Dentistry, Sendai 980–8575, Japan*

*riki@mail.tains.tohoku.ac.jp

Abstract. In this study, we investigated the effects of DNA methyltransferases inhibitor zebularine (ZEB) and histone deacetylases inhibitor suberoylanilide hydroxamic acid (SAHA) on the apoptosis induced by cisplatin (CDDP) or 5-fluorouracil (5-FU) in human oral squamous cell carcinoma (HSC)-3 cells. HSC-3 cells were incubated with CDDP (5 μg/ml) or 5-FU (250 μg/ml) with or without ZEB (120 μM) and/or SAHA (1.5 μM). CDDP or 5-FU alone induced apoptosis in about 30% of cells. The combination of CDDP/SAHA or CDDP/ZEB led to a significant increase in apoptotic cells up to 80% after 48 h incubation, and the triple combination of CDDP/SAHA/ZEB showed a synergetic effect on apoptosis induction. Although the combination of 5-FU/SAHA showed a moderate increase in apoptosis after 72 h, the combination of 5-FU/ZEB inhibited apoptosis rather than that of 5-FU alone. These results indicate that epigenetic active agents (ZEB and SAHA) could sensitize HSC-3 cells to apoptosis induced by these anti-cancer drugs, which may be an important characteristic of solid cancer treatment.

Key words. epigenetics, apoptosis, zebularine, SAHA, oral squamous cell carcinoma

1 Introduction

Epigenetic alterations, including the histone acetylation and DNA methylation, play an important role in the regulation of gene expression associated with cell cycles and apoptosis that may affect the chemosensitivity of cancers. Inhibitors of DNA methyltransferases and histone deacetylases can reactivate epigenetically silenced genes for tumor suppression and thereby decreasing tumor cell growth in vitro and in vivo [1, 2]. However, the precise mechanism of their inhibitors is little known in terms of drug susceptibility and apoptosis induction of oral cancers. Furthermore, chemotherapeutic potential of cisplatin (CDDP) and 5-fluorouracil (5-FU) that are widely used for chemotherapy of oral cancers is low, and remains unsatisfactory. In this study, we investigated the effects of zebularine (ZEB) and suberoylanilide hydroxamic acid (SAHA) on the enhancement of susceptibility to anti-cancer agents in oral squamous cell carcinoma cells.

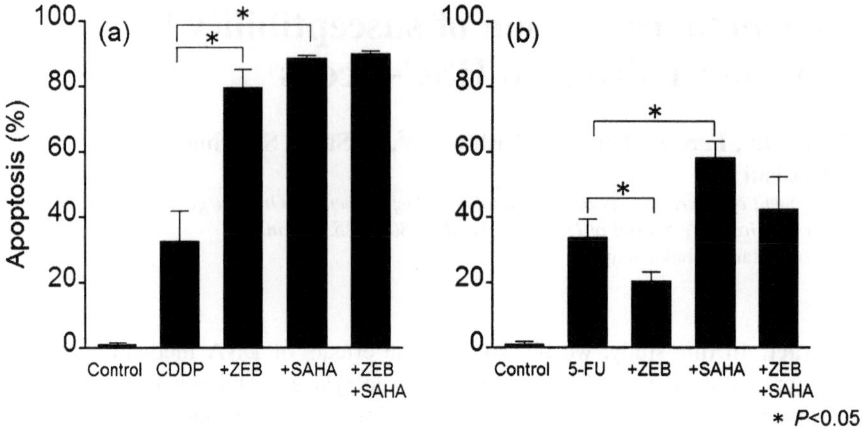

Fig. 1. Apoptosis induced by the combination of anti-cancer drugs with suberoylanilide hydroxamic acid (*SAHA*) or zebularine (*ZEB*). Cells (2×10^4 cells/well) were seeded in a 12-well flat bottomed culture plate and incubated in RPMI-1640 medium with 10% *FBS* (fetal bovine serum) for 24 h. Cells were treated with ZEB (120 μM) for 48 h, followed by treatment with cisplatin (CDDP, 5 μg/ml) for 48 h or 5-fluorouracil (*5-FU*, 250 μg/ml) for 72 h, or with SAHA (1.5 μM) concomitantly with CDDP or 5-FU. Apoptotic cells induced by CDDP (**a**) or 5-FU (**b**) in combination with SAHA or ZEB were determined by a *TUNEL* (terminal deoxynucleotidyl transferase mediated dUTP Mick End Labeling) assay. *HSC*, human oral squamous cell carcinoma

2 Results and conclusions

The combination of CDDP/SAHA or CDDP/ZEB led to a significant increase in apoptotic cells up to 80% after 48 h incubation (Fig. 1a). The combination of 5-FU/SAHA showed a moderate increase in apoptosis, whereas the combination of 5-FU/ZEB showed a decrease in apoptosis compared with the treatment of 5-FU alone after 72 h (Fig. 1b). Although the triple combination of CDDP/SAHA/ZEB showed synergetic effect on apoptosis induction, the combination of 5-FU/SAHA/ZEB decreased apoptosis rather than that of 5-FU/SAHA. These results indicate that SAHA could sensitize human oral squamous cell carcinoma (HSC)-3 cells to apoptosis induced by both anti-cancer drugs (CDDP, 5-FU). The action of ZEB in combination with 5-FU may be complex and associated with the drug metabolism cascade, which is distinct from combination with CDDP.

References

1. Rikiishi H, Shinohara F, Sato T, et al (2007) Chemosensitization of oral squamous cell carcinoma cells to cisplatin by histone deacetylase inhibitor, suberoylanilide hydroxamic acid. Int J Oncol 30:1181–1188
2. Sato T, Suzuki M, Sato Y, et al (2006) Sequence-dependent interaction between cisplatin and histone deacetylase inhibitors in human oral squamous cell carcinoma sells. Int J Oncol 28:1233–1241

Interface Oral Health Science 2007: 281–282

Histamine amplifies proinflammatory signaling cascade in human gingival fibroblasts

T. Minami[1,2*], T. Kuroishi[1], A. Ozawa[2], Y. Endo[1], H. Shimauchi[2], and S. Sugawara[1]
[1]*Division of Oral Immunology;* [2]*Division of Periodontology and Endodontology, Tohoku University Graduate School of Dentistry, Sendai 980–8575, Japan*
*takumi@mail.tains.tohoku.ac.jp

Abstract. As histamine is an important mediator in immune responses, histamine, inflammatory cytokines, and bacterial components released in inflamed periodontal tissues may be synergistically involved in inflammatory processes. The present study showed that human gingival fibroblasts (HGF) express histamine receptors (Rs) H1R and H2R, and responded to histamine to produce interleukin (IL)-8. The stimulation of HGF with tumor necrosis factor-α, IL-1α, and lipopolysaccharide markedly induced IL-8 production, and the IL-8 production was synergistically augmented in the presence of or pretreatment with histamine. The histamine response and the synergistic effect were reproduced by an H1R agonist. Selective inhibitors of mitogen-activated protein kinases (MAPKs), nuclear factor (NF)-κB, and phospholipase C (PLC) significantly inhibited the synergistic effect. These results indicate that HGF are capable of secreting IL-8 in response to histamine through H1R, and that histamine synergistically augments the inflammatory stimuli by the amplification of the MAPK and NF-κB pathway through H1R-linked PLC.

Key words. histamine, fibroblasts, inflammation, MAPK, NF-κB

1 Introduction

Periodontitis is caused by gram-negative periodontopathic bacteria, and inflamed gingival epithelial cells, as well as infiltrated lymphocytes, appear to express several inflammatory cytokines, interleukin (IL)-1, IL-6, IL-8, and tumor necrosis factor (TNF)-α. It is reported that histamine is more released in the inflamed gingiva of periodontitis patients, but it is unclear whether periodontal tissues express histamine receptors and are able to respond to histamine. Therefore, we used TNF-α, IL-1α, and lipopolysaccharide (LPS) as major inflammatory stimuli and measured the secretion of IL-8, one of the major chemokines, from human gingival fibroblasts (HGF) in response to the inflammatory stimuli with or without histamine. We also used specific signaling inhibitors to elucidate the signaling pathway.

2 Materials and methods

After incubation for 1 day, confluent HGF were washed with medium three times, and test stimulants were added for the time indicated. For the inhibition experiments, HGF were preincubated with inhibitors for 30 min–1 h at 37°C and were then stimulated with test stimulants at 37°C. After the incubation, the levels of IL-8 in the supernatants were determined with OptEIA human IL-8 enzyme-linked immunosorbent assay kit (BD Biosciences, San Jose, CA, USA).

3 Results and discussion

The present study showed that HGF express H1R and H2R and responded to histamine to produce IL-8. The IL-8 levels induced by inflammatory stimuli, TNF-α, IL-1, and LPS, with histamine were comparable with those induced by a tenfold concentration of the inflammatory stimuli alone. The result indicates that histamine augments the sensitivity tenfold to the inflammatory stimuli at the site of inflammation. H1R are linked Gαq/11 protein, and Gαq/11 stimulate NF-κB and mitogen-activated protein kinases (MAPKs), and signaling through TNFR, IL-1R, and the LPS receptor also activates both cascades. This study showed that histamine activates MAPK and NF-κB signaling cascades via H1R using specific inhibitors and histamine receptor agonists. The synergistic IL-8 production was also significantly suppressed by the inhibition of NF-κB and MAPKs. These observations indicate that the amplification of MAPKs and NF-κB are equally involved in the synergism in HGF. The principal mechanism of H1R activation is through Gαq/11, resulting in the activation of phospholipase C (PLC). The inhibition of PLC suppressed the production of IL-8 induced by histamine and TNF-α to the levels of TNF-α alone. The results indicate that histamine activates PLC through H1R, and consequently amplifies the MAPK and NF-κB pathway induced by the inflammatory stimuli. In conclusion, the present study showed that HGF secrete IL-8 in response to histamine through H1R, and that histamine synergistically augments IL-8 secretion induced by TNF-α, IL-1α, and LPS by the amplification of the MAPK and NF-κB pathway through H1R-linked PLC. It was suggested that the histamine participated in the amplification of the inflammatory reaction in periodontitis. Therefore, control of histamine receptors at inflammatory sites might be beneficial in the regulation of periodontitis. Details were shown in our report [1].

Reference

1. Minami T, Kuroishi T, Ozawa A, et al (2007) Histamine amplifies immune response of gingival fibroblasts. J Dent Res (in press)

Interface Oral Health Science 2007: 283–284

An antibacterial protein CAP18/LL-37 enhanced production of hepatocyte growth factor in human gingival fibroblast cultures

Hitomi Maeda[1,2,3]*, Akiko Uehara[1], Takashi Saito[3], Hideaki Mayanagi[2], Isao Nagaoka[4], and Haruhiko Takada[1]
[1]*Department of Microbiology and Immunology; and* [2]*Division of Pediatric Dentistry, Graduate School of Dentistry;* [3]*Dentistry for Disabled, Dental Hospital, Tohoku University, Sendai 980–8575;* [4]*Department of Biochemistry, Juntendo University School of Medicine, Tokyo 113–8421; Japan*
*mail-to-ekht@mail.tains.tohoku.ac.jp

Abstract. Human cationic antibacterial protein CAP18/LL-37 exhibits bactericidal and various immunobiological activities. The constitutive expression of CAP18/LL-37 in oral epithelial cells was demonstrated, and that CAP18/LL-37 activated human gingival fibroblasts to enhance production of hepatocyte growth factor, which has been shown to exert multiple biological activities. These findings might be related to restoration and regeneration of periodontal tissues.

Key words. CAP18/LL-37, hepatocyte growth factor, oral epithelial cells, gingival fibroblasts, innate immunity

1 Background

Human cationic antibacterial protein CAP18/LL-37 belongs to the cathelicidin family, which plays an important role in the innate host defense system; the cathelicidin possesses potent sterilizing activities against Gram-negative and Gram-positive bacteria. CAP18/LL-37 is produced by hematopoietic cells and epithelial cells, and is found in a number of tissues and body fluids such as saliva, plasma, and airway surface liquid [1]. Fibroblasts are capable of producing hepatocyte growth factor (HGF), and we have found that human gingival fibroblasts produce HGF upon stimulation with cytokines such as interleukin (IL)-1α [2]. HGF has been shown to exert multiple biological activities as a mitogen, a motogen and a morphogen for various cells [3].

2 CAP18/LL-37 enhanced production of HGF in human gingival fibroblasts

We demonstrated the constitutive expression of CAP18/LL-37 in oral epithelial cells. Therefore, we examined whether CAP18/LL-37 regulated HGF production in human gingival fibroblasts. We used IL-1α as a positive control according to our

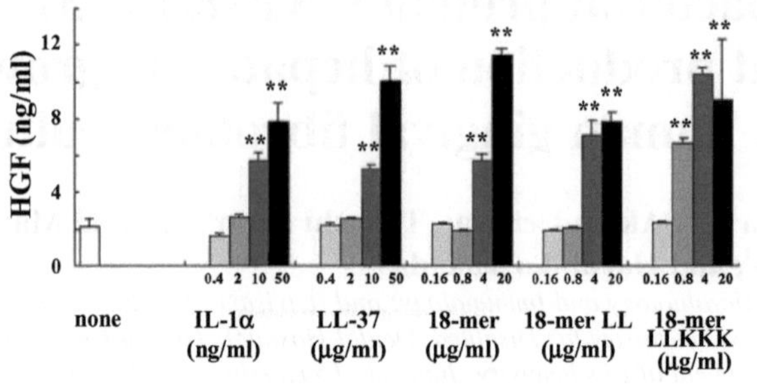

Fig. 1. Enhanced hepatocyte growth factor (*HGF*) production in human gingival fibroblasts by *LL-37* and its *18-mer* derivatives. * $P < 0.05$, ** $P < 0.01$

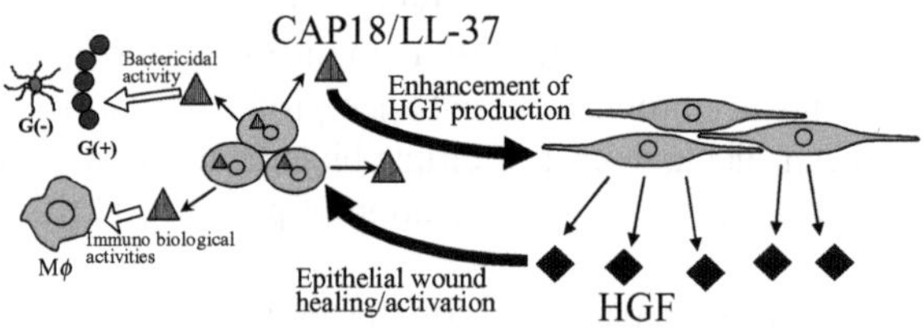

Fig. 2. Possible interaction between *oral epithelial cells* and *gingival fibroblasts* through *CAP18/LL-37* and *HGF*

previous report [2]. We found that human gingival fibroblasts exhibited enhanced HGF production upon stimulation with CAP18/LL-37 in a dose-dependent manner (Fig. 1). Furthermore, we revealed that 18-mer derivatives of CAP18/LL-37, especially 18-mer LLKKK exhibited stronger HGF production than CAP18/LL-37.

3 Conclusion

Human oral epithelial cells produced CAP18/LL-37, and CAP18/LL-37 enhanced HGF production in human gingival fibroblasts. HGF might be involved in restoration and regeneration of periodontal tissues (Fig. 2).

References

1. Bowdish DME, Davidson DJD, Hancodk REW (2006) Curr Top Microbiol Immunol 306:27–66
2. Tamura M, Arakaki N, Tsubouchi H, et al (1993) J Biol Chem 268:8140–8145
3. Ohnishi T, Daikuhara Y (2003) Arch Oral Biol 48:797–804

Interface Oral Health Science 2007: 285–286

Proinflammatory cytokine production and leukocyte adhesion molecule expression of endothelial cells in response to *Abiotrophia defectiva* infection

Shihoko Tajika[1]*, Minoru Sasaki[1], Sachimi Agato[1], Rikako Harada-Oikawa[1], Shigeyuki Hamada[2], and Shigenobu Kimura[1]
[1]*Department of Oral Microbiology, Iwate Medical University School of Dentistry, Morioka 020–8505;* [2]*Advanced Research Institute for the Sciences and Humanities, Nihon University, Tokyo 102–0073; Japan*
*stajika@iwate-med.ac.jp

Abstract. *Abiotrophia defectiva*, one of the oral streptococci, possessed a relatively higher adhesive ability to the cultured human umbilical vein endothelial cells (HUVEC) as well as fibronectin and vitronectin than other oral streptococci tested. Further, *A. defectiva* induced HUVEC to produce IL-8 and TNF-α, and subsequently to express E-selectin, ICAM-1 and VCAM-1. Thus, *A. defectiva* entering into blood streams could adhere to endothelial cells and induce proinflammatory responses through cytokine productions and leukocyte adhesion molecule expressions.

Key words. *Abiotrophia defectiva*, infective endocarditis, proinflammatory cytokine, leukocyte adhesion molecule, endothelial cells

Abiotrophia defectiva and the closely related bacteria, *Granulicatella adiacens*, inhabit human oral cavity as well as intestinal and genitourinary tracts as normal flora, and can be one of the major causes of infective endocarditis, accounting for 5 to 6% of the occurrences [1]. Since these bacteria are nutritionally variant and thus fastidious growing, the organisms have been presumed to be responsible for many incidences of 'culture negative' endocarditis [2]. Although it was reported that *A. defectiva* and *G. adiacens* were isolated with similar frequencies from the blood specimens of the patients with infective endocarditis, the frequency of oral colonization of *A. defectiva* was much lower than that of *G. adiacens* in healthy volunteers (11.8 and 87.1%, respectively) [3]. Thus the adhesive characteristics of these two species to endothelial cells could differ, and that may partly account for the virulence of these microbes in infective endocarditis. In this study, the adhesive abilities of *A. defectiva* to various extracellular matrix proteins (ECM) and HUVEC were examined in comparison to those of *G. adiacens* as well as other oral streptococci.

The binding of *A. defectiva* to HUVEC was significantly higher than that of the oral streptococci, *G. adiacens*, *G. paraadiacens*, *G. elegans*, *G. balaenopterae*, *Streptococcus mutans* and *S. sanguinis*. The bindings of *A. defectiva* to immobilized fibronectin and vitronectin, but not collagen, were also higher than those of the oral streptococci tested.

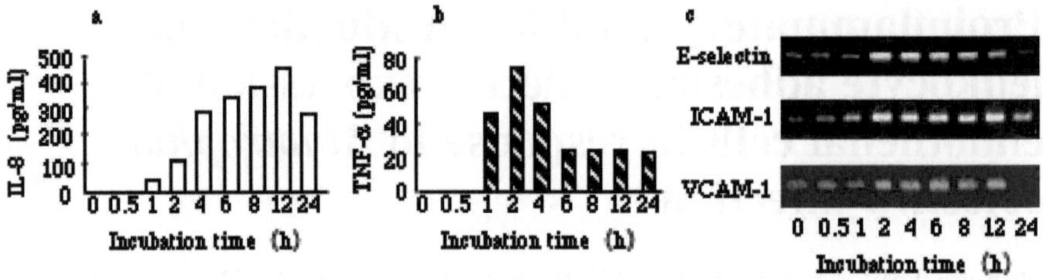

Fig. 1. *Abiotrophia defectiva*-induced cytokine production and leukocyte adhesion molecules expression of HUVEC. HUVEC (1×10^5 cells) were stimulated with *A. defectiva* ATCC 49176 (10^7 CFU) at 37°C. After incubation, IL-8 (**a**) and TNF-α (**b**) in the supernatants were analyzed by ELISA. The expression of the leukocyte adhesion molecules (*E-selectin*, *ICAM-1* and *VCAM-1*) specific mRNA in *A. defectiva*-stimulated HUVEC were detected by RT-PCR (**c**).

To elucidate further the pathogenic ability of *A. defectiva* in infective endocarditis, *A. defectiva*-induced proinflammatory cytokine productions and leukocyte adhesion molecule expressions were examined. The results indicate that *A. defectiva* induce HUVEC to produce IL-8 and TNF-α (Fig. 1a, b), and subsequently to express E-selectin, ICAM-1 and VCAM-1 (Fig. 1c). Furthermore, the stimulation with *A. defectiva* induced IκB degradation in HUVEC. Thus, NF-κB activation could be involved in *A. defectiva*-induced activation of HUVEC.

The present findings indicate that *A. defectiva* possesses a relatively higher adhesive ability to endothelial cells as well as ECM than other oral streptococci, and could induce endothelial cells to produce IL-8 and TNF-α, and subsequently to express E-selectin, ICAM-1 and VCAM-1. Thus, *A. defectiva* entering into blood streams could adhere to endothelial cells and induce proinflammatory responses through cytokine productions and leukocyte adhesion molecule expressions, which may account for the potential pathogenic traits of the organisms in infective endocarditis.

References

1. Ruoff KL (1991) Nutritionally variant streptococci. Clin Microbiol Rev 4:184–190
2. Bouvet A (1995) Human endocarditis due to nutritionally variant streptococci: *Streptococcus adjacens* and *Streptococcus defectivus*. Eur Heart J 16 (Suppl B):24–27
3. Ohara-Nemoto Y, Tajika S, Sasaki M, et al (1997) Identification of *Abiotrophia adiacens* and *Abiotrophia defectiva* by 16S rRNA gene PCR and restriction fragment length polymorphism analysis. J Clin Microbiol 35:2458–2463

Interface Oral Health Science 2007: 287–288

IL-18 expressed in salivary gland cells induces IL-6 and IL-8 in the cells in synergy with IL-17

Azusa Sakai[1,2]*, **Toshinobu Kuroishi**[1], **Yumiko Sugawara**[2], **Takashi Sasano**[2], **and Shunji Sugawara**[1]
[1]*Division of Oral Immunology;* [2]*Division of Oral Diagnosis, Tohoku University Graduate School of Dentistry, Sendai 980–8575, Japan*
*azusa-914@umin.ac.jp

Abstract. Interleukin (IL)-18, an immunoregulatory and proinflammatory cytokine, has been shown to play an important pathogenic role in inflammatory and autoimmune disorders. In the present study, immunohistochemical examination showed that the expression of IL-18 was detected in acinar and ductal epithelial cells in the salivary glands of patients with Sjögren's syndrome (SS) but not in those of healthy subjects. Human salivary gland human parotid gland cell lines (HSY) cells constitutively expressed mRNA of IL-18 and caspase-1. Receptors for IL-18 and IL-17 were expressed on the cell surface. IL-18 induced the secretion of IL-6 and IL-8 in the presence of low amount of IL-17, a T cell-derived proinflammatory cytokine. These results suggest that IL-18 expressed in salivary gland cells is associated with pathogenesis of SS in microenvironment of salivary glands in synergy with IL-17.

Key words. IL-18, IL-17, salivary gland, Sjögren's syndrome

1 Introduction

SS is a chronic autoimmune disease of the exocrine glands with infiltration of lymphocytes. It is reported that IL-18 actively contributes to the modulation of immune regulation in the exocrine glands. Recently, over expression of IL-18 in salivary glands with keratin 5 promoter in mice results in SS-like massive cell infiltration into salivary glands, and atrophy of glandular cells was observed. Furthermore, the expression of IL-17 mRNA in the salivary glands was prominent by DNA microarray analysis in the mice. Therefore, it is supposed that the infiltrating $CD4^+$ T cells into the salivary glands of SS patients secrete IL-17, and that there are some relations between IL-17 and IL-18 in pathogenesis of SS. In the present study, we examined the expression of IL-18 in salivary glands and IL-18-induced production of proinflammatory mediators, IL-6 and IL-8, in synergy with IL-17 using human salivary gland cells in culture.

2 Materials and methods

Labial salivary gland and parotid saliva were extirpated and gathered, respectively, from patients with SS and healthy volunteers as controls. All subjects were informed and consented with this study. The Ethical Review Board of Tohoku University Graduate School of Dentistry approved the experimental procedures. Expression of IL-18 was analyzed by immunohistochemistry assay. Human salivary gland cell lines (HSY) was used to analyses the expression of IL-18 by RT-PCR and Western blotting, secretion of IL-18 and other cytokine by ELISA, and manifestation of IL-18 receptor (IL-18R) and IL-17 receptor (IL-17R) by Flow cytometry.

3 Results and discussion

The expression of IL-18 was detected in acinar and ductal epithelial cells in the salivary glands of SS patients, and slightly detected in some ducts of the salivary glands in normal. No IL-18 expression was detected in infiltrating mononuclear cells in the salivary glands of SS patients, indicating that activated macrophages or antigen-presenting dendritic cells were not infiltrated in the field. These results indicated that expression of IL-18 in the salivary gland epithelial cells in acinar is correlated with SS.

HSY cells constitutively expressed IL-18 and caspase-1 mRNA. Western blotting showed that the cells constitutively expressed a precursor form of IL-18 but not an 18-kDa active form in the cells. Incubation of the proIL-18 containing cell lysate of HSY cells with caspase-1 converted to an 18-kDa mature form, indicating that the IL-18 expressed in salivary gland cell was properly processed in the presence of caspase-1. Elevation of intracellular Ca^{2+} by A23817 significantly induced IL-18 secretion in a dose-dependent manner in HSY cells.

Flow cytometric analyses showed that HSY cells expressed IL-18R and IL-17R on the cell surface. The results indicate that IL-18 secreted by salivary gland cells is able to bind own cells and to activate themselves in an autocrine manner, and that IL-18 and IL-17 may synergistically activate salivary gland cells. Stimulation of HSY cells with IL-18 alone did not induce secretion of IL-6 and IL-8. However, in the presence of IL-17 at 10 mg/ml or 1 ng/ml, IL-18 induced the secretion of IL-6 or IL-8 in a dose-dependent manner. These results indicate that IL-18 synergy with IL-17 is involved in pathology of salivary gland disorder.

4 Conclusion

It is suggested by the present study that IL-18 could play an important role in occurrence, formation, and promoting inflammation in salivary gland, cooperating with IL-17 by inducing proinflammatory cytokines and chemokines.

Interface Oral Health Science 2007: 289–290

Infiltration of immune cells in salivary gland by IL-18 overexpression in mice

K. Sato[1,2]*, T. Kuroishi[2], T. Nishioka[1,2], Y. Sugawara[1], T. Hoshino[3], T. Sasano[1], and S. Sugawara[2]
[1]*Division of Oral Diagnosis, Department of Oral Medicine and Surgery and* [2]*Division of Oral Immunology, Department of Oral Biology, Tohoku University Graduate School of Dentistry, Sendai 980–8575;* [3]*Kurume University School of Medicine, Kurume 830–0011; Japan*
*sato-kyon@umin.ac.jp

Abstract. Sjögren's syndrome (SS) is a chronic autoimmune disease of the exocrine glands with infiltration of lymphocytes and destruction of glandular cells. We previously reported that interleukin (IL)-18, an inflammatory cytokine that is involved in autoimmune diseases, is produced in salivary gland (SG) of SS patients. In this study, histological changes and lymphocytes subpopulations in SG of keratin 5 (K5)/IL-18 transgenic (Tg) mice overexpressing mature IL-18 using human K5 promoter were examined. Histological analysis revealed severe infiltration of lymphocytes and atrophy of glandular duct cells in SG. Flow cytometric analyses showed that T, B, NK cells and Macrophages are infiltrated in SG of the mice. These results indicated that overexpression of IL-18 with K5 promoter induced SS-like feature.

Key words. IL-18, Sjögren's syndrome, immune cells

1 Background

Sjögren's syndrome (SS) is a chronic autoimmune disease of the exocrine glands. Extensive lymphocytic infiltration predominantly with $CD4^+$ T cells is detected in salivary and lachrymal glands from SS patients. A predominant T helper cell type 1 (Th1) pattern of cytokines was expressed in minor salivary glands from patients with primary SS. Dryness of the mouth and eyes results from destruction of the salivary and lachrymal glands. Interleukin (IL)-18 is a multifunctional regulator of innate and acquired immune response. IL-18 is identified not only in immune cells such as activated macrophages and dendritic cells, but also in non-immune cells such as keratinocytes and epithelial cells of various organs. IL-18 is expressed in duct cells and mononuclear cells infiltrated in periductal area of SS salivary glands. Serum IL-18 levels are significantly higher in SS patients than in healthy subjects, indicating that IL-18 is critically involved in SS pathogenesis. In this study, we examined the alteration of SG in K5/IL-18 transgenic (Tg) mice.

2 Materials and methods

Mice: female K5/IL-18 Tg mice (C57BL/6 background, 2–12 month old), n = 13.

Histological analysis: formalin-fixed samples were embedded in paraffin and stained with hematoxylin and eosin.

Flow cytometry: cells were purified from SG, submandibular lymph nodes (SML), thymus and spleen. Cells were stained with antibodies for various immune cell markers, and flow cytometric analyses were performed using FACSCalibur cytometer (BD Biosciences, San Jose, CA, USA).

3 Results and discussion

Hypertrophy of SG, SML and thymus was observed in 6–8-month-old K5/IL-18 Tg mice with female predominance. Histological analysis revealed severe infiltration of lymphocytes and atrophy of glandular duct cells in SG of K5/IL-18 Tg mice but not in wild type mice. Flow cytometric analyses with 12-month-old female K5/Il-18 Tg mice showed that the percentages of $CD3^+$ T cells were 15.0% ($CD3^+CD4^+$, 5.0%; $CD3^+CD8^+$, 5.5%) and 22.2% ($CD3^+CD4^+$, 8.4%; $CD3^+CD8^+$, 12.7%) in SG and SML, respectively. On the other hand, the percentages of $CD3^+$ T cells were similar between thymus and spleen, 38.9% ($CD3^+CD4^+$, 21.7%; $CD3^+CD8^+$, 15.2%) and 29.3% ($CD3^+CD4^+$, 17.3%; $CD3^+CD8^+$, 10.0%), respectively. No $CD4^+CD8^+$ double positive T cells were detected in thymus. The percentages of $CD19^+$ B cells were higher in SG (67.6%) and SML (56.3%) than in thymus (41.3%) and spleen (46.0%). The percentages of $NK1.1^+$ cells were higher in SG (6.0%) than in other tissues (SML, 1.5%; thymus, 1.9%; spleen, 1.5%). The percentages of $F4/80^+$ cells were higher in SG (14.3%) and spleen (5.7%) than in SML (0.2%) and thymus (1.0%).

4 Conclusion

These results suggest that overexpression of IL-18 in SG showed SS-like feature with infiltrations of immune cells and alterations of glandular cells in SG in vivo. Therefore, the mice could be beneficially used for further analysis of precise mechanisms of SS pathogenesis.

Interface Oral Health Science 2007: 291–292

Gelatinase activity in human saliva and its fluctuation in the oral cavity

Yoshitada Miyoshi[1]*, Makoto Watanabe[1], and Nobuhiro Takahashi[2]
[1]Division of Aging and Geriatric Dentistry; [2]Division of Oral Ecology and Biochemistry, Tohoku University Graduate School of Dentistry, Sendai 980–8575, Japan
*miyoshi@mandible.dent.tohoku.ac.jp

Abstract. This study aimed to determine the activities of gelatinase in whole saliva before and after a meal. Paraffin-stimulated whole saliva was collected from seven healthy volunteers (male, age: 27.3 ± 1.3) at 30 min before a meal and 30 min after a meal. Gelatinase activity and collagenase activity were measured by the Gelatinase assay kit and the Collagenase assay kit, respectively. Furthermore, the saliva samples were incubated at 37°C for 2 h and the activities were measured again. All the saliva samples before a meal had gelatinase (1.07 ± 0.28 U/ml) and collagenase (0.11 ± 0.14 U/ml) activities. The gelatinase activity decreased to 10 ± 13% after a meal ($P < 0.05$). At 30 min after a meal, the activity increased again and reached 56 ± 46% of the activity before a meal. The present study confirmed that whole saliva contains gelatinase and collagenase activities. The gelatinase activity in the saliva samples, especially samples obtained after a meal, increased during a 2-h incubation. It was suggested that whole saliva has an activating system for gelatinase, which may activate a latent type of gelatinase in saliva.

Key words. gelatinase activity, whole saliva

Introduction

Human proteases contribute to the growth and the turnover of human tissues by degrading extracellular matrices, while the proteases are also involved in inflammatory diseases and tumorous diseases. Recent studies suggest that the proteases in the oral cavity are related to the etiology of dentin caries [1] and periodontitis [2]. However, their activity and fluctuation in the oral cavity have not been investigated well.

Therefore, this study aimed to investigate (1) the activities of gelatinase and collagenase in whole saliva, (2) the fluctuation of the activity of gelatinase in whole saliva before and after a meal, and (3) the activation of gelatinase activity in whole saliva.

Measurement of gelatinase and collagenase activities

After informed consent was obtained, whole saliva was collected from seven healthy volunteers (male, age: 27.3 ± 1.3) by chewing a sheet of paraffin film for 1 min at 30 min before a meal and 30 min after a meal. Furthermore, the saliva samples were incubated at 37°C for 2 h and the activities were measured again.

Gelatinase and collagenase activities were measured by the Gelatinase assay kit (LL-20002, Life Lab., Co., Japan) and the Collagenase assay kit (LL-20001, Life Lab., Co., Japan), respectively. Degraded fluorogenic gelatin or collagen produced by the enzymatic reaction was determined quantitatively by the fluorescence spectrophotometer (model 650, Hitachi Co., Ltd., Japan) at an excitation wavelength of 495 nm and an emission wavelength of 520 nm. Enzyme activity was calculated using a standard gelatinase (MMP-9, 0.5 U/ml, Life Lab., Co.) or collagenase (MMP-1, 0.5 U/ml, Life Lab., Co.) as control.

The statistical analyses were conducted with the SPSS statistical package (SPSS Version 14, SPSS Japan Inc., Japan). Mann–Whitney *U* test was used to determine the difference between activities of gelatinase and collagenase. ANOVA and Bonferroni were used to determine the difference among activities of gelatinase in whole saliva before and after a meal.

Gelatinase activity and its fluctuation

(1) All the saliva samples collected before a meal had both gelatinase (1.07 ± 0.28 U/ml) and collagenase (0.11 ± 0.14 U/ml) activities. The activity of gelatinase was significantly higher than that of collagenase ($P < 0.05$).
(2) The gelatinase activity decreased to 10 ± 13% after a meal ($P < 0.05$). At 30 min after a meal, the activity tended to return and reached 56 ± 46% of the activity before a meal.
(3) The gelatinase activities of saliva samples collected before a meal from four subjects increased to 181 ± 51% during a 2-h incubation at 37°C, while those from the other three subjects decreased to 54.2 ± 24%. On the other hand, the gelatinase activities of saliva samples collected after a meal during the incubation increased to 764 ± 1190% in all the subjects except in one subject, although the increased activities were lower than those activities before a meal.

Whole saliva was confirmed to contain gelatinase and collagenase activities, and the activity of gelatinase was higher than that of collagenase. The gelatinase activity in whole saliva was fluctuated by a meal. It decreased during a meal and tended to return after a meal. The gelatinase activity was basically increased by incubation, especially when the activity was low, like in the saliva after a meal.

It is suggested that the decrease in gelatinase activity in whole saliva is due to the depletion of gelatinase secretion from salivary glands. The gelatinase might be produced constantly and stored in salivary glands. In addition, it is possible that whole saliva has an activating system for gelatinase, which may activate a latent type of gelatinase in saliva.

References

1. Tjaderhane L, Larjava H, Sorsa T, et al (1998) J Dent Res 77:1622–1629
2. Makela M, Salo T, Uitto VJ, et al (1994) J Dent Res 73:1397–1406

Interface Oral Health Science 2007: 293–294

Genome-wide gene expression analysis of human myelomonocytic cell line THP-1 exposed to lipopolysaccharide (LPS)

Masayuki Taira[1]*, Minoru Sasaki[2], Shigenobu Kimura[2], and Yoshima Araki[1]
[1]*Departments of Dental Materials Science;* [2]*Oral Microbiology, Iwate Medical University School of Dentistry, Morioka 020–8505, Japan*
*mtaira@iwate-med.ac.jp

Abstract. DNA microarray analysis clarified that lipopolysaccharide stimulation caused THP-1 cells to exclusively up-regulate inflammation- and immunity-related genes, in which several genes in the route of the Toll-like receptor signaling pathway were up-regulated.

Key words. myelomonocytic cell line THP-1, lipopolysaccharide, DNA microarray, inflammation, Toll-like receptor

When activated with lipopolysaccharide (LPS), monocytes could produce a variety of cytokines and chemokines. The objectives of this study were (i) to clarify genes of human monocytic cell line THP-1 whose expressions were significantly increased by LPS [1], and (ii) to analyze gene expressions in apoptosis and Toll-like receptor signaling pathways.

THP-1 cells (1×10^8) in 10 ml of RPMI 1640 supplemented with 10% fetal bovine serum were cultured at 37°C in a humidified 5% CO_2 atmosphere for 1 day. The cells were stimulated with *Escherichia coli* O26 LPS (concentration = 1 μg/ml) for additional 4 h. Total RNAs were extracted with Trizol (Invitrogen, Carlsbad, CA, USA). Gene expressions were evaluated by DNA microarray [$n = 2$ for LPS (–) and LPS (+)], using 55k human genome U133 Plus 2.0 array (Affymetrix, Santa Clara, CA, USA) and GeneSpringGX (Agilent Technologies, Santa Clara, CA, USA). Gene expressions in apoptosis and Toll-like receptor signaling pathways were analyzed with pathway database [2].

1. It was confirmed that 1,632 genes were significantly up-regulated (more than 1.14-fold, $P < 0.05$) by LPS stimulation, of which 1,523 and 1,603 genes were the inflammation-related and immunity-related ones, respectively. Nine genes (TNFAIP6, CCL4, CXCL10, IL1β, THBS1, MCP-1, MAFB, IL-8, CCL3) were up-regulated more than 50-fold. Thirty-nine genes were up-regulated more than 20-fold (Table 1).
2. It was suggested in apoptosis signaling pathway analysis that by LPS stimulation, endoplasmic reticulum stress might cause apoptosis of THP-1 cells while (counter) DNA survival system simultaneously took effect.

It was also confirmed that LPS stimulation up-regulated genes on the route in the Toll-like receptor signaling pathway such as those of TRL4, MyD88, IRAK1, TAB2, IKKα, and NF-κB, thereby leading to up-regulation of many down-stream inflammatory cytokine genes such as those of TNF-α, IL-1β, IL-6 [2], and co-stimulatory molecule genes such as those of CD40 and CD80.

Table 1. Thirty-nine genes of THP-1 cells whose expression was more than 20-fold up-regulated by lipopolysaccharide stimulation

Fold change	Gene symbol	GenBank number
971	TNFAIP6	AW188198
885	CCL4: ACT2: G-26: LAG1: MIP1 B: SCYA2: SCYA4: AT744.1: MGC104418: MGC126025: MGC126026: MIP-1-beta	NM_002984
761	TNFAIP6: TSG6	NM_007115
628	CXCL10: C7: IFI10: INP10: IP-10: crg-2: mob-1: SCYB10: gIP-10	NM_001565
216	LOC285628	AL389942
136	IL1B: IL-1: IL1F2: IL1-BETA	NM_000576
130	THBS1: TSP: TSP1	NM_003246
99	MCP-1: HC11: MCAF: MCP1: SCYA2: GDCF-2: SMC-CF: HSMCR30: MGC9434: GDCF-2: HC11: CCL2	S69738
87	MAFB: KRML: MGC43127	NM_005461
80	IL1B: IL-1: IL1F2: IL1-BETA	M15330
76	IL8: K60: NAF: GCP1: IL-8: LECT: LUCT: NAP1:3-1OC: CXCL8: GCP-1: LYNAP: MDNCF: MONAP: NAP-1: SCYB8:	NM_000584
75	NUP62	AI859620
73	IL8	AF043337
53	CCL3: MIP1A: SCYA3: GOS19-1: LD78ALPHA: MIP-1-alpha	NM_002983
47	CXCL1: GRO1: GROa: MGSA: NAP-3: SCYB1: MGSA-a: MGSA alpha	NM_001511
43	THBS1	AV726673
40	CCL8	AI984980
38	SOD2	BF575213
37	ICAM1	AI608725
36	ICAM1: BB2: CD54: P3.58	NM_000201
35	CD80: LAB7: CD28LG: CD28LG1	BC042665
34	TNFAIP3	AI738896
34	GJB2: HID: KID: PPK: CX26: DFNA3: DFNB1: NSRD1	M86849
33	SLAMF7	AL121985
32	SOD2	W46388
32	MAML2	AI148006
30	SLC2A6: GLUT6: GLUT9: HSA011372	NM_017585
29	SGK: SGK1	NM_005627
29	PTX3: TSG-14: TNFAIP5	NM_002852
28	MAFB	AW135013
28	EBI3	NM_005755
27	TNFAIP3: A20: TNFAIP2: MGC104522	NM_006290
27	GBP2	BF509371
26	MIHC: AIP1: API2: CIAP2: HAIP1: HIAP1: MALT2: RNF49	U37546
25	IER3: DIF2: IEX1: PRG1: DIF-2: GLY96: IEX-1: IEX-1L	NM_003897
22	BTG2: PC3: TIS21: MGC126063: MGC126064	NM_006763
20	MAML2	AU147805 AI684439

References

1. Kyoto Encyclopedia of Genes and Genomes (2007) Toll-like receptor signaling pathway: *Homo sapiens* (human). http://www.kegg.jp/kegg/pathway/hsa/hsa04620.html
2. Sharif O, Bolshakov VN, Raines S, et al (2007) Transcriptional profiling of the LPS induced NF-κB response in macrophages. BMC Immunol 8:1 (17 pp)

Interface Oral Health Science 2007: 295–296

CD14-dependent and independent B-cell activations by stimulation with lipopolysaccharide from *Porphyromonas gingivalis*

Yu Shimoyama[1], Yuko Ohara-Nemoto[2], Arisa Yamada[3], Hirohisa Kato[3], Shihoko Tajika[1], and Shigenobu Kimura[1]*
[1]*Departments of Oral Microbiology;* [3]*Dental Pharmacology, Iwate Medical University School of Dentistry, Morioka 020–8505;* [2]*Division of Oral Molecular Biology, Nagasaki University Graduate School of Biomedical Sciences, Nagasaki 852–8588; Japan*
*kimuras@iwate-med.ac.jp

Abstract. The functional role of CD14 and Toll-like receptor (TLR) in the *Porphyromonas gingivalis* lipopolysaccharide (Pg-LPS)-induced activation of B cells was assessed, using CD14-, TLR2- and TLR4-overexpressed murine B cells (CH12.LX). After stimulation with Pg-LPS, CD14- and TLR4-transfected, but not TLR2-transfected, CH12.LX showed higher induction of nuclear factor (NF)-κB activation. Although Pg-LPS induced control CH12.LX to proliferate accompanying by up-regulations of TGF-β and IL-6 mRNA, CD14-transfected cells showed the up-regulation of TGF-β mRNA, but not IL-6 mRNA nor proliferative responses to Pg-LPS. Thus, Pg-LPS could induce B cell activation both in CD14-dependent and independent pathways: In a CD14-dependent pathway, TGF-β production could be induced through TLR4 and NF-κB activation. In contrast, Pg-LPS induced proliferation and IL-6 production in the CD14-independent pathway.

Key words. *Porphyromonas gingivalis*, lipopolysaccharide, B cell activation, CD14, TLR

Porphyromonas gingivalis, a black-pigmented gram-negative anaerobic rod, has been implicated as a major pathogen of human chronic (adult) periodontitis. Since periodontitis lesions are characterized as infiltration with increased numbers of B cells/plasma cells, the lipopolysaccharide (LPS) has been considered to stimulate immunocompetent cells, especially B cells located in the gingival tissues.

We have previously reported that membrane CD14 could be involved in some cellular activation events of B cells elicited by *Escherichia coli* LPS [1]. To assess the functional role of CD14 and TLR in the *P. gingivalis* (Pg)-LPS-induced activation of B cells, in this study, a murine B cell line, CH12.LX, was transfected with a nuclear factor (NF)-κB-dependent E-selectin promoter luciferase reporter gene, with or without an expression plasmid of mouse CD14, TLR2 and TLR4. Further, tyrosine phosphorylation in B cells in relation to the Pg-LPS-induced B cell proliferation was investigated, because a trigger signal by Pg-LPS could be transduced to initiate tyrosine phosphorylation in B cells [2].

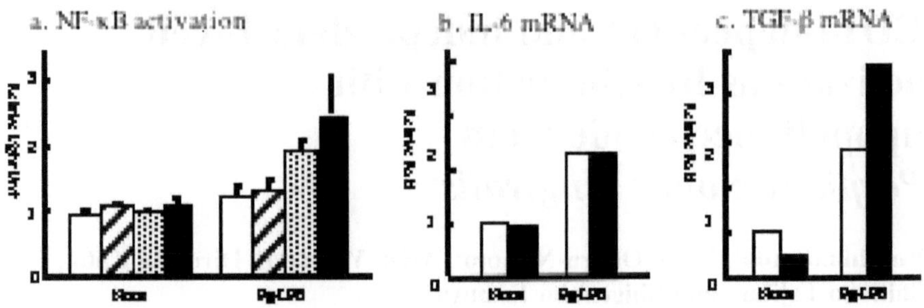

Fig. 1. *Porphyromonas gingivalis* lipopolysaccharide (*Pg-LPS*)-induced nuclear factor (*NF*)-κB activation and accumulation of IL-6 and TGF-β mRNA in CD14- (*closed bar*), TLR2- (*hatched bar*), and TLR4-transfected (*dotted bar*), and control (*open bar*) CH12.LX. NF-κB activation was detected by the luciferase activities in CH12.LX transfected with the CD14 expression plasmid and/or the TLR expression plasmid, together with pRL-TK and pGL3-mELAM-1 luciferase reporter plasmids for 48 h, either untreated or stimulated with Pg-LPS (1 μg/ml) for 6 h before harvest. Relative quantifications of IL-6 and TGF-β mRNA in CH12.LX were detected by real-time PCR

After stimulation with Pg-LPS, CD14- and TLR4-transfected, but not TLR2-transfected, CH12.LX showed higher induction of NF-κB activation, suggesting that Pg-LPS could induce B cell activation in a CD14-dependent pathway through TLR4 and NF-κB activation (Fig. 1a). Pg-LPS also induced cell proliferation of control (a parent vector-transfected) CH12.LX, accompanied by up-regulations of TGF-β and IL-6 mRNA, and the moderate induction of luciferase activity. However, the overexpression of membrane CD14 resulted in enhancement of the up-regulations of TGF–β mRNA (Fig. 1c), but not IL-6 mRNA (Fig. 1b) nor proliferative responses (data not shown) to Pg-LPS. Furthermore, the treatment of B cells with an MAP kinase inhibitor, herbimycin A, abrogated the Pg-LPS-induced proliferation as well as the tyrosine phosphorylation.

Thus, the present findings suggest that Pg-LPS could induce B cell activation in both CD14-dependent and independent pathways. In the CD14-dependent pathway, TGF-β production could be induced upon stimulation with Pg-LPS through NF-κB activation. The CD14-independent pathway mediated by tyrosine phosphorylation could also exist in the Pg-LPS-induced activation of B cells, leading to cell proliferation and IL-6 production.

References

1. Kimura S, Tamamura T, Nakagawa I, et al (2000) CD14-dependent and independent pathways in lipopolysaccharide-induced activation of a murine B-cell line, CH12.LX. Scand J Immunol 51:392–399
2. Kimura S, Koga T, Fujiwara, T, et al (1995) Tyrosine protein phosphorylation in murine B lymphocytes by stimulation with lipopolysaccharide from *Porphyromonas gingivalis*. FEMS Microbiol Lett 130:1–6

Interface Oral Health Science 2007: 297–298

Priming effects of microbial or inflammatory agents in metal allergies

N. Sato[1,3], M. Kinbara[1], T. Kuroishi[1], H. Takada[2], K. Kimura[3], S. Sugawara[1], and Y. Endo[1]*

[1]*Departments of Molecular Regulation;* [2]*Microbiology;* [3]*Fixed Prosthodontics, Graduate School of Dentistry, Tohoku University, Sendai 980–8575, Japan*

*endo@pharmac.dent.tohoku.ac.jp

Abstract. Although T-cells are thought to play central roles in Ni-allergy, this idea is based largely on in vitro studies, because in vivo studies have been limited by the paucity of adequate murine models of Ni-allergy (and indeed it has been said that it is difficult to induce Ni-allergy in mice). However, we recently found that a lipopolysaccharide (LPS) of *Escherichia coli* acted as a potent adjuvant, sensitizing mice to Ni. LPS also sensitized mice to other metals (Cr, Co, Pd, and Ag). Here, we report that in addition to LPS, a variety of microbial or inflammatory substances sensitize mice to Ni. Our findings suggest that a microbial or an inflammatory milieu is an important factor leading to metal allergies.

Key words. metal allergy, nickel allergy, adjuvant, inflammation, infection

1 Introduction

Metals are thought to cause various types of allergic reactions (including dermatitis, lichen planus, palmoplantar pustulosis, asthma, and rhinitis), and even to cause carcinomas. Ni, a constituent of many alloys, is the most frequent contact allergen. Unlike classical haptens, metal ions form geometrically highly defined, but reversible, coordination complexes with partner molecules. Thus, the host may recognize metal ions in complicated ways. The knowledge that the partner molecules are intact self-proteins led us to speculate that metal allergies might actually be forms of autoimmune disease. Although T-cells are thought to play central roles in Ni-allergy, this idea is largely based on in vitro studies, because the in vivo studies have been limited by the paucity of adequate murine models of Ni-allergy (and indeed it has been said that it is difficult to induce Ni-allergy in mice). However, we recently found that an *Escherichia coli* lipopolysaccharide (LPS) [a ligand of Toll-like receptor 4 (TLR4)] acted as a potent adjuvant, sensitizing mice to Ni, and that the Ni(+LPS)-allergy was fully induced even in nude mice lacking T-cells. LPS also sensitized mice to other metals (Cr, Co, Pd, and Ag). Here, the effects of a number of inflammatory agents derived from various sources, including microbial and chemical substances, were examined.

2 Materials and Methods

An equivolume mixture of 1 mM $NiCl_2$ and a solution of a test substance was injected intraperitoneally (0.1 ml/10 g body weight) into mice (BALB/c, C3H/HeN, or C3H/HeJ). Then, 10 days later, 5 mM $NiCl_2$ was delivered as a challenge by intradermal injection into the left and the right pinnas, near the root of the ear (20 μl each ear). Mice were anesthetized with ethyl ether just before the challenging injection. Ear swelling was measured at a site 2–3 mm distant from the challenge-site at the indicated times (using a Peacock dial thickness gauge, Ozaki MFG Co. LTD, Tokyo, Japan), and the induced difference (versus before the challenge) was recorded.

3 Results

The following microbial or inflammatory substances sensitized BALB/c mice to Ni: (a) an LPS of *Prevotella intermedia* (*P. int*) [prepared by the phenol–chloroform–petroleum ether extraction method (Galanos's method)], (b) a mannan of *Saccharomyces cerevisiae* (a putative TLR2 and/or TLR4 ligand), (c) a muramyl dipeptide (a cell-wall component of gram-positive bacteria, an NOD-2 ligand), (d) a double-stranded RNA (polyIpolyC, a TLR3 ligand), (e) concanavalin A (a T-cell mitogen), and (f) alendronate (an inflammatory bisphosphonate). *P. int* LPS (but not *E. coli* LPS) was effective even in C3H/HeJ mice (mice with a TLR4 mutation) as well as in their controls (C3H/HeN mice).

4 Discussion

These results suggest that a microbial or an inflammatory milieu (irrespective of the types of TLR present) is an important factor leading to metal allergies. Unexpectedly, the LPS preparation of *P. int* (an oral black-pigmented gram-negative bacterium) used in the present study exhibited adjuvant activity even in C3H/HeJ mice. This result—which suggests that the major adjuvant effect of our *P. int* LPS preparation is independent of TLR4 [unlike that of *E. coli LPS*]—raises the possibility that a contaminant substance(s) might be responsible for the adjuvant activity of the *P. int* LPS preparation. It would be of interest to identify this substance(s) in the *P. int* bacterium. Incidentally, *Porphyromonas gingivalis* (another oral black-pigmented gram-negative bacterium) contains a lipoprotein that strongly stimulates TLR2, making it a potentially interesting subject for study.

Interface Oral Health Science 2007: 299–300

Dental examinations for oral health promotion in a rural town

Naoko Tanda[1]*, Masaki Iwakura[4], Kyoko Ikawa[1], Jumpei Washio[1,3], Ayumi Kusano[1], Kazutaka Amano[1], Yuhei Ogawa[1], Yudai Yamada[1], Yoshiko Shigihara[1], Yoshiro Shibuya[1], Megumi Haga[4], Ken Osaka[2], and Takeyoshi Koseki[1]

[1]*Division of Preventive Dentistry, Department of Oral Health and Development Sciences;* [2]*Division of International Oral Health, Department of Oral Health and Development Sciences;* [3]*Division of Oral Ecology and Biochemistry, Department of Oral Biology, Tohoku University Graduate School of Dentistry, Sendai 980–8575;* [4]*Department of Human Health and Nutrition, Faculty of Comprehensive Human Sciences, Shokei Gakuin College, Natori 981–1295; Japan*

*ntanda@mail.tains.tohoku.ac.jp

Abstract. We carried out residents-attractive dental health examinations by increasing the learning contents with systemic health examination of a rural town in Japan. Town staffs or some residents who had finished training programs participated in the dental examinations as staffs. Oral malodor was measured to motivate oral hygiene of examinees. We prepared questionnaires and asked the impression of examinees about "the dental examination with learning" 7 months after the examination. Questionnaires showed that 85% of dental examinees recognized it as satisfactory. This fact indicates the possibility that the new dental examination can be a model for oral health promotion in the community.

Key words. dental examination, learning contents, oral health promotion

1 Introduction

The rate of examinees of dental health check-ups organized by municipal health departments is usually less than 10%, because residents recognize it as less necessary common oral diseases to be diagnosed and to be recommended to visit dental offices. The aim of this study was to develop more residents-attractive dental health examinations by increasing the learning contents and to contribute to oral health promotion of rural community.

2 Dental examinations

The new dental examinations were held in 2003, 2004 with systemic health examinations of a rural town in Japan. Town staffs or some residents who had finished training programs participated in the dental examinations as staffs. First, we measured oral odor of most of the examinees with sulfide monitors, Breathtron™ [1]. Second, dentists examined the oral condition of the examinee and explained the needs of oral hygiene, based on the value of Breathtron™. Third, examinee learned how to brush teeth with resident staffs, according to examiner's advice.

We prepared questionnaires and asked the impression of examinees about "the dental examination with learning", 7 months after the examination in 2004. We asked examinees of 2004 the impression of the dental examination, including the learning opportunity of tooth brushing, and oral malodor measurement.

We also asked non-examinees of dental examination 2004 the reason of not joining the dental examination and the impression about oral malodor measurement.

3 Review of the dental examinations

The questionnaires were gathered from 335 examinees who participated in both 2003 and 2004, 187 examinees who participated only in 2003, 96 examinees who participated only in 2004, and 335 residents who participated in no dental examinations.

In 2004 72% of dental examinees joined the opportunity for learning how to brush teeth, and 88% of them recognized it as satisfactory. Oral malodors in 94% of dental examinees of 2004 were measured with Breathtron™ [1], and 85% of them agreed that measurement of oral malodor promoted a better understanding of oral hygiene. The main reasons for not joining dental examination in 2004 were "Need for haste" and "Consultation with a family dentist". Less than 1% of examinees selected "Unnecessary malodor measuring".

Measurement of oral malodor was not the obstacle to proceed with the dental examination. Since 85% of examinees of 2004 agreed with its usefulness for promoting a better understanding of oral hygiene, measurement of oral malodor seems one of the learning contents in the dental examination.

Questionnaires showed that 85% of dental examinees of 2004 recognized the examination satisfactory. This fact indicates the possibility that the new dental examination can be a model for oral health promotion in the community.

Reference

1. Tanda N, Washio J, Ikawa K, et al (2007) A new portable sulfide monitor with a zinc-oxide semiconductor sensor for daily use and field study. J Dent 35:552–557

Interface Oral Health Science 2007: 301–302

Non-destructive ultrasonic device detects early caries lesions

Yudai Yamada[1], Yuhei Ogawa[1], Kazutaka Amano[1], Sadao Omata[2], and Takeyoshi Koseki[1]*
[1]*Division of Preventive Dentistry, Department of Oral Health and Development Sciences, Tohoku University Graduate School of Dentistry, Sendai 980–8575;* [2]*Department of Electrical and Electronic Engineering, Nihon University, Tokyo; Japan*
*tkoseki@mail.tains.tohoku.ac.jp

Abstract. To diagnose the activities and progressions of dental caries lesions, we developed nondestructive measuring device to detect and quantify the mineral loss in the early stage of caries. This novel device with ultrasonic technology measures acoustic frequency characteristics of the tooth surface. Extracted human tooth were embedded in polymethylmethacrylate and polished. The results of the repeated measurements indicated the sufficient reproducibility of acoustic measurements in same samples. Since this acoustic measurement was influenced by the moist condition of tooth surfaces, tooth surfaces should be air-dried and quickly measured for the reproducibility in clinics. We developed the nondestructive ultrasonic measuring device to detect the mineral loss from tooth surfaces and diagnose the initial change of progression of dental caries lesions in enamel and dentine.

Key words. dental caries, mineral loss, ultrasonic device, nondestructive measurement

1 Introduction

To diagnose the activities and progressions of dental caries, it is important to determine the mineral loss from tooth surface accurately, nondestructively, and readily. However, it is difficult to detect the exact mineral loss by traditional palpation with a probe and a visual inspection. We are still facing the clinical problems on how to estimate the mineral loss in early enamel lesions and root-surface lesions. The aim of this study was to develop nondestructive measuring device to detect and quantify the mineral loss in the early caries lesions.

2 Materials and methods

Novel device with ultrasonic technology consisted of oscillating piezoelectric drive element for sensing and piezoelectric detective element to obtain frequency characteristics of the tooth surface [1]. The positive feedback loop from detective element to drive element resulted in amplifying the frequency shifts. The piezoelectric sensing units were connected to the Tohoku University Constant Load (TUCL) probe to obtain the constant applying force to the surfaces. The positioning and sensing motion of the ultrasonic device were controlled by a computerized XYZ table.

Extracted human teeth were cut by water-cooled saw with a thin blade and the cutting surfaces were polished with wet silicon carbide sandpaper. The tooth samples were embedded in polymethylmethacrylate, and then, the sample surfaces were polished again on a wet abrasive paper (grid 800). The cutting surfaces of the samples were stored in distilled water at 4°C until measuring acoustic property.

3 Results and discussion

The repeated measurements of the same tooth sample indicated the sufficient reproducibility of frequency shift under the same measuring condition. Then, we evaluated the operational tolerance of the acoustic measurement in measuring environments. The results demonstrated that basic oscillating frequency and magnitude of frequency shifts changed depending on the operating ambient temperature. By using the temperature correction formula referring to the basic oscillating frequency, the influence of ambient temperature was eliminated. Since the acoustic measurement was influenced by the moist condition of tooth surfaces, enamel and dentin surfaces should be air-dried and measured in a short time for the reproducibility in clinics. Surface scanning of acoustic properties of embedded tooth demonstrated the demineralization-mapping performance of this acoustic measuring system. Acoustic readouts of the device manipulated by a computer-controlled XYZ table and by hand were identical.

In conclusion, we developed the nondestructive ultrasonic measuring device to detect the mineral loss from tooth surfaces and diagnose the initial change of progression of dental caries lesions in enamel and dentine.

Reference

1. Eklund A, Bergh A, Lindahl OA et al (1999) A catheter tactile sensor for measuring hardness of soft tissue. Med Biol Eng Com 37:618–624

Interface Oral Health Science 2007: 303–304

The TUCL probe, novel constant load periodontal probe for the standardized probing measurements

Takeyoshi Koseki[1]*, Emi Ito[1], Kyoko Ikawa[1], Yudai Yamada[1], Yuhei Ogawa[1], Kazutaka Amano[1], and Hidetoshi Shimauchi[2]
[1]*Division of Preventive Dentistry, Department of Oral Health and Development Sciences;*
[2]*Division of Periodontology and Endodontology, Department of Oral Biology, Tohoku University Graduate School of Dentistry, Sendai 980–8575; Japan*
*tkoseki@mail.tains.tohoku.ac.jp

Abstract. To reduce the errors of the periodontal probing measurements, we developed a novel constant load periodontal probe, Tohoku University-type constant load periodontal probe (TUCL probe), which has a similar shape with the one used commonly. The tip of the TUCL probe buckles at the hinge of the handle by the excess load on the tip. The probing forces performed by six new resident dentists were recorded by using the model tooth with pressure sensor mounted on the dummy jaw. The maximal force probed upper right first incisor showed 28 to 109 g (average 56 g) when using normal CPI probes. When they used TUCL probes, however, their maximal probing force showed 16 to 33 g (average 20 g). This indicated TUCL probe could exclude the exceed load of probing, which depends on the examiner's skill. The novel TUCL probe with constant probing load will standardize the periodontal probing examination.

Key words. periodontal probe, constant force, standardization, probing force

1 Introduction

Periodontal probing is one of the most common examinations for assessing the status of gingival health. To achieve the minimal errors of the probing measurements, examiners are requested to calibrate their probing manipulation. To assess accurately the severity of periodontal diseases by probing, we developed a novel constant force periodontal probe, Tohoku University-type constant load periodontal probe (TUCL probe), which has a similar shape with the one used commonly. The aim of this study is to evaluate the performance and the reproducibility of this novel probe.

2 Materials and methods

The tip of the TUCL probe buckles at the hinge of the handle by the excess load on the tip. When the tip is lifted by touching, compressed built-in spring gives back constant force on the tip, which is 25 g on the Williams type of the TUCL probe

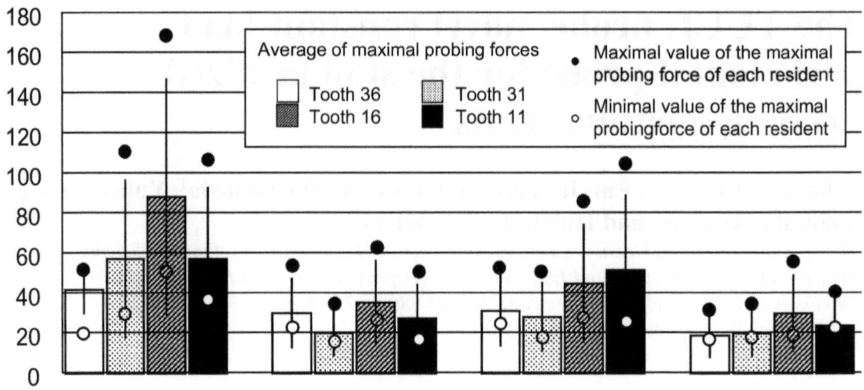

Fig. 1. Maximal probing force measured by new resident dentists. The test tooth on dummy jaw was probed by new resident dentists. The average, minimal and maximal values of the maximal probing force by new resident dentists were indicated

and 20 g on the CPI type of the TUCL probe.The periodontal probes used in this study were TUCL probe (Shioda Co, Ltd.), Gram probe (YMD Co, Ltd.), Crick probe (HerrHawe Co. Ltd), and CPI probe (CPITN-C, YMD Co, Ltd.). The dummy jaw with test tooth was set on the dental chair to simulate the probing measurement in clinic. The probing forces were determined by force sensor connected to the test teeth and recorded with A/D converter. The test teeth were 16 and 11 in upper jaw, and 36 and 31 in lower jaw. The test tooth on dummy jaw was probed by six new resident dentists.

3 Results and discussion

In dummy jaw examination by the new residents, the maximal force probed upper right first incisor showed 28 to 109 g (average 56 g) when using normal CPI probes (Fig. 1). When they used TUCL probes, however, their maximal probing force showed 16 to 33 g (average 20 g). This indicated TUCL probe could exclude the exceed force of probing, which depends on the examiner's skill. In conclusion, the accuracy and reproducibility of probing operation by using TUCL probe were confirmed. The novel TUCL probe with constant probing force will standardize the periodontal probing examination.

Interface Oral Health Science 2007: 305–306

The TUCL probe for easy learning of probing manipulation

Emi Ito[1], Emiko Kato[2], Yoko Sato[2], Kyoko Ikawa[1], Yudai Yamada[1], Yuhei Ogawa[1], Kazutaka Amano[1], Hidetoshi Shimauchi[3], and Takeyoshi Koseki[1]*

[1]*Division of Preventive Dentistry, Department of Oral Health and Development Sciences;* [3]*Division of Periodontology and Endodontology, Department of Oral Biology, Tohoku University Graduate School of Dentistry, Sendai 980-8575;* [2]*Miyagi Advanced Dental Hygienist College, Sendai 980-0803; Japan*

*tkoseki@mail.tains.tohoku.ac.jp

Abstract. It takes time to learn accurate probing skills, because it is difficult to minimize inherent probing measurement error. Thus, we invented the Tohoku University-type constant load periodontal probe (TUCL probe) to standardize the probing measurements. The TUCL probe gives constant probing force by itself. In this study, we reported the application of the TUCL probe in the training of the probing operation. The probing forces of 60 new dental hygienist students were recorded before and after the practice by using conventional probe and weighing scale, or TUCL probe. For learning the force for probing, 36% of students reported that the TUCL probe was an easier learning tool, compared to the weighing scale. In conclusion, the TUCL probe is a significantly useful instrument, not only for standardizing probing measurement, but also for probing operation training.

Key words. periodontal probe, constant force, probing, learning, dental hygienist

1 Introduction

Periodontal probing is one of the most common techniques for assessment of periodontal conditions. Probing measurements, however, have inherent measurement errors caused by a wide range of sources, such as instrument, examiner, and disease status. For training the probing operation in dental hygienist school, it takes time to learn probing the periodontal pockets accurately, because it is difficult to avoid inherent probing measurement error. Thus, we invented the Tohoku University-type constant load periodontal probe (TUCL probe; Fig. 1) to standardize the probing measurements. In this paper, we reported the application of the TUCL probe in the training of the probing operation.

2 Materials and methods

Sixty first-year dental hygienist students, who have not learned the periodontal instrumentation, were divided into two practice groups A and B. Before practice,

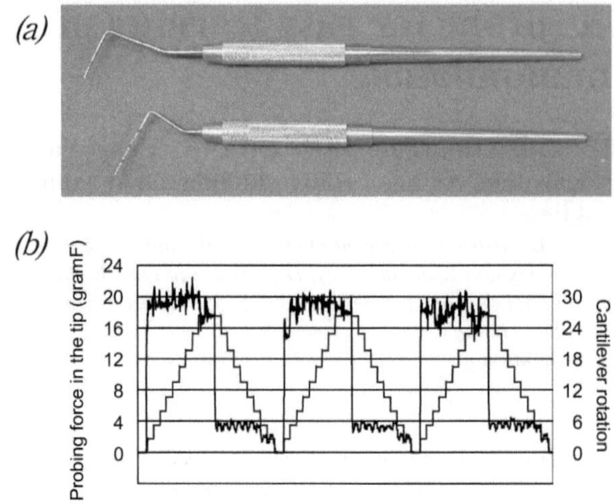

Fig. 1. The TUCL probe, which produces constant probing force on the probe tip.
The TUCL probe (**a**) gives constant probing force by itself. When tip is lifted by touching, compressed built-in spring gives back constant force on the tip (**b**)

the probing forces of all students were recorded by using model tooth. The students in practice A group first learned probing operation by using conventional perioprobe and weighting scale, whereas in practice B group first learned by using TUCL probe. After practice, the probing forces were recorded again. After the training, the students filled the questionnaire concerning about the probing operation and their skills.

3 Results and discussion

Almost all dental hygienist students reported the difficulty of the sounding force of probing and the insert direction of probe tip. After the probing practice, 86% of hygienist students reported the easy learning of probing force by using TUCL probe, and 66% of hygienist students supported it by using weighing scale. Though 57% of the students felt some difficulty in using TUCL probe, 83% of students reported that the learning time of handling TUCL probe is less than 30 min. For learning the sounding force of probing, 36% of students supported that the TUCL probe was an easier learning tool, compared to weighting scale (supported by 9%). In conclusion, the TUCL probe is significant not only for standardizing probing measurement, but also a useful training tool for probing operation.

Interface Oral Health Science 2007: 307–308

Psychological characterization of halitosis patients by using Egogram and the Halitosis Scale Questionnaires

Ayumi Kusano[1], Masaki Iwakura[1], Kyoko Ikawa[1], Naoko Tanda[1], Jumpei Washio[2], Yuhei Ogawa[1], Yudai Yamada[1], and Takeyoshi Koseki[1]*
[1]*Division of Preventive Dentistry, Department of Oral Health and Development Sciences;*
[2]*Division of Oral Ecology and Biochemistry, Department of Oral Biology, Tohoku University Graduate School of Dentistry, Sendai 980-8575; Japan*
*tkoseki@mail.tains.tohoku.ac.jp

Abstract. Halitosis is a common oral concern and patients often seek help from dentists. The purpose of this study is to evaluate psychological characteristics of halitosis patients and their changes with treatment. The subjects consisted of 150 halitosis patients and 64 non-halitosis patients. They were requested to complete Egogram and the halitosis scale questionnaires, before and after treatment. The results indicated that halitosis patients improved in the objective recognition of their malodor level, their obsession with halitosis, and interpersonal communication after treatment. Therefore, the ideal dentist attitude for halitosis treatment is to help patients overcome their obsession by examining their breath level repeatedly and encourage them to develop social skills. However, these subjects had a tendency for social phobia, and its removal was not easy. Therefore, it would be essential that dentists continue to accept their complaints and relapse of their obsession with bad breath sincerely.

Key words. halitosis, psychology, questionnaires, Egogram

1 Introduction

Halitosis is a common oral concern and patients often seek help from dentists. The 75% of the patients who visited our halitosis clinic in Tohoku University Hospital had no bad breath at first examination [1]. Moreover, 74% of patients presume their halitosis based upon others' behavior such as opening windows, rubbing noses, placing hand over the mouth, and turning away the face. It is known that some halitosis patients have certain psychological characteristics such as social phobia, sensitiveness, personal inferiority, obsession, and low self-esteem [2]. They often receive dental treatment for removing source of bad breath such as tongue-coat cleaning. If they still feel their bad breath, they will visit dentist many times or go to another doctor. The purpose of this study is to clarify the psychological characteristics of halitosis patients and their change with treatment.

2 Materials and methods

The subjects consisted of 150 halitosis patients and 64 control patients. Before and after treatment, they were asked to complete Egogram (transactional analysis) and the Halitosis Scale Questionnaires.

3 Results and discussion

Before treatment, halitosis patients had obvious awareness, presumption based upon others' attitudes, obsession with their bad breath, and a tendency of social phobia. In Egogram, the FC (free child) score of halitosis patients was significantly lower than that of the control patients. This suggests that halitosis patients have negative interpersonal communication. After treatment, halitosis patients perceived their bad breath level objectively, and also their obsession with halitosis tended to decrease. Simultaneously, improvement of their interpersonal communication was found in Egogram; however, social phobia was not improved. The halitosis patients tend to remind their bad breath when communicating with another person. This indicates that even if a patient is free from halitosis, their symptom might relapse.

From our results of the Halitosis Scale Questionnaires and Egogram, the ideal dentists' attitude in halitosis treatment is to help halitosis patients overcome their obsession by objectively examining their breath level, as long as they ask, and to encourage them to develop social skills. Simultaneously, it is important to observe the degree of recovery from obsession with their bad breath and poor interpersonal communication. In addition, halitosis patients had a tendency of social phobia, and its removal was not easy. Therefore, it is essential that dentists should sincerely continue to accept their complaints and relapse of obsession with bad breath.

Reference

1. Iwakura M, Yasuno Y, Shimura M, et al (1994) J Dent Res 73:1568–1574
2. Eli I, Baht R, Kozlovsky A, et al (1996) Psychosomatic Medicine 58:156–159

Section III:
Biomaterial interface

Interface Oral Health Science 2007: 311–316

Released ions and microstructures of dental cast experimental Ti–Ag alloys

Masatoshi Takahashi*, Yukyo Takada, Masafumi Kikuchi, and Osamu Okuno
Division of Dental Biomaterials, Tohoku University Graduate School of Dentistry, Sendai 980-8575, Japan
*m-tak@mail.tains.tohoku.ac.jp

Abstract. This study is an examination of the released ions from dental cast experimental Ti–Ag alloys by an immersion test. Ti–Ag alloys (5–25mass%Ag) and pure titanium (control) were cast into magnesia molds; the hardened surface layer was then removed. After each specimen was immersed in aerated 0.9% NaCl or 1% lactic acid solution at 37°C for 7 days, released ions were analyzed using inductively coupled plasma. Cast Ti–Ag alloys with Ag ≤ 20% formed a single α structure. Ti and Ag ions were not detected from the alloys and pure titanium in the NaCl solution. The microstructures of cast Ti–Ag alloys with 22.5% Ag and 25% Ag consisted of α + intermetallic compounds (Ti_2Ag or Ti_2Ag + TiAg). A small amount of Ti and Ag ions was detected from some of the 22.5% Ag and 25% Ag specimens in the NaCl solution. The preferential dissolution of parts of the intermetallic compounds was observed in the specimens after the test. In the lactic acid solution, a significantly smaller amount of Ti ions was released from all the Ti–Ag alloys than from pure titanium. Ag ions were not detected. The intermetallic compounds remained on the specimen surfaces after the test in the lactic acid solution.

Key words. Ti–Ag alloys, released ions, microstructure, corrosion

1 Introduction

As part of our studies developing a new dental titanium alloy with better mechanical properties and machinability than unalloyed titanium [1–7], Ti–Ag alloys were investigated. Our studies [1, 5] showed that some Ti–Ag alloys had mechanical properties superior to those of pure titanium. The yield strength of 10% Ag and 20% Ag alloys matched that of type 4 hardened dental casting alloys or dental base metal casting alloys such as Co–Cr alloys (in this article, percentages are given as mass percent). The mechanism underlying the improved properties could be attributed to the solid-solution strengthening of α titanium and the fine precipitation of Ti_2Ag. The machinability of titanium alloys was evaluated from the grindability test using the constant load as a screening. Some of the Ti–Ag alloys had better grindability than pure titanium [2, 5]. The grindability of the alloys could be attributed mainly to a decrease in the elongation caused by the precipitation of small

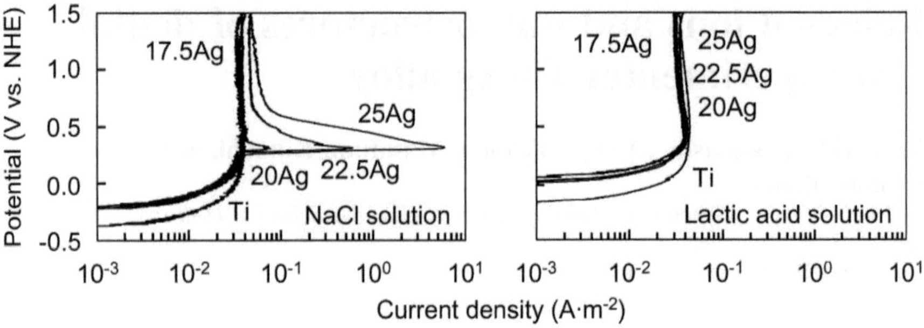

Fig. 1. Anodic polarization curves of the Ti–Ag alloys

amounts of fine intermetallic compounds. The Ti-20%Ag alloys are particularly good candidates for dental CAD/CAM alloys.

In earlier studies [8, 9], anodic polarization tests for the Ti–Ag alloys were performed in 0.9% NaCl or 1% lactic acid solution (Fig. 1). The alloys up to 17.5% Ag had excellent corrosion resistance similar to that of pure titanium in both solutions. Alloys with 20%–25% Ag may be used as dental alloys, because they passivated again immediately after preferential dissolution of Ti_2Ag in the NaCl solution. In order to evaluate the corrosion resistance of the Ti–Ag alloys in detail, it is also important to know the amount of ions released from the alloys.

A series of Ti–Ag alloys with 5%–25% Ag was made in the present study. The immersion test was performed in 0.9% NaCl or 1% lactic acid solution to characterize the relationship between the released ions and the microstructures of the dental cast Ti–Ag alloys.

2 Materials and methods

2.1 Preparation of alloys

Experimental Ti–Ag alloys with 5%, 10%, 17.5%, 20%, 22.5%, and 25% Ag were examined. The characteristics of the Ti–Ag system are shown in Table 1. The desired amounts of titanium sponge (>99.8%, grade S-90, Sumitomo Titanium Corp., Amagasaki, Japan) and pure silver (>99.99%, Ishifuku Metal Industry, Tokyo, Japan) were melted into one 15 g button for each alloy in an argon-arc melting furnace (TAM-4S, Tachibana Riko, Sendai, Japan).

2.2 Preparation of specimens

Each alloy was cast into a magnesia mold (Selevest CB, Selec, Osaka, Japan) using a dental casting unit (Castmatic-S, Iwatani, Osaka, Japan); the hardened surface layer was then removed, producing specimens measuring 15 mm × 10 mm × 1.9 mm.

Table 1. Characteristics of the Ti–Ag system

Reaction type	Reaction	Composition range (mass%)	Reaction point (mass%)	Temperature (°C)
Peritectic	(βTi) + L ↔ TiAg	29.2 to 97	68	1,020
Peritectoid	(βTi) + TiAg ↔ Ti_2Ag	24 to 68	52.9	940
Eutectoid	(βTi) ↔ (αTi) + Ti_2Ag	10.0 to 52.9	15.6	855

2.3 Microstructural observation

The specimen surfaces, which were polished and etched with 0.5 ml HF, 1.0 ml HNO_3, and 300 ml H_2O, were observed using an optical microscope. Selected specimens were also observed before and after an immersion test using a scanning electron microscope (SEM) (JSM-6060, JEOL, Tokyo, Japan).

2.4 X-ray diffractometry

X-ray diffraction (XRD) was performed at room temperature using copper Kα radiation generated at 30 kV and 10 mA in an X-ray diffractometer (Miniflex CN2005, Rigaku, Tokyo, Japan). The experimental conditions were: 2θ range 20°–90° at 0.02° per step and 6 s-photon counting time per step. The peaks on the XRD patterns were indexed to X-ray polycrystalline powder diffraction files.

2.5 Immersion test

Each specimen ($n = 7$) was immersed in 10 ml of 0.9% NaCl or 1% lactic acid solution with saturated dissolved oxygen (6 mg l^{-1}) at 37°C for 7 days. After each specimen was removed from the solution, the released ions were analyzed qualitatively and quantitatively using inductively coupled plasma (ICP) (IRIS-AP, Japan Jarrell Ash, Kyoto, Japan). The data were analyzed by ANOVA/Scheffé ($\alpha = 0.05$).

3 Results and discussion

3.1 Microstructures and XRD

The microstructures of the Ti–Ag (5%–17.5% Ag) alloys (Fig. 2) consisted of α titanium grains, which resembled those of pure titanium. The acicular structure in the Ti–Ag alloys changed to an equiaxed structure as the concentration of Ag

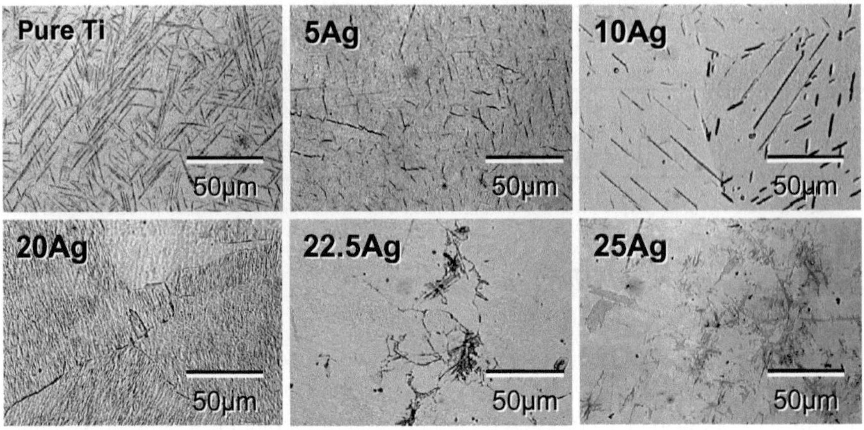

Fig. 2. Microstructures of the etched Ti–Ag alloys

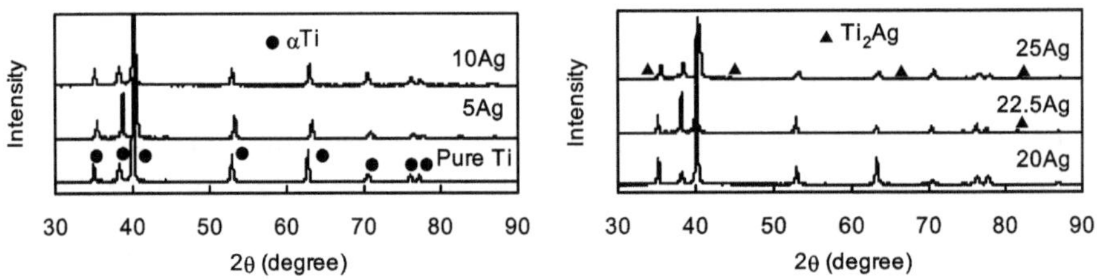

Fig. 3. X-ray diffraction patterns of the Ti–Ag alloys

increased. Only the α titanium peaks were present in the pure titanium and in the 5%–20% Ag alloys in the XRD patterns (Fig. 3).

A few needle-like precipitates were observed in the 20% Ag alloys. With an increase in the Ag concentration, the amount of the precipitates increased (Fig. 2). Both α titanium and Ti_2Ag peaks appeared in the 22.5% and 25% Ag alloys (Fig. 3). The XRD revealed that the precipitates were Ti_2Ag.

3.2 Released ions in 0.9% NaCl solution

The amount of ions released from the Ti–Ag alloys in 0.9% NaCl solution is shown in Fig. 4a. Ti and Ag ions were not detected from Ti–Ag alloys with Ag $\leq$ 20%. No structural changes were found in the alloys. Regardless of the Ag content, the α solid solution in the Ti–Ag alloys released no ions in the NaCl solution, which was similar to the case of pure titanium.

A small amount of Ti and Ag ions were detected from some of the 22.5% Ag and 25% Ag specimens. The amount of Ag ions from the specimens was 1/300 or

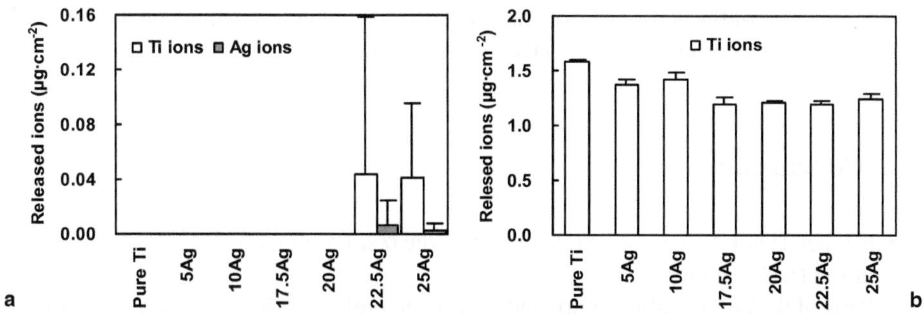

Fig. 4. Amount of ions released from the Ti–Ag alloys and pure titanium. **a** In 0.9% NaCl solution with saturated dissolved oxygen at 37°C for 7 days. **b** In 1% lactic acid solution with saturated dissolved oxygen at 37°C for 7 days

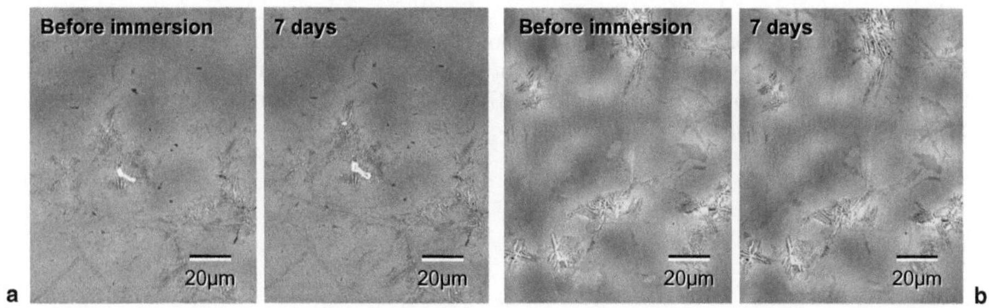

Fig. 5. Changes in surface structure of Ti-25%Ag alloy after immersion. **a** In 0.9% NaCl solution with saturated dissolved oxygen at 37°C. **b** In 1% lactic acid solution with saturated dissolved oxygen at 37°C

less than that from Ag–Pd–Au alloy in an immersion test under the same conditions [10]. The microstructures of the corroded alloy surfaces indicated the deterioration of parts of the intermetallic compounds (Fig. 5a). The dissolved structures were probably TiAg, which was surrounded with Ti_2Ag precipitated due to the peritectoid reaction.

3.3 Released ions in 1% lactic acid solution

The amount of ions released from the Ti–Ag alloys in 1% lactic acid solution is shown in Fig. 4b. A significantly smaller amount of Ti ions ($P < 0.05$) was released from all the Ti–Ag alloys than from pure titanium. As the concentration of Ag increased, the amount of Ti ions for the 5%–17.5% Ag alloys tended to decrease. There was no significant difference ($P > 0.05$) in the amount of Ti ions among the 17.5%–25% Ag alloys. Ag ions were not detected in the Ti–Ag alloys. Differently from the case of the NaCl solution, intermetallic compounds remained on the specimen surfaces after testing in the lactic acid solution (Fig. 5b).

4 Conclusion

- Neither Ti nor Ag ions were released from the α solid solution in the Ti–Ag alloys in the NaCl solution.
- Parts of the intermetallic compounds dissolved preferentially in the NaCl solution and released a small amount of Ti/Ag ions.
- The amount of ions released from the Ti–Ag alloys in the lactic acid solution was smaller than that from pure titanium.

Ti–Ag alloys up to 25% Ag have potential for dental alloys.

Acknowledgments. The authors gratefully acknowledge the Sumitomo Titanium Corporation for supplying the high-purity titanium sponge used in this study. A portion of this study was supported by a Grant-in-Aid for Young Scientists (B) No.18791418 from The Ministry of Education, Culture, Sports, Science and Technology of Japan.

References

1. Takahashi M, Kikuchi M, Takada Y, et al (2002) Mechanical properties and microstructures of dental cast Ti-Ag and Ti-Cu alloys. Dent Mater J 21:270–280
2. Kikuchi M, Takahashi M, Okabe T, et al (2003) Grindability of dental cast Ti-Ag and Ti-Cu alloys. Dent Mater J 22:191–205
3. Kikuchi M, Takahashi M, Okuno O (2003) Mechanical properties and grindability of dental cast Ti-Nb alloys. Dent Mater J 22:328–342
4. Takahashi M, Kikuchi M, Okuno O (2004) Mechanical properties and grindability of experimental Ti-Au alloys. Dent Mater J 23:203–210
5. Takahashi M, Kikuchi M, Takada Y, et al (2005) Grindability and mechanical properties of experimental Ti-Au, Ti-Ag and Ti-Cu alloys. In: Watanabe M, Takahashi N, Takada H (eds) Interface oral health science. Elsevier, Amsterdam, pp 326–327
6. Sato H, Kikuchi M, Komatsu M, et al (2005) Mechanical properties of cast Ti-Hf alloys. J Biomed Mater Res 72B:362–367
7. Kikuchi M, Takahashi M, Sato H, et al (2006) Grindability of cast Ti-Hf alloys. J Biomed Mater Res 77B:34–38
8. Takada Y, Nakajima H, Okuno O, et al (2001) Microstructure and corrosion behavior of binary titanium alloys. Dent Mater J 20:34–52
9. Takahashi M, Kikuchi M, Takada Y, et al (2006) Electrochemical behavior of cast Ti-Ag alloys. Dent Mater J 25:516–523
10. Takada Y, Okuno O (1996) Reversing potentials and increasing amount of released ions with contact corrosion between stainless steels and Au-Ag-Pd alloys. J J Dent Mater 15:525–531

Interface Oral Health Science 2007: 317–322

Induction of octacalcium phosphate by surface modification of TiO_2 film prepared by electron cyclotron resonance plasma oxidation

Yusuke Orii[1,3], Hiroshi Masumoto[2], Takashi Goto[2], Yoshitomo Honda[3], Takahisa Anada[3], Keiichi Sasaki[1], and Osamu Suzuki[3]*
[1]*Division of Advanced Prosthetic Dentistry;* [3]*Division of Craniofacial Function Engineering, Tohoku University Graduate School of Dentistry, Sendai 980-8575;* [2]*Institute for Materials Research, Tohoku University, Sendai; Japan*
*suzuki-o@mail.tains.tohoku.ac.jp

Abstract. TiO_2 (titania) films on titanium substrates were fabricated by electron cyclotron resonance (ECR) plasma oxidation. A calcification experiment of the Ti substrates treated with the ECR plasma was carried out by immersing in a supersaturated solution with respect to hydroxyapatite (HA) and octacalcium phosphate (OCP), but slightly saturated with respect to dicalcium phosphate, dihydrate (DCPD). After the immersion, OCP precipitated predominantly on TiO_2 film, whereas formation of HA, the most stable phase under physiological condition, was inhibited. The calcification ability was affected by the substrate temperature during the ECR plasma oxidation. The surface roughness increased with increasing the oxidation temperature. These results indicate that the surface oxidation by ECR plasma modulates the surface characteristics to enhance OCP precipitation.

Key words. titania film, octacalcium phosphate, osteoconductivity, electron cyclotron resonance plasma oxidation

1 Introduction

The present study was designed to investigate whether the state of surface oxide film of the titanium (Ti), created by electron cyclotron resonance (ECR) plasma oxidation, affects nucleation and growth of octacalcium phosphate (OCP), a precursor of biological apatite crystals in bone and tooth [1], and a scaffold to actively induce new bone demonstrated by us [2], on the surface of Ti substrate. ECR plasma allows the oxide film to form at a low temperature and a short time [3].

2 Materials and methods

Figure 1 shows the ECR plasma apparatus. ECR plasma was used for oxidizing Ti substrates. The microwave power was set to 900 W. A magnetic field (8.75×10^{-2} T) was applied to the plasma chamber to satisfy the ECR condition. A mirror-type

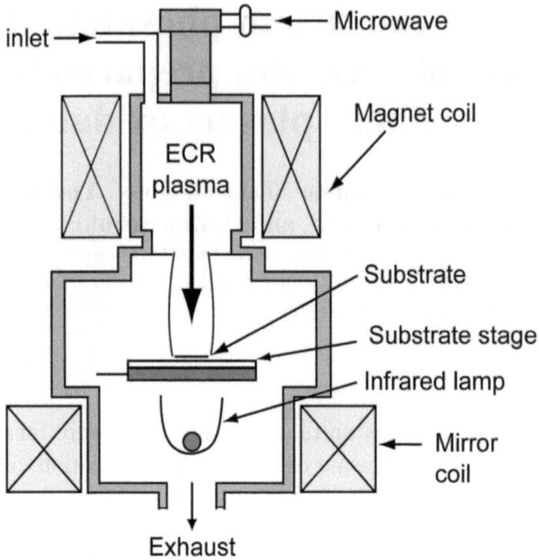

Fig. 1. ECR plasma apparatus

magnetic field (4.5×10^{-2} T at the substrate stage) was applied in order to raise a plasma density. The oxygen gas pressure during ECR oxidation was from 3.3×10^{-3} to 9.3×10^{-1} Pa, and the oxidation time was 5–60 min. The substrates (10 mm × 10 mm × 1 mm) of commercially pure Ti plate (CP-Ti, JIS Grade 2) were used. An infrared lamp was used for controlling the substrate temperature in the range from room temperature (RT) to 600°C. The substrate temperature rose to about 180°C by the ECR plasma irradiation without intentionally heating by the infrared lamp. After the ECR plasma treatment, surface roughness was examined by a three-dimensional non-contact form measuring instrument (NH-3, Mitaka Kohki, Tokyo, Japan)

Calcification experiment was conducted by immersing the TiO_2 films deposited on the Ti substrates in phosphate-buffered saline at room temperature. The solution used was supersaturated slightly with respect to dicalcium phosphate dihydrate (DCPD) but supersaturated sufficiently with respect to both OCP and HA, respectively, which potentially promotes the nucleation and calcification of these calcium phosphate crystals [4]. After the calcium phosphate was deposited on the substrate, the substrate was washed with pure water, dried in an oven and examined by X-ray diffraction (XRD: Mini Flex; Rigaku Electrical, Tokyo, Japan) and MARUTO scanning electron microscope (e-SEM, Tokyo, Japan) to identify the crystalline phase of the deposits.

3 Results and discussion

Figure 2 shows surface photographs of TiO_2 films on Ti substrates after calcification of pure Ti and oxidized at RT to 600°C. Substrate surface was covered with a small amount of calcium phosphate precipitates in the substrate oxidized at RT to 300°C.

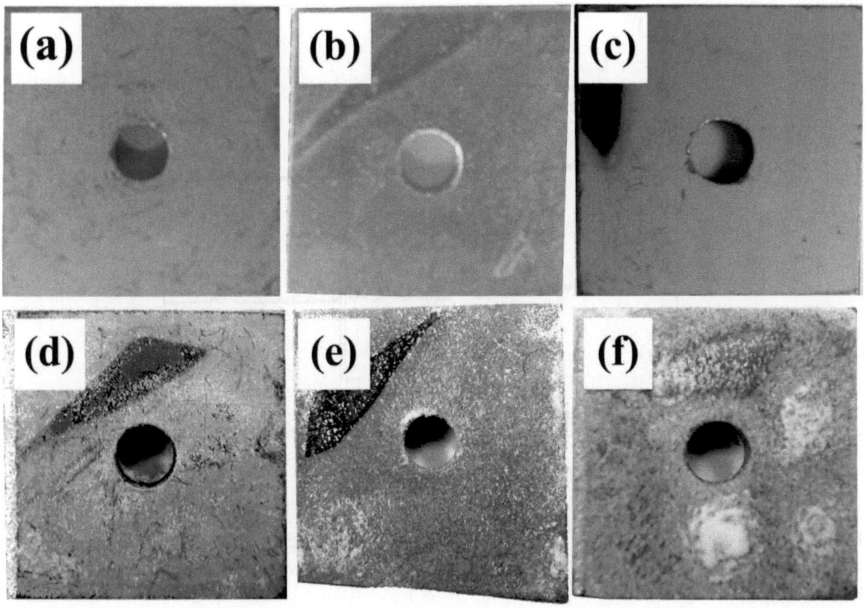

Fig. 2. Surface photographs of the TiO_2 films after calcification. [Pure Ti (**a**), and oxidized at RT (**b**), 300°C (**c**), 400°C (**d**), 500°C (**e**) and 600°C (**f**)]. (Reproduced from Masumoto et al. [5], with permission from Trans Tech Publications)

Many precipitates of the calcium phosphate were observed in the substrate oxidized over 400°C. The quantity of precipitates increased with increasing substrate-oxidized temperature.

Figure 3 shows the crystalline phase of TiO_2 films after calcification oxidized at RT, 400°C, 600°C. Rutile-type TiO_2 films were obtained by ECR plasma oxidation. OCP and DCPD were identified in the substrate oxidized over 400°C. The intensity of the OCP and DCPD peaks increased with increasing substrate-oxidized temperature.

Table 1 shows surface roughness of TiO_2 films oxidized at RT, 300°C, 450°C, 600°C. Roughness of TiO_2 surface changed depending on the substrate temperature. The surface roughness increased with increasing the oxidation temperature, and reached the maximum at 600°C.

The surface oxidation on titanium substrate by ECR plasma modifies the surface characteristics, which enhance OCP precipitation. The results suggest that surface structure of TiO_2 film may be a factor to control calcification ability. OCP has been shown to be rapidly replaced with new bone rather than a biodegradable β-tricalcium phosphate (β-TCP) ceramic [6]. This is not inconsistent with the order of solubility of OCP and β-TCP in an in vitro physiological environment [7]. The biodegradability was coupled with appearance of tartrate-resistant acid phosphatase (TRAP)-positive osteoclast-like cells [8]. It seems likely that calcification ability to form OCP on the substrate is favorable from the point of view of fixation in the

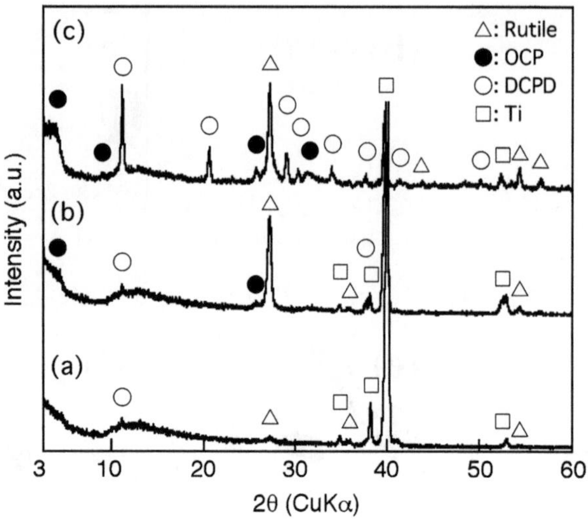

Fig. 3. X-ray diffraction (XRD) patterns of TiO_2 films after calcification oxidized at RT (**a**), 400°C (**b**), 600°C (**c**). (Reproduced from Masumoto et al. [5], with permission from Trans Tech Publications)

Table 1. Surface roughness (Ra (μm)) of TiO_2 films

	Substrate temperature			
Oxidation time	RT	300°C	450°C	600°C
30 min	0.178	0.198	0.243	0.384
60 min	0.164	0.173	0.212	1.314

Ti implant by newly formed bone. Further experiment is under way to establish a linkage between the surface structure and the calcification ability.

4 Conclusion

The surface conditions, obtained by various ECR plasma oxidation conditions, control the induction of calcium phosphate nucleation and the growth. Estimation of osteoconductivity is under way to establish the linkage between ECR plasma condition and biocompatibility of the Ti substrate covered by OCP precipitate.

Acknowledgments. The ECR treatment was performed using a multi-controlled chemical vapor deposition apparatus at Advanced Research Center of Metallic Glasses, Institute for Materials Research, Tohoku University. This work was partly supported by Ministry of Education, Culture, Sports, Science, and Technology of Japan: Grants-in-Aid (No. 16656197, 17076001, 18659567).

References

1. Brown W, Smith J, Frazier A, et al (1962) Crystallographic and chemical relations between octacalcium phosphate and hydroxyapatite. Nature 196:1050–1055
2. Suzuki O, Kamakura S, Katagiri T, et al (2006) Bone formation enhanced by implanted octacalcium phosphate involving conversion into Ca-deficient hydroxyapatite. Biomaterials 27:2671–2681
3. Masumoto H, Goto T, Masuda Y, et al (1991) Preparation of $Bi_4Ti_3O_{12}$ films on a single-crystal sapphire substrate with electron cyclotron resonance plasma sputtering. Appl Phys Lett 58:243
4. Honda Y, Kamakura S, Sasaki K, et al (2007) Formation of bone-like apatite enhanced by hydrolysis of octacalcium phosphate crystals deposited in collagen matrix. J Biomed Mater Res B Appl Biomater 80:281–289
5. Masumoto H, Goto T, Honda Y (2007) Preparation of titania films on implant titanium by electron cyclotron resonance plasma oxidation. Key Eng Mater 330–332:565–568
6. Kamakura S, Sasano Y, Shimizu T, et al (2002) Implanted octacalcium phosphate is more resorbable than β-tricalcium phosphate and hydoxyapatite. J Biomed Mater Res 59:29–34
7. Brown WE, Mathew M, Tung MS (1981) Crystal chemistry of octacalcium phosphate. Prog Cryst Growth Charact 4:59–87
8. Imaizumi H, Sakurai M, Kashimoto O, et al (2006) Comparative study on osteoconductivity by synthetic octacalcium phosphate and sintered hydroxyapatite in rabbit bone marrow. Calcif Tissue Int 78:45–54

Interface Oral Health Science 2007: 323–328

Biomaterials based on mineralised collagen—an artificial extracellular bone matrix

Michael Gelinsky[1]*, Anne Bernhardt[1], Marlen Eckert[1], Thomas Hanke[1], Ulla König[1], Anja Lode[1], Antje Reinstorf[1], Corina Vater[1], Anja Walther[1], Atsuro Yokoyama[2], and Fumio Watari[2]

[1]*Max Bergmann Center of Biomaterials, Institute of Materials Science, Technical University Dresden, 01069 Dresden, Germany;* [2]*Biomedical, Dental Materials and Engineering, Graduate School of Dental Medicine, Hokkaido University, Sapporo 060-8586, Japan*

*gelinsky@tmfs.mpgfk.tu-dresden.de

Abstract. Extracellular matrix (ECM) of bone tissue consists of a highly organised nanocomposite made of fibrillar collagen type I and calcium phosphate mineral phase hydroxyapatite. We have developed a process to synthesise a material, mimicking bone ECM, and produced several biomaterials out of this mineralised collagen, suitable for use in oral medicine and maxillofacial as well as in general surgery. Synthesis of the nanocomposite, development of the different types of scaffolds, some of their properties, and possible applications are discussed.

Key words. scaffold, collagen, hydroxyapatite, biomimetic, nanocomposite

1 Introduction

Bone biomineralisation is a complex and multi-step process, starting with collagen type I biosynthesis, followed by precipitation of hydroxyapatite nanocrystals. Calcium phosphate crystallisation is induced by expression of the enzyme alkaline phosphatase (ALP) and controlled by several non-collagenous proteins such as osteocalcin and osteopontin, secreted by osteoblasts. The resulting mineralised extracellular bone matrix can be interpreted as a highly organised nanocomposite material with not only unique mechanical, but also biological properties [1]. In our lab we have been working on synthetically mineralised collagen as biomimetic implant material, scaffold for bone tissue engineering, and in vitro model for bone biomineralisation and remodelling for several years. In this study, a review of the biomaterials based on mineralised collagen and those suitable for use in oral medicine and maxillofacial as well as in general surgery is given. This includes membrane-like materials for covering of bone defects as well as non-mineralised collagen membranes for tissue engineering of oral mucosa, porous three-dimensional scaffolds, and biphasic implant materials for the therapy of osteochondral defects and also resorbable calcium phosphate bone cements, functionalised with mineralised collagen as fibre reinforcement.

2 Materials and methods

Preparation of mineralised collagen

Preparation of mineralised collagen fibrils was described in detail elsewhere [2]. Briefly, a collagen type I solution in hydrochloric acid is combined with calcium and sodium chloride, TRIS- and phosphate buffer (final pH = 7.0). Warming up the mixture to 37°C starts the fibril reassembly process which occurs coevally with precipitation of nanocrystalline hydroxyapatite (HAP). Here, the collagen acts as a template for mineral deposition. Finally, a homogenous nanocomposite consisting of about 30% collagen and 70% mineral is formed which can be isolated by centrifugation as a soft, wax-like material.

Biomaterial development

Using the mineralised collagen as a basic material, several types of scaffolds have been developed. Applying a vacuum filtration process, followed by chemical cross-linking with *N*-(3-dimethylaminopropyl)-*N*′-ethylcarbodiimide (EDC), a flat, membrane-like material ("tape") was achieved [3, 4]. Treatment of these mineralised tapes with acidic TRIS buffer led to dissolution of the HAP nanocrystals and therefore to the formation of a pure collagen membrane, suitable for tissue engineering of oral mucosa. By freeze–drying mineralised collagen suspensions, followed by EDC crosslinking, porous 3D scaffolds were obtained [5, 6]. By combining these with a layer of a non-mineralised collagen hyaluronic acid composite, joint freeze–drying and crosslinking, biphasic, but monolithic scaffolds for therapy of osteochondral defects were developed [7]. The modification of a resorbable calcium phosphate bone cement with mineralised collagen fibrils was described in detail in previous studies [8, 9].

Cell culture and animal experiments

Cell seeding and analysis of the cell matrix constructs were carried out under the same conditions as described elsewhere [4]. Briefly, bone marrow derived human mesenchymal stromal cells (hBMSC) were seeded on membranes of mineralised as well as demineralised collagen. For this experiment, collagen, isolated from calf skin (Collaplex 1.0, GfN, Waldmichelbach, Germany) was used. Adhesion of cells was investigated by fluorescence microscopy. In addition, proliferation and osteogenic differentiation of hBMSC, growing on mineralised collagen membranes (collagen isolated from bovine tendon, kindly provided by Syntacoll, Saal/Donau, Germany), was studied. Cells were cultivated in DMEM with 10% foetal calf serum (FCS) without (–OS) and with osteogenic supplements (dexamethasone, β-glycerophosphate and ascorbic acid 2-phosphate; +OS). Cell number was determined by measurement of lactate dehydrogenase (LDH) activity, and osteogenic differentiation was monitored by ALP activity quantification. Porous 3D scaffolds made of mineralised collagen were tested in an animal model. The material was implanted in a defect made in rat femur. Details of the procedure were published by Yokoyama et al. [6].

3 Results

Synchronous collagen fibril reassembly and mineralisation with nanocrystalline HAP led to a homogenous composite material which mimics extracellular matrix (ECM) of healthy bone tissue [2]. Using these mineralised collagen fibrils, several different types of scaffolds have been developed.

Membranes made of mineralised or demineralised collagen

Densification of resuspended mineralised collagen by means of vacuum filtration leads to a flat, membrane-like material ("tape") [3]. After stabilisation by chemical crosslinking of the collagen with carbodiimide derivative EDC, the tapes can be used as a model substrate for osteogenic differentiation of hBMSC [4] and for in vitro studies on bone matrix remodelling by co-culturing osteoblasts and osteoclasts [10]. A scanning electron microscopy (SEM) image of the microstructure of the tape is given in Fig. 1a, showing the close interaction between the reconstituted collagen fibrils and the nanocrystalline HAP phase and the microporosity of the material.

Proliferation of hBMSC seeded on tapes was investigated by measurement of LDH activity after cell lysis at different time points of culture. An increase in the number of living cells could be observed in the presence (+OS) as well as in the absence (–OS) of osteogenic supplements (Fig. 2a). However, "–OS" cells showed higher proliferation rates in comparison with "+OS" cells. Thus, a more than sevenfold increase was detected for non-induced cells over a period of 21 days compared to a fourfold increase of osteogenically induced cells. To analyse the osteogenic differentiation of hBMSC, ALP activity in the same cell lysates were determined and related to the cell number. Specific ALP activity of the "+OS" cells

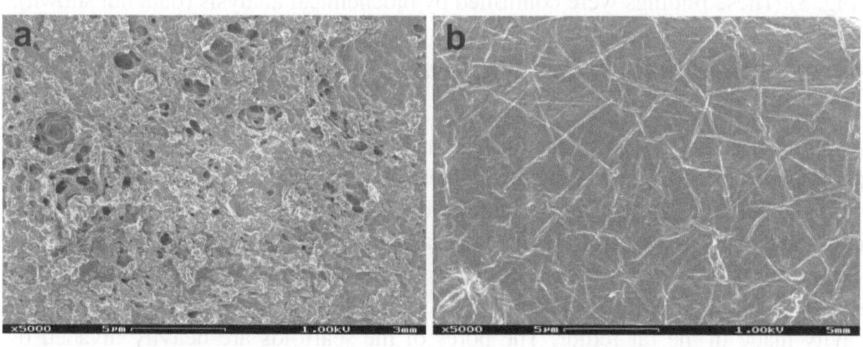

Fig. 1. SEM micrographs of the surface of **a** mineralised collagen membrane ("tape") and **b** demineralised collagen membrane. ×5,000

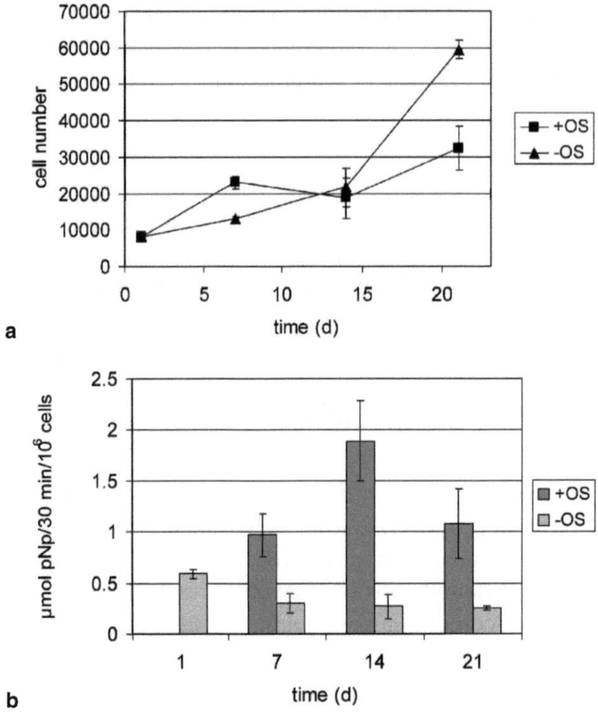

Fig. 2. Comparison between cell culture without ("–OS") and with osteogenic supplements ("+OS"). **a** Proliferation of hBMSC, and **b** specific activity of alkaline phosphatase (ALP) of hBMSC, growing on membranes of mineralised collagen

raised over the cultivation period of 21 days with a maximum on day 14 (Fig. 2b). In contrast, ALP activity of non-induced cells was not increased.

To utilise the tapes for non-mineralised tissues like oral mucosa too, they were demineralised by storage in acidic 0.1 M TRIS-buffer (pH 2) for 2 days. After dissolution of the mineral phase the surface of the tape is smooth and the micropores had disappeared (Fig. 1b). Both materials support hBMSC attachment: there were no obvious differences between mineralised and demineralised collagen. At early stages of culture many cells had already attached to both types of membrane and showed initial spreading. After 24 h most of the cells were attached and exhibited the characteristic fibroblast-like morphology known for mesenchymal stromal cells (Fig. 3). These findings were confirmed by biochemical analysis (data not shown).

Porous three-dimensional scaffolds

Applying freeze–drying and chemical crosslinking, porous 3D scaffolds can be prepared out of mineralised collagen which exhibit interconnecting pores with diameters of about 200 µm [5]. Due to their elastic properties in the wet state the material is suitable to act as a scaffold for cell culturing under cyclic mechanical stimulation. The material already has been tested successfully in an animal model [6]. Figure 4 shows a histological image taken 2 weeks after implantation of the material in a cavity made in the rat femur. The pores of the scaffolds are heavily invaded by osteoblasts which produce new mineralised matrix, deposited directly on the inner

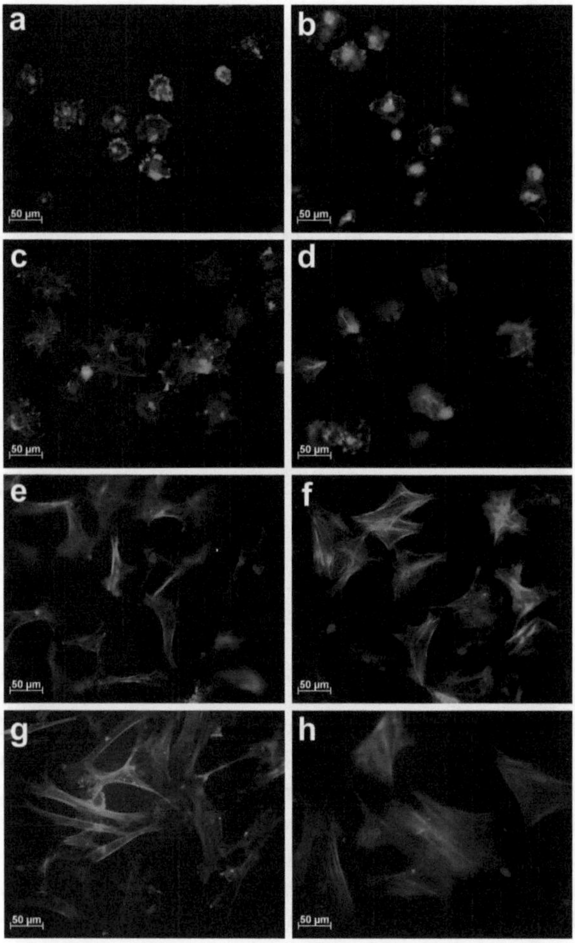

Fig. 3. Adhesion of hBMSC, 30 min, 1, 4, and 24 h after seeding. **a**, **c**, **e**, **g** (*left row*) mineralised collagen membrane, **b**, **d**, **f**, **h** (*right row*) demineralised collagen membrane. **a**, **b** 30 min, **c**, **d**: 1 h, **e**, **f**: 4 h, **g**, **h**: 24 h after seeding. Fluorescence micrographs; samples were fixed and stained with Alexa Fluor488 phalloidin for cytoskeleton and DAPI for nuclei; *scale bars* = 50 µm

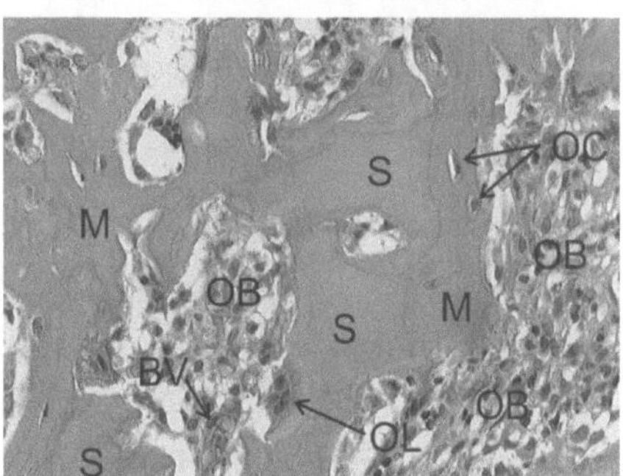

Fig. 4. Histological image of a porous 3D scaffold made of mineralised collagen, 2 weeks after implantation in a bone cavity of rat femur. HE stain, ×100. *S* scaffold, *M* newly deposited bone matrix, *OB* osteoblasts, *OC* osteocytes, *OL* osteoclast, *BV* blood vessel

surface of the scaffold. In the newly deposited matrix some osteocytes are embedded. Remaining scaffold material is resorbed by osteoclasts (proven by TRAP staining; images not shown). In the pores already blood capillaries can be found—showing that the interconnecting pore system is suitable for fast vascularisation.

Combining the mineralised collagen in the liquid (suspended) state with a (non-mineralised) collagen hyaluronic acid composite, followed by joint freeze–drying and chemical crosslinking, biphasic, but monolithic scaffolds for the therapy of osteochondral defects could be achieved [7].

Calcium phosphate bone cements, functionalised with mineralised collagen

Addition of mineralised collagen fibrils to the solid precursor phase of a hydraulic calcium phosphate bone cement leads to a cement paste with better cohesion and an improved performance in vitro [8, 9] and in vivo [11]. Collagen acts as a fibre reinforcement of the brittle cement phase and improves the fracture toughness after hardening of the material. Presently, the cement is tested in a critical size defect model in the lower jaw bone of mini pig.

4 Discussion

Applying different methods for scaffold fabrication, we were able to develop several biomaterials out of mineralised collagen fibrils—a material which mimics ECM of healthy bone tissue. A review on other materials based on collagen HAP composites was recently given by Wahl et al. [12]. The scaffolds were proven to be biocompatible and showed good results in cell culture as well as in animal experiments. By co-culturing osteoblasts and osteoclasts on artificial ECM of bone in vitro models for bone remodelling can be established [10].

References

1. Rho JY, Kuhn-Spearing L, Zioupos P (1998) Med Eng Phys 20:92–102
2. Bradt JH, Mertig M, Teresiak A, et al (1999) Chem Mater 11:2694–2701
3. Burth R, Gelinsky M, Pompe W (1999) Tech Text 8:20–21
4. Bernhardt A, Lode A, Boxberger S, et al (2007) J Mater Sci Mater Med (in press). doi:10.1007/s10856-006-0059-0
5. Gelinsky M, König U, Sewing A, et al (2004) (in German). Materialwiss Werkstofftech 35:229–233
6. Yokoyama A, Gelinsky M, Kawasaki T, et al (2005) J Biomed Mater Res B Appl Biomater 75B:464–472
7. Gelinsky M, Eckert M, Despang F (2007) Int J Mater Res 98:749–755
8. Knepper-Nicolai B, Reinstorf A, Hofinger I, et al (2002) Biomol Eng 19:227–231
9. Hempel U, Poppe M, Reinstorf A, et al (2004) J Biomed Mater Res B Appl Biomater 71B:130–143
10. Domaschke H, Gelinsky M, Burmeister B, et al (2006) Tissue Eng 12:949–958
11. Schneiders W, Reinstorf A, Pompe W, et al (2007) Bone 40:1048–1059
12. Wahl D, Czernuszka JT (2006) Eur Cell Mater 11:43–56

Interface Oral Health Science 2007: 329–334

Osteoclast-mediated bone remodeling in guided bone regeneration with sintered bone grafts

Yoshinaka Shimizu[1]*, Keisuke Okayama[1,2], Mitsuhiro Kano[1], Hiroyasu Kanetaka[3], and Masayoshi Kikuchi[1]
[1]*Division of Oral and Craniofacial Anatomy, Department of Oral Function and Morphology;*
[2]*Division of Advanced Prosthodontics, Department of Oral Function and Morphology;*
[3]*Division of Orthodontics and Dentofacial Orthopedics, Department of Oral Health and Development Science, Tohoku University Graduate School of Dentistry, Sendai 980-8575; Japan*
*shimizu@anat.dent.tohoku.ac.jp

Abstract. This study examined the effects of graft material on osteoclast-mediated bone remodeling in guided bone regeneration (GBR). Sintered rabbit bone particles were used as the graft material. A polytetrafluoroethylene membrane was molded into a dome and anchored to the frontal bone in 16 male rabbits. The space under the membrane was filled with a blood clot (control group) or sintered bone particles (experimental group). Animals were killed 2, 4, 8, and 12 weeks after operation. The resected samples were fixed in 4% paraformaldehyde and demineralized. Paraffin-embedded histological sections were stained with hematoxylin and eosin and underwent a histochemical assay to determine tartrate-resistant acid phosphate (TRAP) activity. The proportions of newly formed bone and graft particles and the numbers and densities of osteoclasts and multinucleated giant cells (MGCs) were calculated. The proportion of newly formed bone increased up to 4 weeks in both the control and experimental groups. Subsequently, the proportion decreased in the control group, but did not change significantly in the experimental group. Osteoclast density on newly formed bone was higher in the control group than that in the experimental group. We conclude that the use of a sintered bone graft inhibits bone resorption by osteoclasts.

Key words. guided bone regeneration, sintered bone, bone remodeling, osteoclast, histomorphometry

Introduction

Guided bone regeneration (GBR) has been performed at implant-recipient sites that lack sufficient bone to increase the volume and quality of bone, and thereby enhance implant stability and long-term outcomes. Various devices and surgical techniques have been developed to augment bone. Good clinical outcomes require an understanding of the biologic mechanisms and temporal dynamics of newly formed bone in GBR.

In GBR, bone is newly formed over time in a secluded space containing a blood clot [1]. Initially, woven bone is rapidly formed. This bone is then immediately remodeled to a mature bone with lamellar structure. Although the maintenance of newly formed bone is a prerequisite to creating a biomechanical and clinical environment conducive to long-term implant stability, mature bone gradually decreases in response to various extrinsic factors [2], subsequently entering the marrow spaces [3]. Previous studies have suggested important differences in the pattern of new bone formation in a secluded space containing graft materials [4]. Such materials apparently contribute to the prolonged maintenance of newly formed bone during GBR.

Osteoclasts have important roles in the maintenance and resorption of newly formed bone during bone remodeling. The effects of graft materials on osteoclasts remain unclear. We therefore histomorphometrically investigated the effects of graft materials on the recruitment or localization of osteoclasts and multinucleated giant cells (MGCs) in a model of GBR.

Experimental model of GBR

Sixteen male Japanese white rabbits weighing about 3 kg each were used in this study. General anesthesia was induced by injecting pentobarbital sodium salt (Tokyo Kasei, Tokyo, Japan) at a dose of 0.5 mg/kg body weight into an ear vein. In addition, about 2 ml of local anesthesia (1% lidocaine, Astra Zeneca, Osaka, Japan) was injected subcutaneously into the operation site. The frontal bone was exposed via a midsagittal incision through the skin and periosteum. Cortical bone defects (3×15 mm^2) were made in the external cortical plate of the right and left frontal bones. An expanded polytetrafluoroethylene (e-PTFE) membrane reinforced with a thin titanium mesh (Gore-Tex, WL Gore, AZ, USA) was molded into a dome ($10 \times 5 \times 5$ mm^3) and filled with venous blood from the rabbit's ear in the control group, and sintered bone particles in the experimental group (Fig. 1). Two membranes (control and experimental groups) were placed over the defects and anchored to the bone surface by means of four mini-screws (Ti-SIS pins, SIS-System Trade, Klagenfurt, Austria). The skin was sutured over the membranes.

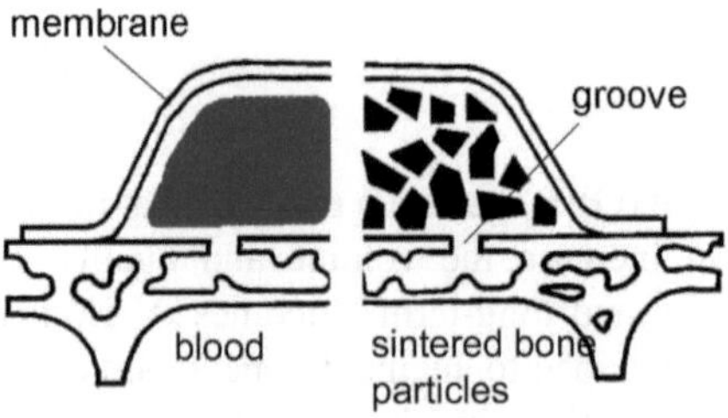

Fig. 1. Schema of the experimental site (frontal section)

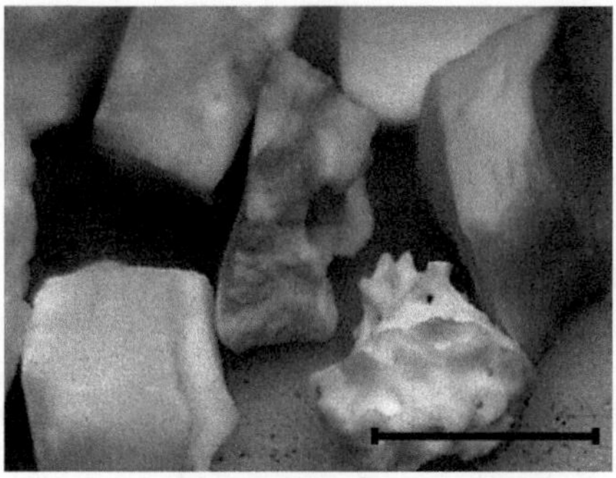

Fig. 2. Stereomicroscopic image of pulverized and separated sintered bone particles. *Bar* = 500 μm

Sintered bone particles

The sintered bone particles used in this study were derived from natural rabbit bone. This type of sintered bone is classified as a xenogenic material, with biocompatible and osteoconductive properties [5]. The sintered bone was prepared as follows: boiled rabbit cortical bone was immersed in a mixture of 1% NaOH and 1% H_2O_2 (1:1) to remove proteins on the bone surface for 1 h. The mixture was then neutralized with 1 N HCl. The bone was placed in an electric furnace and sintered at 600°C for 3–5 h and 1,100°C for 3.5 h. The sintered cortical bone was pulverized in a bone mill and sorted into particles of 300–500 μm by means of a standard sieve. The sintered bone particles were white and irregularly shaped on examination with a stereomicroscope (Fig. 2). The calcium and phosphate contents of the sintered bone particles were Ca 42.5%, P 18.2%, and Ca/P ratio 2.34.

Histomorphometry

The rabbits were anesthetized and killed 2, 4, 8, and 12 weeks after operation. For histological examination, the resected samples were demineralized in 10% EDTA and embedded in paraffin. Histological sections were sliced and stained with hematoxylin and eosin and underwent a histochemical assay to determine tartrate-resistant acid phosphate (TRAP) activity. After histological examination by light microscopy, the digitized images were photographed, and the following variables were measured: bone and graft particles volumes, the proportions of bone and graft particles (The bone proportion in the experimental group was calculated as a proportion in the space remaining after exclusion of the graft particles.), and the numbers and the densities of osteoclasts and MGCs. Totally 58 sites of right and left frontal bones in 18 rabbits were used for statistical analyses (two-way analysis of variance and Tukey's test).

Bone formation and sintered bone particles in GBR

Initially, woven bone, lined by a dense layer of cuboidal osteoblasts, proliferated around the groove in the control and experimental groups. The space under the membrane was filled with newly formed bone, fibrous tissue, and adipose tissue at 12 weeks in the control group. Mature bone with a lamellar structure was found from 4 weeks onward. The sintered bone particles were surrounded by newly formed bone, fibrous tissue, and adipose tissue, without inflammatory cell infiltration. The height and extent of newly formed bone were greater in the control group than in the experimental group throughout the experiment (Fig. 3).

Bone volume in the control group increased rapidly for up to 4 weeks ($P < 0.01$) and then decreased significantly ($P < 0.01$). In the experimental group, bone volume increased gradually up to 4 weeks and then did not change significantly. Bone volume in the experimental group was larger than that in the control group from 8 weeks onward (Fig. 4a). The proportion of newly formed bone in the experimental group did not differ from that in the control group at 4 weeks. At 8 and 12 weeks, the proportion of newly formed bone in the experimental group was significantly larger than that in the control group ($P < 0.01$) (Fig. 4b). The proportion of graft particles did not change significantly (Fig. 4c).

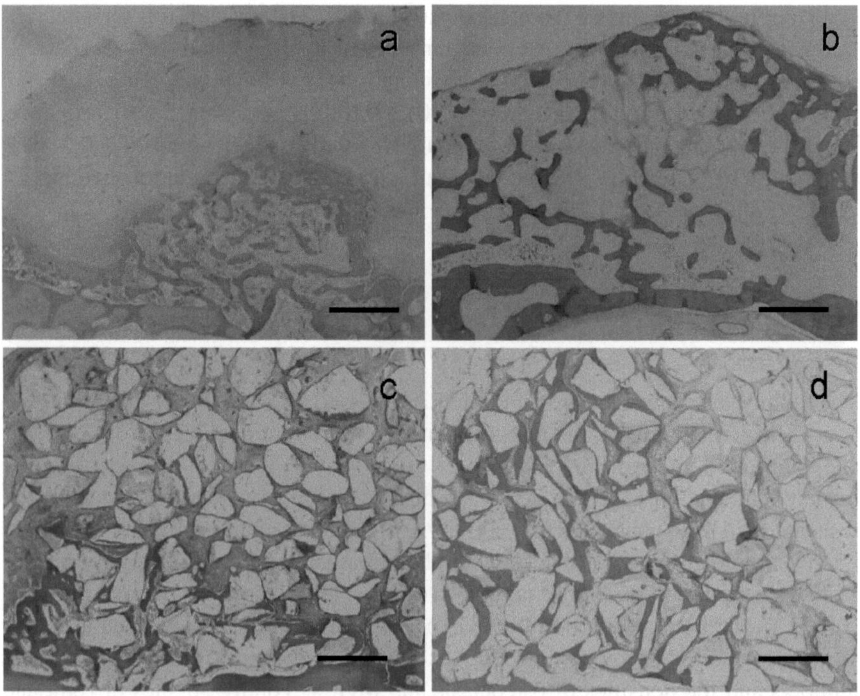

Fig. 3. Histological findings at low magnification (hematoxylin and eosin stain). **a** At 4 weeks in the control group. **b** At 12 weeks in the control group. **c** At 4 weeks in the experimental group. **d** At 12 weeks in the experimental group. *Bars* = 500 μm

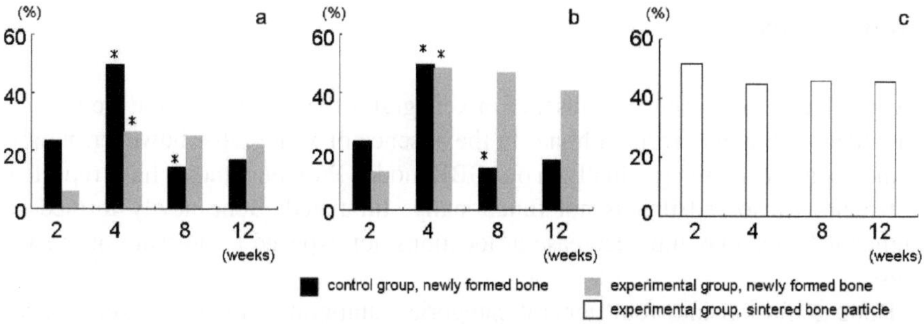

Fig. 4. **a** The volume of newly formed bone. **b** The proportion of newly formed bone in the space remaining after exclusion of the sintered bone particles. **c** The proportion of sintered bone particles. $*P < 0.05$

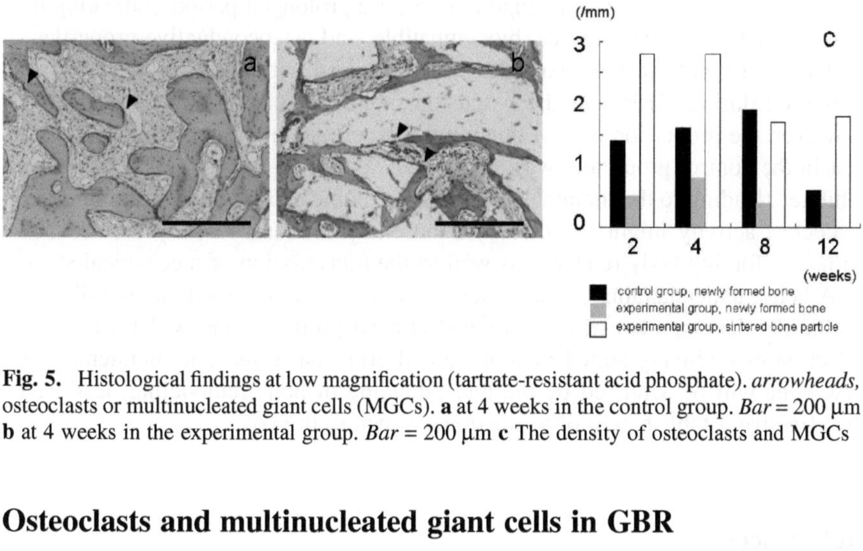

Fig. 5. Histological findings at low magnification (tartrate-resistant acid phosphate). *arrowheads,* osteoclasts or multinucleated giant cells (MGCs). **a** at 4 weeks in the control group. *Bar* = 200 μm **b** at 4 weeks in the experimental group. *Bar* = 200 μm **c** The density of osteoclasts and MGCs

Osteoclasts and multinucleated giant cells in GBR

In both the control and experimental groups, TRAP-positive osteoclasts were found on newly formed bone from 2 weeks onward. At 12 weeks, a few pale osteoclasts were sporadically detected on newly formed bone trabeculae. In the experimental group, TRAP-positive MGCs were detected on the surfaces of sintered bone particles from 2 to 12 weeks (Fig. 5a, b).

The numbers of osteoclasts and MGCs were greatest at 4 weeks in the control and experimental groups. Throughout the experimental period, the number of osteoclasts in the experimental group was lesser than that in the control group. The densities of osteoclasts and MGCs were highest at 8 weeks in the control group and at 4 weeks in the experimental group (Fig. 5c). The densities of osteoclasts and MGCs were lower in the experimental group than in the control group at 2, 4 and 12 weeks. The number of TRAP-positive MGCs increased at 2 weeks and decreased at 8 weeks. The density of TRAP-positive MGCs showed no significant change at 4 weeks and decreased at 8 weeks (Fig. 5c).

Conclusions

Irrespective of the presence or absence of xenografts, the space under the membrane was filled with newly formed bone. In the absence of xenografts, however, newly formed bone decreased gradually in our GBR model. Previous studies have reported that bone is resorbed if it is not functionally stimulated. Bone newly formed at augmented sites may thus decrease at locations not exposed to pressure or tensile stress.

Bone grafts fall into four general categories: autografts, allografts, xenografts, and alloplasts. Sintered bone is classified as a xenograft and considered biocompatible and osteoconductive. Numerous studies of human bone augmentation have shown that bovine-derived particles depleted of organic components are associated with successful bone regeneration. The sintered bone particles had good bone-to-graft contact and showed no change in size for a prolonged period, indicating that this material has nonabsorbable, biocompatible, and osteoconductive properties.

The use of sintered bone particles was associated with no disturbance of bone formation during GBR, but the extension of newly formed bone was delayed. However, the proportion of new bone in the experimental group was equivalent to that in the control group and was maintained. The recruitment of osteoclasts was inhibited, leading to the maintenance of bone volume and proportion. Inhibition of osteoclast activity might be related to the encapsulation of newly formed bone without a foreign body reaction, as well as the transmission of mechanical stress.

A better understanding of the biologic characteristics of new bone in GBR will enable the development of more refined clinical protocols and will facilitate the selection of optimally suited membrane and graft materials. The maintenance of new bone will be achieved by the understanding of bone remodeling, leading to the effective establishment of a functional structure in vivo.

References

1. Nishimura I, Shimizu Y, Ooya K (2004) Effects of cortical bone perforation on experimental guided bone regeneration. Clin Oral Implants Res 15:293–300
2. Asai S, Shimizu Y, Ooya K (2002) Maxillary sinus augmentation model in rabbits: effect of occluded nasal ostium on new bone formation. Clin Oral Implants Res 13:405–409
3. Xu H, Shimizu Y, Onodera K et al (2005) Long-term outcome of augmentation of the maxillary sinus using deproteinized bone particles experimental study in rabbits. Br J Oral Maxillofac Surg 43:40–45
4. Okazaki K, Shimizu Y, Xu H et al (2005) Blood-filled spaces with and without deproteinized bone grafts in guided bone regeneration. A histomorphometric study of the rabbit skull using non-resorbable membrane. Clin Oral Implants Res 16:236–243
5. Matsuda M, Kita S, Takekawa M et al (1995) Scanning electron and light microscopic observations on the healing process after sintered bone implantation in rats. Histol Histopathol 10:673–679

Interface Oral Health Science 2007: 335–340

Expression of bone matrix proteins and matrix metalloproteinases during repair of rat calvarial bone defects

Tomoko Itagaki[1,2], Takahiro Honma[1,2], Megumi Nakamura[2], Ichiro Takahashi[3], Seishi Echigo[1], and Yasuyuki Sasano[2]*
[1]*Division of Oral Surgery;* [2]*Division of Craniofacial Development and Regeneration;* [3]*Division of Orthodontics and Dentofacial Orthopedics, Tohoku University Graduate School of Dentistry, Sendai 980-8575, Japan*
*sasano@anat.dent.tohoku.ac.jp

Abstract. Little information has been available on repair of bone defects, whereas numerous studies have been reported on that of bone fractures. The present study was designed to investigate the repairing process of bone defects focusing on the bone healing rate and the cellular activity of extracellular matrix production and degradation using the standardized rat experimental model. Our results indicated that osteoblasts and osteocytes decline bone formation and extracellular matrices (ECM) remodeling resulting in bone healing being ceased within 24 weeks regardless of completion of bone defect repair.

Key words. bone healing, MMP, ECM, real-time PCR, in situ hybridization

1 Introduction

Repair of bone defects depends on a size of the defect, i.e., a bone defect larger than a certain size (a critical size) does not heal completely [1, 2]. There have been few reports on repair of bone defects, whereas numerous studies have investigated that of bone fractures [3–5]. It has not been known how and why bone formation ceases in the course of healing of the large bone defect. Bone formation during development involves extensive remodeling of extracellular matrices (ECM), which is achieved by both production and degradation of ECM [6–9]. Our previous study suggested that osteoblasts and osteocytes secrete matrix metalloproteinases (MMPs) 2, 8, and 13 and play a role in ECM degradation as well as ECM production during bone development [7, 10]. We investigated the process of bone healing in the critical size defect focusing on the bone healing rate and the cellular activity of ECM production and degradation using the standardized rat calvarial bone defect model in the present study.

2 Morphometric analysis of repair of the standardized rat calvarial bone defect

Twelve-week-old male Wistar rats were used. A full-thickness standardized trephine defect, 8.8 mm in diameter, was made in the rat parietal bone under anesthesia. The rats were fixed by perfusion through the aorta in weeks 1, 2, 3, 4, 6, 12, 18, 24, and 36. The resected calvaria were radiographed for morphometric analysis of bone matrix apposition per week as previously described [11]. Then the specimens were processed for in situ hybridization (ISH) as described below.

Bone healing proceeded gradually and almost ceased in week 24 with about 25% of the defect unrepaired (Figs. 1, 2a). The bone healing rate, i.e., the rate of bone matrix apposition per week was the largest in the fourth week and decreased thereafter (Fig. 2b). Little bone was apposed in the 36th week leaving the defect unrepaired [11].

The bone healing or the bone matrix apposition may cease within 24 weeks in the calvarial bone defect, regardless of completion of the defect repair.

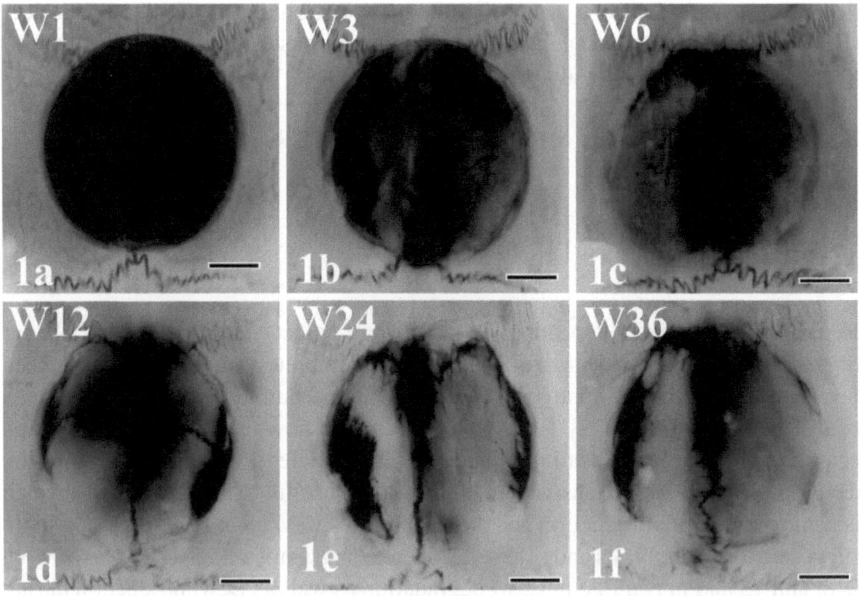

Fig. 1. Radiographs of bone healing in 8.8 mm defects in weeks (*W*) 1 (**a**), 3 (**b**), 6 (**c**), 12 (**d**), 24 (**e**), 36 (**f**). The defect is not completely repaired with bone in week 36. *Scale bars* = 2.2 mm. (From [11] with modification, with permission)

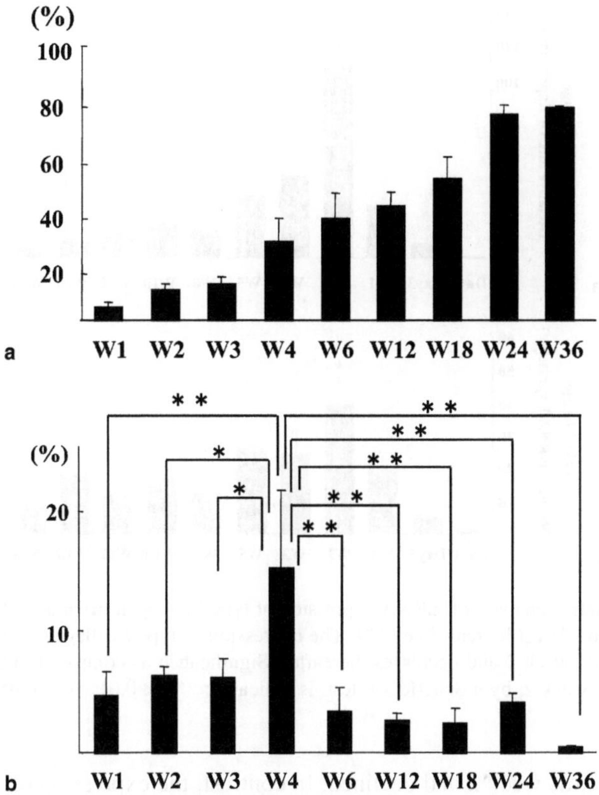

Fig. 2. Quantitative radiographic analysis for the total bone healing (**a**) and the bone healing rate (**b**) in 8.8 mm defects. Bone healing proceeds gradually and almost ceases in week 24 with about 25% of the defect unrepaired (**a**). The bone matrix apposition per week is the largest in the fourth week and decreases thereafter (**b**). Little bone formation is identified in the 36th week. Significance as compared to week 4, by a Scheffé's *F* test, is indicated: $^{**}P < 0.01$, $^{*}P < 0.05$. (From [11] with modification, with permission)

3 Quantitative analysis of mRNA expression of bone matrix proteins and MMPs during repair of the bone defect

RNA was extracted from tissue that filled the original bone defect in days 1, 3 and weeks 1, 2, 3, 5, 8, 10, 12, 18, and 24 and processed for quantitative analysis of expression of type I collagen, osteocalcin, and MMPs 2, 8, and 13 using real-time PCR.

The expression of type I collagen and osteocalcin as well as MMP 2 increased towards week 2 and decreased thereafter. Similarly, the expression of MMP13

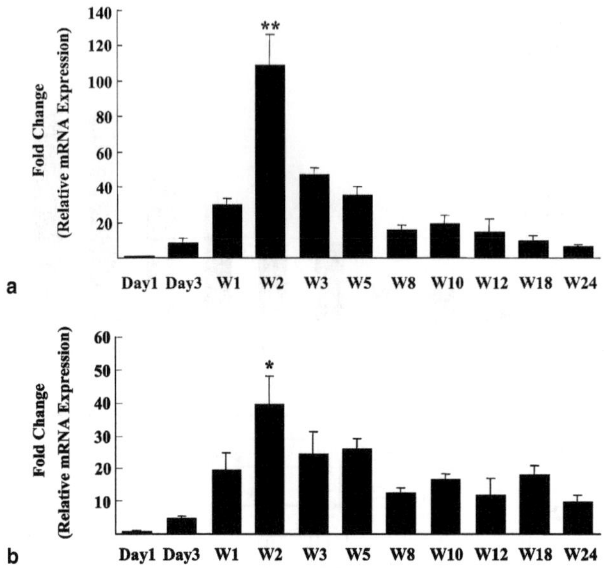

Fig. 3. Quantitative analysis of mRNA expression of type I collagen (**a**) and MMP 2 (**b**) during repair of the bone defect by real-time PCR. The expression of type I collagen as well as MMP 2 increases towards week 2 and decreases thereafter. Significance as compared to any time point before and after week 2, by a Scheffé's F test, is indicated: $^{**}P < 0.01$, $^{*}P < 0.05$

increased towards week 2 and declined. In contrast, the expression of MMP 8 was the highest in day 1 and then decreased. Expression of the bone matrix ECM proteins and MMPs was no longer identified in week 24.

The results indicate that ECM production and degradation is the most noticeable around week 2 and declines with time (Fig. 3a, b).

4 Localization of mRNA transcripts of bone matrix proteins and MMPs during repair of the bone defect

The resected calvaria were decalcified, embedded in paraffin, and then processed for in situ hybridization for type I collagen, osteocalcin, and MMPs 2, 8, and 13.

The mRNA transcripts of type I collagen and osteocalcin were localized in osteoblasts and osteocytes in week 2 (Fig. 4a). Some of those cells expressed MMPs 2, 8, and 13 (Fig. 4b). Expression of the bone matrix ECM proteins and MMPs was no longer identified in week 24.

The results demonstrate that osteoblasts and osteocytes express MMPs 2, 8, and 13 as well as type I collagen and osteocalcin and remodel bone ECM during repair of the bone defect. These osteogenic cells cease bone healing, i.e. bone ECM production and degradation, within 24 weeks in the experimental model.

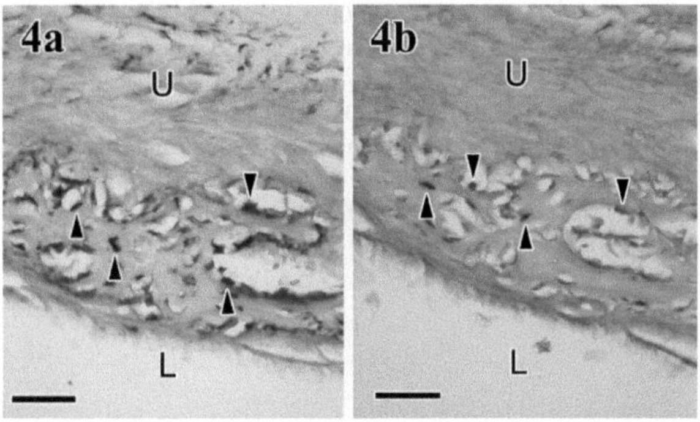

Fig. 4. Localization of mRNA transcripts of type I collagen (**a**) and MMP 2 (**b**) in week 2 demonstrated by in situ hybridization. The mRNA transcripts of type I collagen are localized in osteoblasts and osteocytes. Some of those cells also express MMP 2. *Arrowheads* ostoblasts and osteocytes. *Scale bars* = 50 μm, *U* upper surface of the parietal bone, *L* lower surface of the parietal bone

5 Conclusions

The present study suggests that osteoblasts and osteocytes cease bone formation and ECM remodeling within a limited period regardless of completion of defect repair.

References

1. Schmitz JP, Hollinger JO (1986) The critical size defect as an experimental model for craniomandibulofacial nonunions. Clin Orthop Relat Res 205:299–308
2. Schmitz JP, Schwartz Z, Hollinger JO, et al (1990) Characterization of rat calvarial nonunion defects. Acta Anat (Basel) 138:185–192
3. Precious DS, Hall BK (1994) Repair of fractured membrane bones. In: Hall BK (ed Bone, vol. 9) Differentiation and morphogenesis of bone. CRC Press, Boca Raton, pp 145–163
4. Dimitriou R, Tsiridis E, Giannoudis PV (2005) Current concepts of molecular aspects of bone healing. Injury 36:1392–1404
5. Einhorn TA (2005) The science of fracture healing. J Orthop Trauma 19(10 Suppl):S4–S6
6. Werb Z, Chin JR (1998) Extracellular matrix remodeling during morphogenesis. Ann NY Acad Sci 857:110–118
7. Sasano Y, Zhu JX, Tsubota M, et al (2002) Gene expression of MMP8 and MMP13 during embryonic development of bone and cartilage in the rat mandible and hind limb. J Histochem Cytochem 50:325–332
8. Ortega N, Behonik D, Stickens D, et al (2003) How proteases regulate bone morphogenesis. Ann N Y Acad Sci 995:109–116

9. Nakamura M, Sone S, Takahashi I, et al (2005) Expression of versican and ADAMTS1, 4, and 5 during bone development in the rat mandible and hind limb. J Histochem Cytochem 53:1553–1562
10. Maruya Y, Sasano Y, Takahashi I, et al (2003) Expression of extracellular matrix molecules, MMPs and TIMPs in alveolar bone, cementum and periodontal ligaments during rat tooth eruption. J Electron Microsc 52:593–604
11. Honma T, Itagaki T, Nakamura M, et al (2007) Bone formation in rat calvaria ceases within a limited period regardless of completion of defect repair. Oral Dis (in press)

Interface Oral Health Science 2007: 341–342

Mold filling of wedge-shaped Ti–Hf alloy castings

Hideki Sato[1]*, Masafumi Kikuchi[2], Masashi Komatsu[1], Osamu Okuno[2], and Toru Okabe[3]
[1]*Division of Operative Dentistry;* [2]*Division of Dental Biomaterials, Tohoku University Graduate School of Dentistry, Sendai 980-8575, Japan;* [3]*Department of Biomaterials Science, Baylor College of Dentistry, Texas A&M University System H. S. C., Dallas, TX 75246, USA*
*yamasa@mail.tains.tohoku.ac.jp

Abstract. An object of this study was to examine the mold filling capacity of some Ti–Hf alloys. Regardless of the edge angle, the Ti–Hf alloys behaved similar to commercially pure (CP) Ti due to their isomorphous nature of solidifying congruently over the ranges of the alloys examined.

Key words. mold filling, wedge, Titanium alloy, Hafnium

1 Introduction

In previous studies, we investigated the mechanical properties and the corrosion behavior of as-cast Ti–Hf alloys. In those studies, it was found that the strengths of the alloys increased with the Hf concentration, and the alloys with more than 20%Hf were significantly stronger than commercially pure (CP) Ti. The electrochemical evaluation indicated that there were no significant differences between these alloys and CP Ti or pure Ti (prepared from sponge used to make the Ti–Hf alloys). Because of these encouraging findings, we investigated other characteristics needed for dentistry. In this study, we evaluated the mold filling characteristics of these Ti–Hf alloys using a technique established in other studies by casting the metals into wedge-shaped cavities.

2 Materials and methods

Titanium alloys with 10–40 mass% Hf were made with sponge Ti (Toho Titanium, Japan) and sponge Hf (Tohotec, Japan) in an argon-arc melting furnace. Molds were prepared using magnesia investment (Selevest CB, Selec, Japan) in two wedge-shaped acrylic patterns with either 30° or 15° angles. Each alloy was cast in a centrifugal casting machine (Ticast Super R, Selec). The castings were cut into sections perpendicular to the angled edge to create 13 specimens; the surfaces of the sections were then photographed (50×). Mold filling was evaluated as the missing length (μm) between the angled edge of the casting and the theoretical edge (mold filling

Table 1. Mold filling index for each metal [(μm)(S.D.)]

Angle	Commercially pure	Sponge Ti	10 Hf	20 Hf	30 Hf	40 Hf
30°	118 (39)	114 (27)	132 (32)	106 (21)	117 (50)	109 (37)
15°	181 (34)	176 (65)	204 (57)	203 (75)	197 (59)	208 (39)

index). CP Ti and titanium made from sponge Ti (PT) were used as controls. The data were analyzed using the ANOVA/Student–Newman-Keuls test ($\alpha = 0.05$).

3 Results and discussion

Binary Ti–Hf alloys are in an isomorphous system in which the two metals are mutually soluble at all proportions. Thus, we thought that the mold filling capacity of our series of Ti–Hf alloys would probably be similar to that of CP Ti. No statistical differences were found in the mold filling index values for both the 15° and 30° angles of a series of Ti–Hf alloys compared to CP Ti or pure Ti (Table 1). However, the mold filling values for the Ti–Hf (particularly for the 15° specimens) were consistently higher than that for the CP Ti or pure Ti. A comparison of the present results and our previous mold filling index values for several titanium alloys showed that the means of each alloy cast at the 15° angle were always higher (worse) than those of the CP Ti or pure Ti. The reduced temperature difference between the liquidus and solidus lines of the alloys does not always explain the results.

Interface Oral Health Science 2007: 343–344

Corrosion characteristics of magnetic assemblies composing dental magnetic attachments

Yukyo Takada*, Noriko Takahashi, and Osamu Okuno
Division of Dental Biomaterials, Tohoku University Graduate School of Dentistry, 4-1 Seiryo-machi, Aoba-ku, Sendai 980-8575, Japan
*takada@mail.tains.tohoku.ac.jp

Abstract. This study electrochemically evaluated corrosion resistance of commercially available cup yoke type magnetic assemblies. Anodic polarization curves in 0.9% NaCl solution at 37°C showed that the magnetic assemblies tested in this study broke down at the range of 0.75–1.3 V because of pitting corrosion in the vicinity of the shield ring of 316L. However, the pitting potentials were significantly higher than that of 316L ($P < 0.05$). EPMA showed increase in Cr content on the surface of shield ring compared to its inside ($P < 0.05$) because the welding bead deeply covered the surface of shield ring. The laser welding resulted in increasing the pitting potentials of magnetic assemblies.

Key words. ferritic stainless steel, magnetic assembly, pitting corrosion, laser welding, corrosion resistance

1 Introduction

Recently, dental magnetic attachments have been widely used for retaining dentures in oral cavities. The retention is supplied by the attractive force between magnetic assemblies and keepers fixed in the denture and root caps, respectively. When the dental magnetic attachment works in an oral cavity, the magnetic assemblies, which are covered with ferritic and austenitic stainless steels, contact with the dental precious alloys in the corrosive environment. In this study, corrosion resistance of commercially available cup yoke type magnetic assemblies was electrochemically evaluated by their pitting potentials obtained from anodic polarization curves.

2 Materials and methods

Commercially available cup yoke type magnetic assemblies, such as Magfit DX800 (Aichi Steel, Nagoya, Japan), GIGAUS D800 (GC, Tokyo, Japan), Hyper Slim, and Hicorex Slim (NEOMAX, Tokyo, Japan), were used in this study. Stainless steels composing yokes and shield rings (SUS444, SUSXM27, SUS447J1, and SUS316L) were also used as controls. Anodic polarization curves of the magnetic assemblies and the stainless steels were measured in 0.9% NaCl solution at 37°C (n = 3). Cathodic polarization curves of a Type 4 gold alloy (PGA-2, Ishifuku, Tokyo, Japan) and a silver alloy (Castwell MC, GC, Tokyo, Japan) were also measured under the same condition. Distribution of elements in laser welding zones between the yoke and the shield ring were analyzed using an electron probe X-ray micro-analyzer (EPMA) (n = 5). The statistical analyses were performed by ANOVA (Tukey's test) at significant level of $\alpha = 0.05$.

3 Results and discussion

Although pitting potentials of the magnetic assemblies were expected to be almost equal to that of 316L composing shield rings, anodic polarization curves gave proof that the magnetic assemblies tested in this study broke down at the range of 0.75–1.3 V, which was significantly higher ($P < 0.05$) than the pitting potential of 316L. The welding zone seemed to corrode preferentially compared with the yokes because pitting corrosion occurred in the vicinity of the welding zone on the shield ring.

Quantitative analyses using EPMA indicated that Cr content in the welding zone was significantly higher ($P < 0.05$) than that in 316L because the welding bead deeply covered the surface of shield ring of 316L. Therefore, the increase in Cr content caused by laser welding possibly contributed to the pitting potential rise, and resulted in improving corrosion resistance of the magnetic assemblies.Corrosion potential and current of each stainless steel in contact with the precious alloys can be electrochemically obtained from the intersection points of the anodic and the cathodic polarization curves. The corrosion potentials of each stainless steel were maintained within passive region in contact with the precious alloys even when their surface area ratio of (stainless steel)/(precious alloy) was 1/1 or 1/10, and were also sufficiently lower than pitting potentials of each magnetic assemblies. These findings indicate that the cup yoke type magnetic assemblies composed of the ferritic stainless steel, such as 444, XM27 or 447J1, showed significantly higher pitting potential than 316L, and that they can maintain passivation and sufficient corrosion resistance even in contact with dental precious alloys used for root caps. However, the shield ring is a very narrow width of about 0.1 mm, and its corrosion resistance is slightly inferior to the yoke against the electrochemical oxidizing environment. In galvanic corrosion, therefore, dental alloys used for root caps should be selected with care of not raising a corrosion potential as much as possible.

4 Conclusion

These magnetic assemblies possibly maintain good corrosion resistance in an oral cavity because their pitting potentials were sufficiently higher than that of 316L used as biomedical stainless steel.

Interface Oral Health Science 2007: 345–346

Elastic properties of experimental titanium alloys

Masafumi Kikuchi*, Masatoshi Takahashi, and Osamu Okuno
Division of Dental Biomaterials, Tohoku University Graduate School of Dentistry, Sendai 980-8575, Japan
*kikuchi@mail.tains.tohoku.ac.jp

Abstract. The Young's moduli of experimental binary titanium alloys with Cu, Ag, Au, Zr, Nb, or Hf (up to 30 mass%) were determined. As the concentration of Cu increased, the Young's modulus monotonically increased. As the concentration of Ag, Au, or Nb increased to 20%, the Young's modulus decreased, followed by a subsequent increase in value. The Young's modulus monotonically decreased with the increase in the Zr concentration. A slight increase in the Young's modulus was present when titanium was alloyed with Hf.

Key words. titanium alloy, Young's modulus, density

1 Introduction

The Young's modulus is one of the most important properties of a dental structural material. A typical Young's modulus for titanium is about one-half that of stainless steel or Co–Cr alloys. A higher or lower modulus than that of titanium is required, depending on the application. The mechanical properties of titanium can be changed through alloying. In the present study, the Young's moduli of experimental binary titanium alloys with Cu, Ag, Au, Zr, Nb, or Hf were determined by the ultrasonic-pulse method in order to investigate the effect of alloying on the elastic property of titanium.

2 Materials and methods

Buttons of titanium and binary titanium alloys with Cu, Ag, Au, Zr, Nb, or Hf (up to 30 mass%) were made in an argon-arc melting furnace. These buttons were cast into magnesia molds by a dental titanium-casting machine. Prior to testing, all the surfaces of the castings were ground to remove the hardened surface layer. An ultrasonic pulser/receiver (5800, Panametrics, Waltham, MA, USA) and transducers (V208 and V156, Panametrics) were used to determine the Young's moduli of the alloys by an ultrasonic-pulse method [1]. The density of each alloy was previously measured using Archimedes' principle. Data were analyzed using ANOVA followed by Scheffé's test ($\alpha = 0.05$).

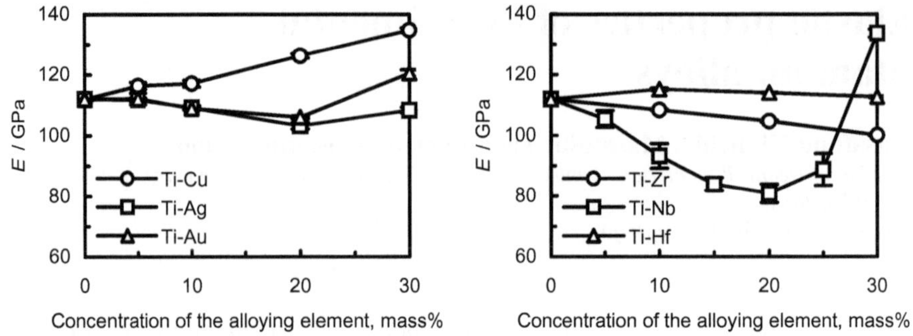

Fig. 1. Young's moduli (*E*) of experimental titanium alloys

3 Results

The densities of all the alloys monotonically increased as the concentration of alloying elements increased. The Young's moduli of the experimental titanium alloys are shown in Fig. 1. As the concentration of Cu increased, the Young's modulus monotonically and significantly increased. As the concentration of Ag, Au, or Nb increased to 20%, the Young's modulus significantly decreased, followed by a subsequent significant increase in value. The Young's modulus monotonically and significantly decreased with the increase in the Zr concentration. There was a slight increase in the Young's modulus when titanium was alloyed with Hf.

4 Discussion

The Young's modulus was dependent on the alloy phases. It is likely that the crystal structure of titanium changed from α to imperfect α by the addition of a small amount of Ag, Au, Zr, or Nb, resulting in a lower Young's modulus. Intermetallic compounds generally have higher Young's modulus than the constituting elements. Although the Young's moduli of the intermetallic compounds Ti_2Cu, Ti_2Ag, and Ti_3Au were not measured in the present study, it is possible that they are higher than those of the constituting elements, and that the increases in the modulus of Ti–Cu, Ti–Ag, and Ti–Au alloys were due to the formation of the intermetallic compounds in the α matrices. The increase in the Young's modulus of the Ti–Nb alloys was caused by the formation of the ω phase in the β matrix. Hf had a smaller effect on the Young's modulus than the other alloying elements.

Reference

1. GE Panametrics (2003) Application notes 930-012. Elastic modulus measurement, GE Panametrics, Waltham, pp 1–4

Interface Oral Health Science 2007: 347–348

Strength of porcelain fused to pure titanium made by CAD/CAM

R. Inagaki[1]*, M. Yoda[1], M. Kikuchi[2], K. Kimura[1], and O. Okuno[2]
[1]*Division of Fixed Prosthodontics and* [2]*Dental Biomaterials, Department of Restorative Dentistry, Tohoku University Graduate School of Dentistry, Sendai 980-8575, Japan*
*inagaki@mail.tains.tohoku.ac.jp

Abstract. This study is an investigation of the bond strength of porcelain to pure titanium, cast and machined by CAD/CAM processes, and a determination of the fracture strength of porcelain fused to pure titanium crown by both methods. The bond strength was evaluated according to the surface treatment conditions. The mean bond strengths were ranged from 36.1 to 49.4 MPa. The surface treatment conditions had a significant effect on the bond strength. The mean fracture strength for the cast frame crowns was 1667 N, and that for the machined frame crowns was 1554 N. There was no significant difference between the two methods. Acceptable bond strength and fracture strength were achieved by both methods.

Key words. CAD/CAM, bond strength, fracture strength, pure titanium

1 Objectives

Recently, computer-aided design and computer-aided manufacturing (CAD/CAM) have been advancing rapidly. These methods may be used to overcome disadvantages associated with porcelain fused to cast pure titanium. However, there are currently no reports on bond strength of porcelain to pure titanium machined by CAD/CAM processes. This study is an investigation into the bond strength of porcelain to cast and CAD/CAM-machined pure titanium in accordance with ISO9693. The fracture strength of porcelain fused to pure titanium crown made by both the methods was also determined.

2 Materials and methods

To evaluate the bond strength, pure titanium (JIS grade 2) specimens were divided into five groups according to the surface treatment condition. Six specimens were tested for each group: Cast and Polished (CP), Cast and Milled by tungsten carbide bur (CM), Machined by CAM and Polished (MP), Machined in the Lengthwise direction (ML), and Machined in the Crosswise direction (MC). Machined specimens (MP, ML, MC) were made with a CAD/CAM system (GN-1, GC, Japan). Commercial porcelain for titanium (VITA-Titankeramik, VITA, Germany) was applied to the specimens. The bond strengths were evaluated in accordance with ISO9693 and statistically analyzed using one-way ANOVA followed by Tukey pair-wise tests ($\alpha = 0.05$).

To evaluate the fracture strength, the crowns were made by the titanium frames prepared with a cast and machined by CAD/CAM processes, and a static load was applied at the incisal edge of the crown. The breakage load was measured at a cross-head speed of 2.5 mm/min and statistically analyzed using a paired *t*-test ($\alpha = 0.05$).

3 Results and discussion

The mean bond strengths for the cast specimens were CP: 36.2(±2.3), CM: 49.4(±1.2) MPa, and for the machined specimens were MP: 36.1(±3.4), ML: 45.5(±3.5), and MC: 46.5(±3.7) MPa. The bond strengths for all groups were above the minimum value specified in ISO9693 (25 MPa) [1]. These means exhibited higher value than the bond strength for gold alloy specimens (30.5 MPa) determined in a previous study [2]. There were no significant differences ($P > 0.05$) between the polished specimens (CP and MP), the machined directions (ML and MC), and the specimens cast/milled by tungsten carbide (CM) and the machined specimens (ML and MC). However, CM had significantly higher bond strength than CP. ML and MC had significantly higher bond strength than MP. ($P < 0.05$)

The mean fracture strength for the crowns made with the cast frames was 1667(±204) N, and that for the crowns made with the machined frames was 1554(±296) N. There was no significant difference between these specimens.

4 Conclusions

These results showed that the surface treatment conditions had a significant effect on the bond strength. There was no significant difference in the fracture strength between cast and machined by CAD/CAM processes. Acceptable bond strength and fracture strength could be achieved by either of these methods.

References

1. ISO9693:1999(E) (1999) Metal-ceramic dental restorative system, 2nd edition. Switzerland, International Organization for Standardization
2. Yoda M, Konno T, Takada U, et al (2001) Bond strength of binary titanium alloys to porcelain. Biomaterials 22:1675–1681

Interface Oral Health Science 2007: 349–350

Preparation of TiO_2 coating on dental metal materials by plasma CVD

R. Marumori[1]*, T. Kimura[2], N. Hayashi[2], M. Yoda[1], K. Kimura[1], and T. Goto[2]

[1]*Division of Fixed Prosthodontics, Department of Restorative Dentistry, Tohoku University Graduate School of Dentistry, Sendai 980-8575;* [2]*Multi Functional Materials Science Laboratory, Institute of Materials Research, Tohoku University, Sendai 980-8577; Japan*

*marumoriita@mail.tains.tohoku.ac.jp

Abstract. In this study, TiO_2 films were prepared by Plasma-enhanced Chemical Vapor Deposition (PECVD) using Ti(O-*i*-Pr)$_2$(dpm)$_2$ precursors. The effects of the deposition conditions on the crystalline phases, microstructures, and color of the deposited films were investigated. At a substrate pre-heating temperature T_{pre} = 623 K, the crystalline phase of the deposited film changed from amorphous to anatase to rutile with increasing microwave power P_M from 0 to 3.0 kW, and the microstructure changed from dense to granular. The deposited film was gray or black but changed to white at a higher T_{pre} (923 K).

Key words. TiO_2 films, PECVD, deposition conditions, microstructure

1 Objectives

Pd–Ag alloys are widely used as a substrate material in resin-veneered dental crowns. Since the adhesion of a dental resin to the Pd–Ag alloy substrate is poor, protuberances are usually formed on the substrate surface to sustain the mechanical retention. Current preferences among patients require that a crown be as similar as possible to the color of human teeth; however, the resin on protuberant substrates sometimes transmits the dark color of the metallic substrate because the resin coating is thin. Therefore, a layer of a ceramic buffer be applied between the Pd–Ag alloy substrate and the resin is proposed. Due to good adhesion of the resin to ceramics and the opaque properties of ceramics [1, 2], the ceramic layer would improve the mechanical properties and control the apparent color of the resin-veneered crown.

2 Materials and methods

The plastic pattern for a square, flat metal casting was made using a plastic board 1.0 mm in thickness and 10.0 mm by 10.0 mm in size. The plastic pattern was then vacuum-invested, and a casting was made with a dental Pd–Ag alloy consisting of

46% silver, 20% palladium, and 12% gold. On the board, TiO_2 films were prepared by Plasma-enhanced Chemical Vapor Deposition (PECVD) using $Ti(O\text{-}i\text{-}Pr)_2(dpm)_2$ precursors. The vaporized temperature of the precursor ranged from 453 to 458 K. Deposition temperature and total pressure were RT to 923 K and 1.2 to 7.0 torr, respectively. The microwave power was 0 to 3.0 kW. The nozzle-substrate distance was 20 to 25 mm. A double tube was used as a source gas nozzle. Ar or O_2 was used as the carrier gas. The Ar gas flow rate was 0 to 100 sccm. The O_2 gas flow rate was 30 to 100 sccm.

3 Results and discussion

At a substrate pre-heating temperature T_{pre} = 623 K, the crystalline phase of the deposited film changed from amorphous to anatase to rutile with increasing microwave power P_M from 0 to 3.0 kW, and the microstructure changed from dense to granular. The deposited film was gray or black but changed to white at a higher T_{pre} (923 K), and the film contained a white area at high P_M (3.0 kW). The black films (0 to 2.0 kW) contained 4 to 8 vol% carbon. Therefore, the carrier gas was changed from Ar gas to O_2 gas.

4 Conclusion

Plasma enhanced the deposition of TiO_2 films at a lower temperature. The films were crystalline with a rutile + anatase mixture phase. The films were granular, and the grain size increased with more plasma power. Black films were obtained at low deposition temperature due to the large amount of carbon in the films. The carbon contents in the films decreased by using oxygen gas as the carrier gas.

References

1. Tanaka T, Hirano M, Matsumura H, et al (1988) Study on ion-coating surface treatment of dental alloys for adhesion part 1. Effect of Cu target on precious alloys (in Japanese). J Jpn Prosthodont Soc 32:181–188
2. Kikuchi T, Yoshida N, Shimakura M, et al (1997) The clinical application of resin facing titanium crown by non-retention method—the influence of heating time on the bond strength (in Japanese). J Jpn Prosthodont Soc 41:481–488

Interface Oral Health Science 2007: 351–352

The possibility to form a new bone by means of using osteogenesis devices placed between bone and periosteum in rabbits

Junichi Hara*, Hitoshi Nei, Zaher Aymach, and Hirosi Kawamura
Division of Maxillofacial Surgery, Department of Oral Medicine and Surgery, Graduate School of Dentistry, Tohoku University, Sendai 980-8575, Japan
*jyjsendai@yahoo.co.jp

Abstract. There are a few reports mentioning that we can expand the healing space gradually with the possibility of inducing a new bone. The aim of this study is to determine the possibility to form a new bone by means of using osteogenesis devices in the healing space.

Key words. a new bone, osteogenesis, periosteum, expansion of the healing space, microangiograph

The bone healing process is known to progress in consecutive stages of impact, induction, inflammation, soft callus formation followed by hard callus formation, and finally, remodeling stage. The findings of the soft callus stage showed a process of revascularization with free osteogenic cells formation. Distraction osteogenesis is also known to expand the soft callus space between the bone parts.

On the other hand, we know that the healing space is created between the periosteum and the bone while ablating the periosteum and suturing it again to its own position.

The aim of this study is to determine the possibility to produce a new bone by means of using osteogenesis devices in the healing space between the periosteum and the bone.

Twenty-five rabbits were used and spread into control and experiment groups; both groups underwent periosteum ablation with placing osteogenesis device in the healing space of control groups, and then the periosteum was sutured in both groups to their own position (Figs. 1, 2; Table 1).

The control groups showed revascularization with a thin new bone formation in the healing space. On the other hand, all the experimental groups showed expansion in the healing space, with more revascularization beside the original bone compared to the side of the periosteum. The first experimental group was observed after a week of expansion cease; the histological findings showed new bone formation over original bone. On 2-week and 4-week groups, the new bone showed maturation much more than the 1-week group (Table 2).

All the groups showed revascularization from the original bone surface into the healing space in addition to formation of new bone which in turn confirms the possibility to produce new blood vessels and bone in the healing space. It can be concluded that expanding the healing space between the periosteum and the bone gradually might introduce a new bone over the original bone surface in the healing space.

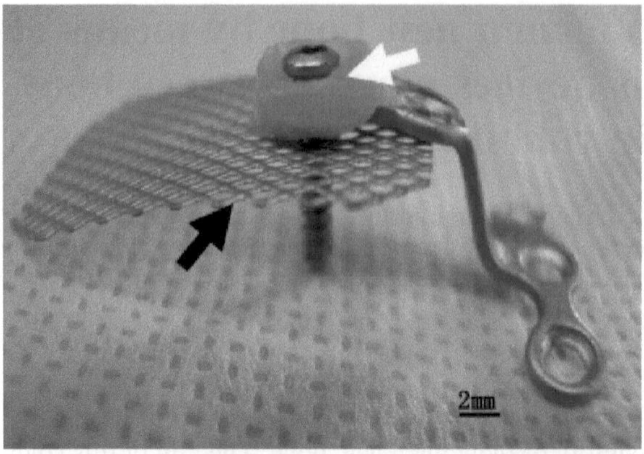

Fig. 1. The healing space expansion device. The titanium mesh (*black arrow*) which faces the periosteum distracting it away from the bone as the screw (*white arrow*) is activated

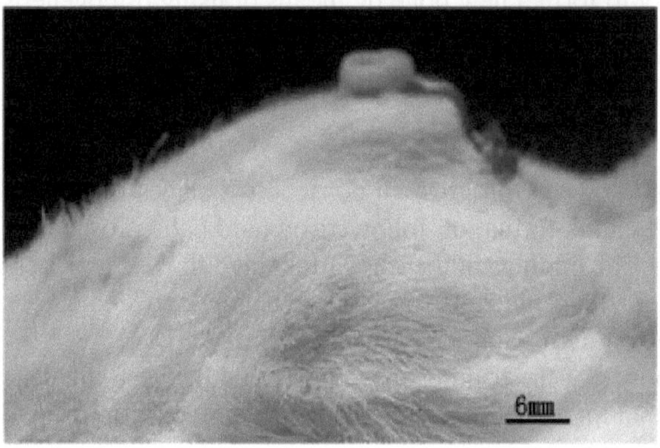

Fig. 2. Photograph of the animal after expansion showing complete healing around the device on the head. The screw has been expanded fully to 5 mm

Table 1. Experiment schedule

experiment	operation→ ←Latency period→	←Expansion period 0.5mm/day→		1week group dead ↓	2week group dead ↓	4week group dead ↓
days	*1*	*7*	*17*	*24*	*31*	*45*
control	← operation			↑ 1week group dead	↑ 2week group dead	↑ 4week group dead

Table 2. New bone height after expansion between the bone and the periosteum

	Postexpansion by (week)			
	1	2	4	Total (mm)**
Control	0.09 ± 0.036	0.11 ± 0.108	0.13 ± 0.017	0.11 ± 0.059
Experiment	2.50 ± 0.539	2.72 ± 1.016	3.40 ± 0.272	2.90 ± 0.753

(mm)**

**Mean value ± standard deviation experimentat week (n = 4), 2, 4 week (n = 5), control (n = 3)

*$P < 0.05$

Interface Oral Health Science 2007: 353–354

The effects of orthopedic forces with self-contained SMA appliance on cranial suture in rat

Sachiko Urayama[1]*, Hiroyasu Kanetaka[1], Yoshinaka Shimizu[2], Akihiro Suzuki[1], Ryo Tomizuka[1], and Teruko Takano-Yamamoto[1]
[1]*Division of Orthodontics and Dentofacial Orthopedics;* [2]*Division of Oral and Craniofacial Anatomy, Tohoku University Graduate School of Dentistry, Sendai 980-8575, Japan*
*sachiko-u@mail.tains.tohoku.ac.jp

Abstract. This study investigated the effect of orthopedic force applied by shape memory alloy (SMA) wire on the cranial bone growth in efforts to develop a self-contained orthopedic appliance. Expansive forces and compressive forces were applied by Ni–Ti SMA wires to interparietal sutures in Wistar rats. All rats were killed at day 14 or 28 post-treatment. Morphological analyses by soft X-ray and micro-computed tomography, and histological observation were performed. Cranial width in the expansive groups was increased significantly compared to the control group ($P < 0.01$). The increase in cranial width in the compressive groups was inhibited. The results suggest that the self-contained SMA appliance has orthopedic application to the cranium.

Key words. orthopedic force, SMA, morphometrical analysis, rat cranium

Introduction

External orthopedic appliances are frequently applied in clinical orthodontics to modify orofacial growth and to obtain harmonious skeletal relations. Current efforts in the development of orthopedic appliances are aimed at creating new internal appliances that are invisible and are automated so as not to require daily activation [1]. We investigated here the effects of orthopedic forces utilizing shape memory alloy (SMA) wire in a self-contained appliance in an experimental rat model of cranial modification through sutural growth control.

Material and methods

Expansive forces (group A, 50 gf; group B, 150 gf) and compressive forces (group C, 30 gf; group D, 90 gf) were applied to interparietal sutures in Wistar rats (male, 6 weeks old) using Ni–Ti SMA wire (0.018 and 0.014 inch Nitinol classic; 3M Unitek, Monrovia, CA, USA) (Fig. 1a). Two-dimensional morphometric analyses

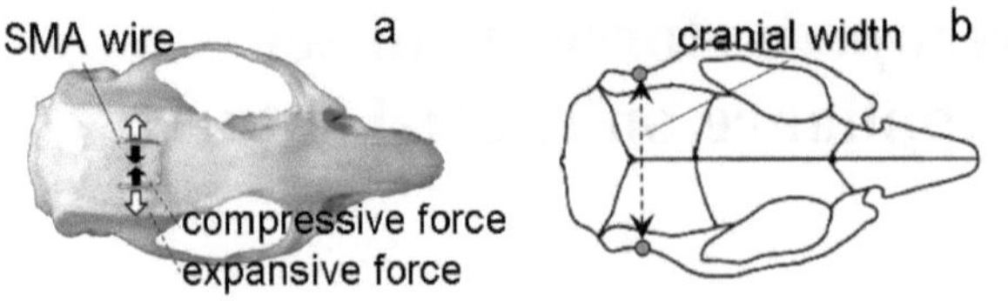

Fig. 1. **a** *SMA wire* applying compressive and expansive forces on the rat cranium. **b** Schema demonstrating the dimension assessed on soft X-ray photographs

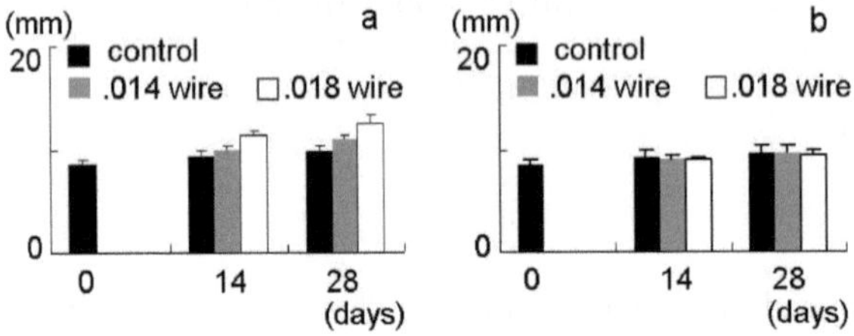

Fig. 2. Time course of cranial width growth in **a** the expansive groups and **b** the compressive groups

to determine cranial width were performed using soft X-ray photographs (Softex, Tokyo, Japan) (Fig. 1b). The interparietal sutures were observed three dimensionally using micro-computed tomography (CT) photographs (ASMX-225CT; Shimadzu Corporation, Kyoto, Japan) and histologically.

Results

Cranial width in the expansive groups was significantly increased compared with the control group ($P < 0.01$, Fig. 2a), whereas cranial width did not significantly differ between the control and compressive groups (Fig. 2b). In the expansive groups, interparietal sutures showed an enlarged opening and thin edges, whereas those in the compressive groups showed a linear shape and thick edges. Three-dimensional observations demonstrated that the interparietal sutures had a more apparent interdigital shape in the expansive groups, compared to a straight form in the compressive groups. Histiologically, new bone formation with osteoblasts on the sutural surfaces was observed in the expansive groups. In contrast, the compressive groups showed osteoclastic resorption and slight hyalinization.

Conclusion

The self-contained orthopedic appliance utilizing SMA wire was found to have orthopedic application to the cranium.

Reference

1. Kanetaka H, Shimizu Y, Hosoda H, et al (2007) Orthodontic tooth movement in rats using Ni-free Ti-based shape memory alloy wire. Mater Trans 48:367–372

Interface Oral Health Science 2007: 355–356

Development of a new ultra-precision-polished pure titanium mirror for dental treatment

Hiroyasu Kanetaka*, Akihiro Suzuki, Ryo Tomizuka, Sachiko Urayama, and Teruko Takano-Yamamoto
Division of Orthodontics and Dentofacial Orthopedics, Tohoku University Graduate School of Dentistry, Sendai 980-8575, Japan
*kanetaka@mail.tains.tohoku.ac.jp

Abstract. The aim of this study was to develop a biosafe and a biocompatible mirror for dental treatment. Mirrors are indispensable manual instruments for dental care and are widely used for intraoral examination and treatment. In this study, light and highly biocompatible mirrors were developed by mirror-polishing of pure titanium and used for dental examination and intraoral photography. The surface roughness and reflectance of the dental mirrors were measured and compared with those of commercially available dental mirrors made of stainless steel. Our results suggested that the pure titanium mirror with a mirror-polished surface was satisfactory for clinical use.

Key words. pure titanium, mirror, dental treatment, biocompatibility, polishing

Introduction

Recently, direct-reflection-type metal mirrors with a mirror-polished metal surface have been used because of several advantages, including distinctness of the reflected image, high safety, and improved operability. However, most metal mirrors are made of stainless steel and contain about 10% nickel. Such mirrors may not be suitable for medical use because they can elicit strong allergic reactions [1]. In addition to this biosafety problem, corrosion of stainless steel mirrors can be caused by chloride disinfectants, containing mainly sodium hypochlorite [2]. In this study, light and highly biocompatible mirrors were developed by mirror-polishing of pure titanium, and the surface roughness and reflectance of the new titanium dental mirrors were compared with those of commercially available dental mirrors made of stainless steel.

Material and methods

Newly developed mirror was made of pure titanium (Daido Steel Co., Ltd., Nagoya, Japan), finished by ultra-precision mirror polishing, and used to produce dental mirrors and mirrors for intraoral photography. These mirrors were compared with a commercially available dental mirrors made of stainless steel (metal mirror, YDM Co., Tokyo, Japan). The surface roughness and reflectance of four newly developed dental mirrors were compared with those of four commercially available dental mirrors made of stainless steel. Surface roughness (measurement range: around 360 μm) was measured using a three-dimensional non-contact-type surface

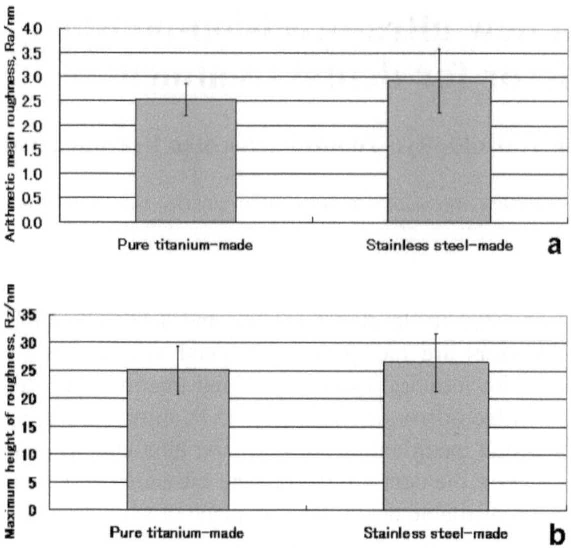

Fig. 1. Surface roughness of metal mirrors

roughness measuring instrument (Talysurf CCI 3000, Taylor Hobson Ltd., Leicester, UK). The reflectance of the mirrors was measured within the visible region (measurement range: 380–780 nm), using a spectrophotometer (U-3120, Hitachi Ltd., Tokyo, Japan). The intraoral mirrors were used to observe teeth and take intraoral photographs.

Results

The arithmetic mean roughness (Ra) of the dental mirror made of pure titanium was 3 nm or less (Fig. 1a), and the maximum height of roughness (Rz) was 30 nm or less (Fig. 1b), comparable to that of the commercially available dental mirror made of stainless steel. Both Ra and Rz showed no significant differences between the two groups.These result suggested that surface of the pure titanium was sufficiently polished by our new original method. The reflectance of the pure titanium mirror was slightly lower than that of the stainless steel mirror, but both provided favorable image.

Conclusion

It was suggested that the pure titanium mirror with mirror-polished surface was considered satisfactory for clinical use.

References

1. Kanerva L, Forstrom L (2001) Allergic nickel and chromate hand dermatitis induced by orthopedic metal implant. Contact Dermatitis 44:103–104
2. Dartar Oztan M, Akman AA, Zaimoglu L, et al (2002) Corrosion rates of stainless-steel files in different irrigating solutions. Int Endod J 35:655–659

Interface Oral Health Science 2007: 357–358

Biodegradable characteristics of octacalcium phosphate combined with collagen implanted in two bony sites

Yuko Suzuki[1], Shinji Kamakura[3], Kouki Hatori[1], Kazuo Sasaki[4], Yoshitomo Honda[2], Takahisa Anada[2], Keiichi Sasaki[1], and Osamu Suzuki[2]*
[1]*Division of Advanced Prosthetic Dentistry;* [2]*Division of Craniofacial Function Engineering, Tohoku University Graduate School of Dentistry, Sendai 980-8575;* [3]*Department of Translational Research, Center for Translational and Advanced Animal Research, Tohoku University School of Medicine, Sendai 980-8574;* [4]*Nippon Meat Packers, Inc., Tsukuba 300-2646; Japan*
*suzuki-o@mail.tains.tohoku.ac.jp

Abstract. Synthetic octacalcium phosphate (OCP) has been shown to be a resorbable bone regenerative material. In this study, OCP combined with collagen (OCP/Col) or collagen was implanted into the critical-sized defects or into the subperiosteal region in rat crania and fixed at 4, 8 and 12 weeks after implantation. The percentage of newly formed bone and the tissue including the implant were determined by a histomorphometric analysis. Activities of tartrate resistant acid phosphatase (TRAP) were stained to examine phagocytotic multinucleated giant cells. The effect of implantation on crystalline phase of OCP in the implant was examined by X-ray diffraction (XRD) and Fourier transform infrared spectroscopy (FTIR). The present study suggests that the bone formation on OCP/Col and biodegradation of OCP/Col is distinct depending on the implantation sites.

Key words. octacalcium phosphate, biodegradation, bone formation, biomaterial

1 Introduction

Octacalcium phosphate (OCP) is thought to be a precursor of the initial mineral crystals of biological apatite in bone and teeth. Previous studies showed that synthetic OCP is an osteoconductive material [1], and that OCP combined with collagen (OCP/Col) enhances bone regeneration accompanied by its biodegradation [2]. The purpose of the present study was to investigate whether OCP/Col is biodegraded and enhances bone regeneration in two separate sites, if implanted.

2 Materials and methods

Male Wistar rats 12-week-old were used. The rats were divided into two groups. In bone defect group, a full-thickness standardized trephine defect, 9 mm in diameter, was made in the calvarium. An OCP/Col or collagen disk (9 mm in diameter,

1 mm thick) was then implanted into the defect. In periosteal group, the periosteum was raised from the bone surface and OCP/Col or collagen disk (9 mm in diameter, 3 mm thick) was implanted into the periosteal pocket. Untreated animals were used as control. Animals were fixed at 4, 8, 12 weeks after implantation.

In subperiosteal group, the maximum thickness of the bone with implant was measured parallel to the calvaria. The structural changes of OCP/Col before and after implantation in subperiosteal region for 4 weeks were examined for X-ray diffraction (XRD) and Fourier transform infrared spectroscopy (FTIR). The specimens were dehydrated and embedded in paraffin, and histological sections were cut. Sections were stained with hematoxylin and eosin (H-E) or tartrate resistant acid phosphatase (TRAP). Light micrographs of the sections stained with H-E were used for the histomorphometric measurement. The percentage of newly formed bone area or remaining implant and surrounding tissue area of implantation area were calculated.

3 Results and conclusions

In the thickness measurement of periosteal region, OCP/Col was relatively retained unresorbed in 4 weeks, but gradually resorbed until 12 weeks up to the control level. XRD and FTIR analyses showed that hydroxyapatite (HA) was formed by the conversion of OCP within collagen matrix by the implantation until 4 weeks. When OCP/Col was implanted into the bone defect, the defects tended to be filled with newly formed bone. In periosteal region, however, inflammatory cell infiltration was observed around the implant surfaces but it was rarely seen in the central region of the implant in week 4. The remaining OCP particle was surrounded by multinucleated giant cells (MNGCs). In week 8, the thickness of OCP/Col was drastically reduced and almost disappeared until week 12. TRAP-positive cells were abundant around OCP/Col in the subperiosteal region rather than that in bone defects over the time. Histomorphometric analysis indicated that OCP/Col in bone defect enhanced bone formation with the biodegrad ation of OCP/Col, whereas OCP/Col in subperiosteal region enhanced little bone formation with the biodegradation of OCP/Col.

In conclusion, the present study suggests that the bone formation on OCP/Col and biodegradation of OCP/Col are distinct depending on the implantation sites.

References

1. Suzuki O, Nakamura M, Miyasaka Y, et al (1991) Bone formation on synthetic precursors of hydroxyapatite. Tohoku J Exp Med 164:37–50
2. Kamakura S, Sasaki K, Honda Y, et al (2006) Octacalcium phosphate combined with collagen orthotopically enhances bone regeneration. J Biomed Mater Res B Apple Biomater. 79:210–217

Interface Oral Health Science 2007: 359–360

New bone formation in β-TCP/MSC complex: effect of osteoblastic differentiation of MSC

Mamoru Kubota, Yoshiyasu Tokugawa, Makoto Nishimura, and Kaoru Igarashi*
Department of Oral Health and Development Sciences, Tohoku University Graduate School of Dentistry, Sendai 980-8575, Japan
*igarashi@mail.tains.tohoku.ac.jp

Abstract. The purpose of this study was to examine the effect of osteoblastic differentiation of mesenchymal stem cell (MSC) on new bone formation in β-TCP/MSC complex in vivo. Bone marrows were collected from beagle and cultured. Cultured cells were implanted into β-TCP. β-TCP/MSC complexes and control materials were surgically implanted into dorsal pouches of beagle. Eight weeks after the surgery, the implanted materials were harvested and evaluated. In all materials, new bone formation was observed. A significant increase in new bone formation was seen in non-differentiation complex and 1-day differentiation complex as compared with the control. However, new bone formation in 7-day differentiation complex was less than the other two complexes. The results suggest that the implantation of MSC intoβ-TCP enhances new bone formation, but the effect of osteoblastic differentiation of MSC is rather inhibitory.

Key words. mesenchymal stem cell, osteoblastic differentiation, β-TCP, new bone formation, beagle dog

Beta tricalcium phosphate (β-TCP) is known to have biocompatible, bioactive, and osteoconductive characteristics, and is one of the most popular biomaterials for bone tissue engineering. However, some studies have suggested that implantation of β-TCP alone is not very effective for bone formation [1].

It is also known that implantation of cells into β-TCP promotes bone regeneration, and bone marrow-derived mesenchymal stem cells (MSCs) are useful for tissue engineering [2]. However, there are considerable differences in the methods of cell implantation among studies, and an optimal condition of the cells is unclear. In the present study, we examined the effect of osteoblastic differentiation of MSC on new bone formation in β-TCP/MSC complex in vivo.

Four female beagle dogs, 1 year old, were used. Bone marrows were collected from the animals' iliac crest, and cultured in DMEM with mesenchymal cell growth supplements. Adherent cells were expanded as MSCs and then implanted into β-TCP. β-TCP/MSC complex was further cultured in the osteoblastic differentiation medium (MSC medium, 80 μg/ml ascorbic acid, 10 mM beta-glycerophosphate and 100 nM dexamethasone) for 0, 1, or 7 days (non-differentiation complex, 1-day differentiation complex, and 7-day differentiation complex). β-TCP without cells served as a control material. The materials were surgically implanted into dorsal

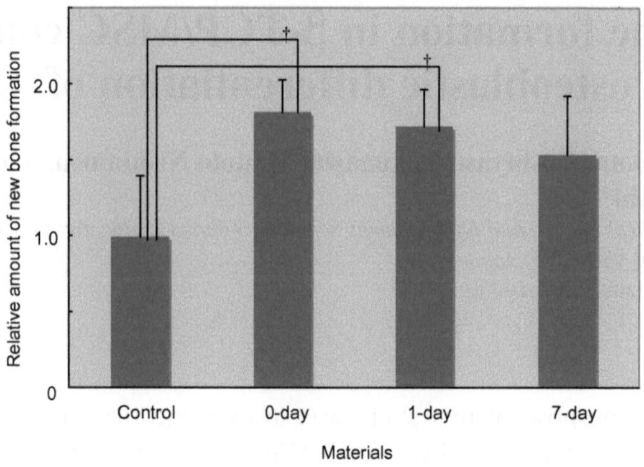

Fig. 1. Effect of osteoblastic differentiation of MSC on new bone formation in β-TCP/MSC complexes. Each *column* represents the mean ± SD (n = 4). †$P < 0.05$ vs. control. *0-day* non-differentiation complex, *1-day* 1-day differentiation complex, *7-day* 7-day differentiation complex

subcutaneous pouches of the animals. Eight weeks after the surgery, the implanted materials were harvested and evaluated. Histomorphometry of sections stained with hematoxylin and eosin was performed to quantitatively determine bone formation using NIH-image. Osteoclasts, defined as multinuclear cells having more than four nucleuses were counted in unit area.

In all materials, the absorption of β-TCP and displacement by new bone was observed. A significant increase in the new bone formation was seen in non-differentiation complex and 1-day differentiation complex as compared with the control (Fig. 1). However, new bone formation decreased as the induction culture of osteoblastic differentiation was prolonged.

In all groups, osteoclasts appeared on the surface of β-TCP. There were no significant differences in the number of osteoclasts among groups.

These results suggest that the implantation of MSC into β-TCP enhances new bone formation, but the effect of osteoblastic differentiation of MSC is rather inhibitory.

References

1. Kurashima K, Kurita H, Wu Q, et al (2002) Biomaterials 23(2): 407–412
2. Kasten P, Vogel J, Luginbuhl R, et al (2005) Biomaterials (29):5879–5889

Interface Oral Health Science 2007: 361–362

Bone regenerative property of synthetic octacalcium phosphate in collagen matrix

Tadashi Kawai[1,2], Takahisa Anada[2], Shinji Kamakura[3], Yoshitomo Honda[2], Aritsune Matsui[1,2], Kazuo Sasaki[4], Seishi Echigo[1], Osamu Suzuki[2]*
[1]*Division of Oral Surgery;* [2]*Division of Craniofacial Function Engineering, Tohoku University Graduate of Dentistry;* [3]*Division of Clinical Cell Therapy, Tohoku University School of Medicine, Sendai;* [4]*Nippon Meat Packers, Inc., Tsukuba, Ibaraki, Japan*
*suzuki-o@mail.tains.tohoku.ac.jp

Abstract. The present study was designed to investigate whether the amount of octacalcium phosphate (OCP) facilitates osteoconductive characteristics of OCP/Collagen in vivo and in vitro. Radiographic and histological examination showed that the quantity of newly formed bone increased with increasing OCP concentration in collagen. The degree of differentiation of osteoblastic cells increased depending on OCP concentration up to day 14 of culture. The present study suggests that the osteoconductive characteristics of OCP/Collagen can be displayed by the intrinsic bone regenerative property of OCP, in concert with collagen matrix.

Key words. octacalcium phosphate, collagen, bone regeneration, osteoconductivity, osteoblasts

1 Introduction

Octacalcium phosphate (OCP) has been suggested to be a precursor of biological apatite crystals in bones and teeth. Our previous studies showed that synthetic OCP, if implanted in murine cranial defects, facilitated bone regeneration and was replaced by newly formed bone, accompanying OCP biodegradation [1]. Furthermore, it was apparent that OCP combined with collagen (OCP/Collagen) enhanced bone regeneration [2]. The aim of this study was to investigate whether the concentration of OCP in collagen matrix influences bone regenerative property of OCP/Collagen in vivo and in vitro.

2 Materials and methods

2.1 Preparation of OCP/Collagen

The OCP was prepared according to the method previously reported [3]. OCP/Collagen discs, including various OCP concentrations, were prepared from pepsin-digested atelocollagen isolated from the porcine dermis and synthetic OCP.

2.2 Implantation procedure

A standardized critical-sized defect was made in the rat calvaria, and each OCP/Collagen disc was implanted into the defect. Five rats were fixed at 4 weeks after implantation, and examined radiographically and histologically.

2.3 Cell culture

Mouse bone marrow derived stromal ST-2 cells were cultured on the dishes pre-coated with OCP/Collagen, having different OCP concentrations. The capacity of proliferation and differentiation was determined up to day 21 of culture.

3 Results and discussion

Radiographic examination showed that the radiopacity of the defect increased with increasing OCP concentration in collagen throughout the defect. Histological examination showed that newly formed bone was observed in relation to OCP granules within collagen matrix. The proliferation of ST-2 cells was inhibited with increasing OCP concentration, whereas alkaline phosphatase activity, a maker for osteoblastic differentiation, increased depending on OCP concentration up to day 14 of culture. The present study showed that the osteoconductive characteristics of OCP/Collagen were facilitated with increasing OCP concentration both in vivo and in vitro. The results suggested that osteoconductive characteristics of OCP/Collagen should be controlled by the intrinsic bone regenerative property of OCP. The study by prolonged implantation of OCP/Collagen is under way to establish the linkage between biodegradable property of this scaffold and the replacement with newly formed bone with time.

References

1. Suzuki O, Kamakura S, Katagiri T, et al (2006) Bone formation enhanced by implanted octacalcium phosphate involving conversion into Ca-deficient hydroxyapatite. Biomaterials 27:2671–2681
2. Kamakura S, Sasaki K, Honda Y, et al (2006) Octacalcium phosphate combined with collagen orthotopically enhances bone regeneration. J Biomed Mater Res B 79:210–217
3. Suzuki O, Nakamura M, Miyasaka Y, et al (1993) Bone formation on synthetic precursors of hydroxyapatite. Tohoku J Exp Med 164:37–41

Interface Oral Health Science 2007: 363–364

Effect of octacalcium phosphate on proliferation and differentiation of bone marrow stromal cell line ST-2

Takashi Kumagai[1,2], Takahisa Anada[2], Yoshitomo Honda[2], Shinji Kamakura[3], Hidetoshi Shimauchi[1], and Osamu Suzuki[2]*
[1]*Divisions of Periodontology and Endodontology;* [2]*Craniofacial Function Engineering, Tohoku University Graduate School of Dentistry, Sendai 980-8575;* [3]*Department of Translational Research, Center for Translational and Advanced Animal Research, Tohoku University School of Medicine Sendai; Japan*
*suzuki-o@mail.tains.tohoku.ac.jp

Abstract. The present study was designed to investigate whether octacalcium phosphate (OCP) affects proliferation and differentiation of mouse bone marrow stromal ST-2 cells, an osteogenic cell line, in vitro. OCP facilitates the differentiation of ST-2 cells rather than hydroxyapatite (HA). It was also shown that OCP tends to convert into HA during the incubation. These results suggest that enhancement of osteoblastic cell differentiation by OCP may be associated with a process of irreversible conversion into HA.

Key words. octacalcium phosphate, hydroxyapatite, bone marrow stromal cells, differentiation

1 Introduction

Our previous studies showed that synthetic octacalcium phosphate (OCP) enhances bone regeneration more than sintered hydroxyapatite (HA) [1] in vivo. Furthermore, in vitro studies suggested that osteoblastic cell differentiation is enhanced by OCP [2] and affected by adsorption of cytokines onto OCP [3]. However, the molecular mechanism of the induction of osteoconductive characteristics by OCP has not yet been elucidated. The present study was designed to investigate whether OCP stimulates differentiation of mouse bone marrow stromal ST-2 cells, an osteoblastic cell line in vitro.

2 Materials and methods

2.1 *Preparation of synthetic OCP and HA*

OCP was prepared according to a method previously reported [4]. HA was also prepared according to the method of Moreno et al. [5]. Synthesized OCP and HA were characterized by X-ray diffraction (XRD). XRD patterns were recorded with

CuK_α X-rays on a diffractometer (Mini Flex; Rigaku Electrical Co., Ltd., Tokyo, Japan) at 30 kV, 15 mA.

2.2 *Cell culture*

ST-2 cells were maintained in α-MEM containing 10% FBS. We prepared OCP or sintered HA coating plates for cell culture. The quantity of OCP and HA coated was adjusted to 3.0 mg in each well of 48-well cell culture plates. ST-2 cells were incubated on OCP or HA plate for 3, 7, 15, and 21 days. The proliferation of cells was determined using a WST-8 assay. Alkaline phosphatase (ALP) activity was measured using a commercially available kit.

3 Results and discussion

The cell numbers on the OCP coating was slightly decreased compared to that of the control at the early period of culture, suggesting the inhibition of cell proliferation. However, the cell numbers reached to be almost same as the control on day 15. The level of ALP activity of OCP coating plates was significantly higher than that of HA coating or the control on day 21. In contrast, ALP activity remained unchanged on HA coating compared to the control. The OCP structure converted to that of HA with the reduction in intensity at $2\theta = 4.8°$ reflection of OCP in XRD. These results suggest that OCP stimulates osteoblastic cell differentiation and that OCP-apatite conversion may be involved in this stimulatory function of OCP.

References

1. Imaizumi H, Sakurai M, Kashimoto O, et al (2006) Comparative study on osteoconductivity by synthetic octacalcium phosphate and sintered hydroxyapatite in rabbit bone marrow. Calcif Tissue Int 78:45–54
2. Suzuki O, Kamakura S, Katagiri T, et al (2006) Bone formation enhanced by implanted octacalcium phosphate involving conversion into Ca-deficient hydroxyapatite. Biomaterials 27:2671–2681
3. Kumagai T, Anada T, Honda Y, et al (2007) Osteoblastic cell differentiation on BMP-2 pre-adsorbed octacalcium phosphate and hydroxyapatite. Key Eng Mater (in press)
4. LeGeros RZ (1985) Preparation of octacalcium phosphate (OCP): a direct fast method. Calcif Tissue Int 37:194–197
5. Moreno EC, Gregoly TM, Brown WE (1968) Preparation and solubility of hydroxyapatite. J Res Natl Bur Stand 72A:773–782

Interface Oral Health Science 2007: 365–366

Fitness of Zirconia all-ceramic crowns with different cervical margin forms

S. Miura*, N. Suto, R. Inagaki, Y. Kaneta, M. Yoda, and K. Kimura
Division of Fixed Prosthodontics, Department of Restorative Dentistry, Tohoku University Graduate School of Dentistry, Sendai 980-8575, Japan
*shoko-m@mail.tains.tohoku.ac.jp

Abstract. The purpose of this study was to evaluate the marginal and internal fitness of zirconia all-ceramic crowns. Three brass dies were prepared with the following cervical margin forms: a shoulder and two types of rounded (curvature radius: 0.2 and 0.5 mm) shoulder preparations. Five crowns for each type of die were fabricated using the CAD/CAM system. The fitness was evaluated with a replica technique using silicon impression materials. The mean gap dimensions and standard deviations at the margins were 36 ± 32 μm for the shoulder, 27 ± 29 μm for the 0.2 mm rounded shoulder, and 41 ± 33 μm for the 0.5 mm rounded shoulder. There were no significant differences among the three groups regarding marginal fitness. The zirconia all-ceramic crowns made using the CAD/CAM system showed clinically acceptable marginal fitness.

Key words. zirconia, all-ceramic crown, fitness, CAD/CAM

1 Objectives

The purpose of this study was to evaluate the marginal and internal fitness of zirconia all-ceramic crowns with three different cervical margin forms.

2 Materials and methods

Three brass dies were prepared with the following cervical margin forms: one shoulder type and two rounded-shoulder types (curvature radius: 0.2 and 0.5 mm). Impressions were made using a vinyl polysiloxane impression material (Exafine regular type: GC, Tokyo, Japan). The dies were fabricated with a type IV stone (New Fujirock: GC, Tokyo, Japan). Fifteen standardized partially sintered zirconia ceramic cores were fabricated using the CAD/CAM system (cercon® smart ceramics: DeguDent, Hanau, Germany) for the three test groups (n = 5). The marginal gap was assessed by measuring the vertical discrepancy between the outer restoration margin and the preparation line. The measurements were performed at eight different points across the entire circumference of each crown. The internal fitness was evaluated with a replica technique with a black silicone impression material

(BITE-CHECKER: GC, Tokyo, Japan) to fill the space between the specimen and the die. A white silicone impression material (FIT CHECKER: GC, Tokyo, Japan) was used to stabilize the black silicone impression material film. The replicas were cut with a scalpel in two axial directions. In this manner, the replica was divided into four pieces. Measurements were performed by the Profile Projector (V16-D: Nikon, Tokyo, Japan) at a magnification of 50×. The measurements of the black silicone layer were made at nine different points. The data were statistically analyzed using one-way ANOVA followed by Tukey's HSD tests ($\alpha = 0.05$).

3 Results and discussion

The mean marginal gaps and standard deviations for the all-ceramic crowns were 36 ± 32 μm for the shoulder type, 27 ± 29 μm for the 0.2 mm rounded-shoulder type, and 41 ± 33 μm for the 0.5 mm rounded-shoulder type. There were no significant differences among the three groups for the marginal gaps.

The mean internal gaps between the dies and the all-ceramic crowns at nine measured points were 33 to 122 μm for the shoulder, 65 to 104 μm for the 0.2 mm rounded-shoulder type, 52 to 129 μm for the 0.5 mm rounded-shoulder type. There were no significant differences among the three groups for the internal gaps.

According to McLean, the clinically acceptable marginal gap limit is not more than 120 μm [1]. In our previous study, the marginal gaps for the zirconia all-ceramic crowns fabricated using the CAM system were about 100 μm [2]. In this study, the mean marginal gaps of the three types of zirconia all ceramic crowns were 36, 27, and 41 μm, respectively. Consequently, the results of this study indicated excellent marginal fitness.

4 Conclusion

The marginal gaps of the zirconia all-ceramic crowns were not affected by the difference in cervical margin forms.

References

1. McLean JW, von Fraunhofer JA (1971) The estimation of cement film thickness by an in vivo technique. Br Dent J 131:107–111
2. Miura S, Inagaki R, Kimura K (2005) A study of the CAM system applied zirconia—material characteristics (in Japanese). Jpn J Esthet Dent 17:147–157

Interface Oral Health Science 2007: 367–368

Periodic changes of marginal adaptation of cervical composite resin restorations

H. Sasazaki* **and M. Komatsu**
Division of Operative Dentistry, Department of Restorative Dentistry, Tohoku University Graduate School of Dentistry, Sendai 980-8575, Japan
*hiromi@ddh.tohoku.ac.jp

Abstract. The purpose of this study was to evaluate the periodic changes of marginal adaptation of cervical composite resin restorations. Class V saucer-type cavities or cervical cavities were prepared. A total of 159 restorations were placed in 40 patients by one operator. Four self-etching bonding systems were used in this study. Clinical findings of these fillings were periodically observed. In order to observe the marginal adaptation, precision replicas were made. These replicas were observed by SEM. In the Fluoro bond group (Shofu, Kyoto, Japan) and the Mega bond group (Kuraray, Okayama, Japan), maximum observation period is 3,121 days. In many cases of enamel and enamel dentin margin cavities, marginal steps were observed after 1 year. The width of steps was extended with time.

Key words. self-etching bonding system, marginal adaptation, clinical performance, replica, SEM

1 Introduction

Self-etching bonding system was now widely used in the world. We reported in the previous study, when the bonding layer was exposed at the margin, it was abraded and marginal step was clearly observed in an early stage. If marginal steps were formed, abrasion and micro fracture of composite resin progressed with time [1]. The purpose of this study was to evaluate the periodical changes of marginal adaptation of cervical composite resin restorations.

2 Materials and methods

Under local anesthesia or without an anesthesia, class V saucer-type cavities or cervical cavities were prepared on vital human teeth. A total of 159 restorations were placed in 40 patients (13 males and 27 females) by an operator. Fluoro bond, Mega bond, Clearfil Photo bond (Kuraray, Okayama, Japan), and Clearfil New bond (Kuraray, Okayama, Japan) were applied according to the manufacturer's instructions. Light-cured type composite resins (Lite fil II A: Shofu, Kyoto, Japan, Clearfil AP-X, Clearfil photo anterior: Kuraray, Okayama, Japan) and Chemical-cured type composite resin (Clearfil F II: Kuraray, Okayama, Japan) were inserted into these cavities and polymerized. Clinical findings of these fillings were periodically observed. In order to observe the marginal adaptation, impressions of these restorations were taken and precision replicas were made. These replicas were observed by the scanning electron microscope.

3 Results

1. In the Fluoro bond group, maximum observation period is 3,121 days. Clinical performance was good. In half the cases of enamel and enamel dentin margin cavities, marginal steps were observed after 1 year.
2. In the Fluoro bond (+phosphoric acid) group, maximum observation period is 2,645 days. Marginal adaptation of early stage was better than Fluoro bond (−phosphoric acid) group. In many cases of enamel dentin margin cavities, marginal steps were observed after 18 months.
3. In the Mega bond group, maximum observation period is 2,505 days. Clinical performance was good. In many cases of enamel-dentin margin cavities, marginal steps were observed after 1 year. The width of steps was extended with time by the observation of replica using scanning electron microscope.

4 Discussion and conclusion

When the bonding layer was exposed at the margin, abrasion progressed from the exposed bonding layer, and then marginal fracture of resin materials occurred. Enamel etching effectively improved the adaptation to enamel.

Reference

1. Sasazaki H, Komatsu M (2002) Short-term clinical assessment of resin restorations treated with self-etching primer. Jpn J Cons Dent 45(2):310–321

Interface Oral Health Science 2007: 369–370

Comparative evaluation of the radiopacity of fiber-reinforced posts

M. Kanehira[1]*, W. J. Finger[2], and M. Komatsu[1]
[1]*Department of Restorative Dentistry, Tohoku University Graduate School of Dentistry, Sendai 980-8575, Japan;* [2]*Department of Preclinical Dentistry, School of Dental Medicine, University of Cologne, 50931 Cologne, Germany*
*kane@mail.tains.tohoku.ac.jp

Abstract. This study evaluated and compared the X-ray opacity of eight marketed fiber-reinforced composite (FRC) post brands by densitometric analysis of radiographs, taken of the separate posts and of posts seated in extracted human canine teeth, respectively. The fiber-reinforced resin posts studied had significantly different radiopacity. Two post brands were hardly detectable on the radiographs, whereas four brands were acceptably radiopaque; and two posts with >2 mm Al equivalent were very satisfactory for unequivocal detection on dental radiographs.

Key words. fiber-reinforced post, radiopacity, X-ray opacity

1 Introduction

Since the introduction of the first carbon fiber post in 1990 [1], a large number of fiber-reinforced posts have been introduced to the European and American markets. Carbon fibers have been substituted by different kinds of glass fibers in order to adjust the post color. The clinical long-term observations indicate that fiber-reinforced composite (FRC) posts have successfully widened the spectrum of endodontic-restorative treatment regimens [2]. One of the clinical disadvantages often encountered with FRC is insufficient or missing X-ray opacity [3]. The aim of this study was to evaluate and to compare by densitometric analyses of radiographs the X-ray opacity of eight marketed fiber-reinforced composite post brands.

2 Materials and methods

Eight commercial fiber post brands were investigated: DT Light Post® (DTL) and DT White Post® (DTW) (RTD, St. Egrève, France), FibreKor® Post (FIK) (Pentron®, Wallingford, CT, USA) FRC Postec® (FPO) and FRC Postec® Plus (FPP) (Ivoclar Vivadent AG, Schaan, Liechtenstein), ER DentinPost (KOE) (Brasseler, Lemgo, Germany), Luscent Anchors™ (LUA) (Dentatus, Hägersten, Sweden) and Para-Post® FiberWhite (PAP) (Coltène, Altstätten, Switzerland). Radiographs were produced on Agfa Dentus M2 Comfort films for determination of the radiographic density of the posts and an eight-step aluminum wedge, exposed for calibration purposes. The photographic densities were determined with a transmission densitometer (X-Rite, Model 301). The optical densities of the eight post radiographs were measured, and the means and standard deviations were then transformed into equivalent aluminum thickness. In the second part, extracted human upper canines were selected. The crowns were removed and the root canals were prepared to at least 15 mm depth. Each of the eight posts was inserted into the root canal preparations and standard radiographs were produced for measurement of the optical densities. Statistical evaluation by ANOVA and Duncan's post-hoc test at $P < 0.05$.

3 Results and discussion

DTW and LUA (Al-equivalent approximately 0.8 mm) are scarcely visible on the radiographs, DTL, FIK, FPO, and PAP (1.0 to 1.6 mm Al) can be identified, whereas KOE (2.3 mm Al) and in particular FPP (2.8 mm Al) are excellently visible. Statistically, the radio densities of the teeth evaluated both without and with each of the eight posts inserted in root canals were significantly different ($P < 0.0001$). DTW and LUA could not be visually differentiated from the control, i.e., the root without inserted post. DTL, FIK, FPO, and PAP with Al equivalents between 4.3 and 4.8 mm are weekly identified, whereas FPP and KOE are clearly seen. This study has shown that the fiber-reinforced resin posts studied have significantly different radiopacity. The FRC Postec Plus post with high X-ray opacity was considered especially adequate for unequivocal detection on dental radiographs.

References

1. Duret B, Reynaud M, Duret F (1990) Un nouvau concept de reconstitution corono-radiculaire: Le Composiposte (1). Chirurg Dent France 540:131–141
2. Ferrari M, Vichi A, Garcia-Godoy F (2000) Clinical evaluation of fiber-reinforced epoxy resin posts and cast post and cores. Am J Dent 13:15B–18B
3. Finger WJ, Ahlstrand WM, Fritz UB (2002) Radiopacity of fiber-reinforced resin posts. Am J Dent 15:81–84

Interface Oral Health Science 2007: 371–372

Quantitative-radiographic and molecular-histological analysis of bone repair in critical and non-critical size rat calvarial bone defects

Takahiro Honma[1,2], Tomoko Itagaki[1,2], Megumi Nakamura[2], Shinji Kamakura[3], Ichiro Takahashi[4], Seishi Echigo[1], and Yasuyuki Sasano[2]*

[1]*Division of Oral Surgery;* [2]*Division of Craniofacial Development and Regeneration;* [4]*Division of Orthodontics and Dentofacial Orthopedics, Tohoku University Graduate School of Dentistry, Sendai 980-8575;* [3]*Division of Clinical Cell Therapy, Department of Translational Research, Center for Translational and Advanced Animal Research (CTAAR), Tohoku University School of Medicine, Sendai 980-8574; Japan*

*sasano@anat.dent.tohoku.ac.jp

Abstract. The present study was designed to investigate bone repair in a critical size defect compared to that in a smaller or a non-critical-size defect. Our original standardized rat calvarial bone defect model was used. The rate of bone formation was examined with X-ray morphometry, and the bone production was assessed by in situ hybridization for type I collagen and osteocalcin. Formation of repaired bone ceased within 24 weeks in both critical- and non-critical-size defects, whereas osteoblasts and osteocytes no longer expressed type I collagen or osteocalcin in week 24. The results suggest that osteoblasts and osteocytes cease bone formation within a limited period regardless of completion of the defect repair.

Key words. bone repair, osteoblasts, ECM, X-ray morphometry, in situ hybridization

1 Introduction

A bone defect that is not repaired with bone completely is designated as a nonunion defect or a critical-size defect. The biological mechanism that regulates the process of bone repair of the critical-size defect is not known. The present study was designed to investigate bone repair in a critical size defect compared to that in a smaller or a non-critical-size defect.

2 Materials and methods

Our original standardized rat calvarial bone defect model was used for the experiment [1]. The rate of bone formation was examined with X-ray morphometry, and the bone production of osteoblasts and osteocytes was assessed by molecular histology with in situ hybridization for type I collagen and osteocalcin [2].

3 Results and discussion

X-ray morphometry data showed that formation of repaired bone ceased within 24 weeks in both critical- and non-critical-size defects, i.e. regardless of completion of the defect repair. In situ hybridization demonstrated that mRNA transcripts of type I collagen and osteocalcin were no longer detected in the 24th week after the defect was created. The results suggested that osteoblasts and osteocytes cease bone formation, and differentiation of osteoblast progenitors declines in 24 weeks. Also, bone repair proceeds from the periosteum on both sides of the parietal bone but not from the surface of the bony edge around the original defect [3].

The results could provide useful information for clinical research on bone repair.

References

1. Kamakura S, Sasano Y, Homma H, et al (1990) Implantation of octacalcium phosphate (OCP) in rat skull defects enhances bone repair. J Dent Res 78:1682–1687
2. Sasano Y, Maruya Y, Sato H, et al (2001) Distinctive expression of extracellular matrix molecules at mRNA and protein levels during formation of cellular and acellular cementum in the rat. Histochem J 33:91–99
3. Honma T, Itagaki T, Nakamura M, et al (2007) Bone formation in rat calvaria ceases within a limited period regardless of completion of defect repair. Oral Dis (in press)

Interface Oral Health Science 2007: 373–374

Immunohistological study on STRO-1 in developing rat molars

Ryuta Kaneko[1,2], Hirotoshi Akita[2], Hidetoshi Shimauchi[1], and Yasuyuki Sasano[2]*
[1]*Division of Periodontology Endodontology;* [2]*Division of Craniofacial Development and Regeneration, Tohoku University Graduate School of Dentistry, Sendai 980-8575, Japan*
*sasano@mail.tains.tohoku.ac.jp

Abstract. STRO-1 is a specific marker of mesenchymal stem cells, but the tissue localization of STRO-1 in developing teeth is not clear. The present study was designed to investigate the immunohistological localization of STRO-1 in developing and erupting rat molars. STRO-1 antigens were found in odontoblasts, dental pulp cells, periodontal ligament cells, cementblasts, periodontal osteoblasts, and blood vessels. The STRO-1 may be involved in hard tissue formation of dentin and alveolar bone.

Key words. STRO-1, mesenchymal stem cells, tooth, dental pulp, periodontium

1 Introduction

STRO-1 is a cell surface antigen of bone marrow stromal cells and suggested to be a specific marker of the cell population containing mesenchymal stem cells [1, 2]. STRO-1 positive cells reside in dental pulp and periodontal ligaments as well as in bone marrow [3]. However, the tissue localization of STRO-1 in developing teeth is not yet clear. The present study was designed to investigate spatiotemporal localization of STRO-1 in developing and erupting rat teeth by immunohistochemistry.

2 Materials and methods

2.1 Tissue preparation

Wistar rats at 2, 3, 6 and 12 weeks were anesthetized and fixed by perfusion through the aorta with 4% paraformaldehyde in phosphate buffer. Mandibles were resected and decalcified in 10% EDTA. After dehydration through a graded series of ethanol solution, the specimens were embedded in paraffin. Sections with 5 μm thickness were cut and processed for immunohistochemistry.

2.2 *Immunohistochemistry*

The sections were deparaffinized and incubated with the monoclonal mouse antibody against human STRO-1 antigens overnight at 4°C and with the secondary antibody for 30 min at room temperature. The immunoreactivity was visualized in diaminobenzidine (DAB) solution. Control sections were processed as above except that non-specific mouse IgM was used as a substitute for the STRO-1 antibody.

3 Results

Immunoreaction for STRO-1 was identified in some bone marrow cells of 2-, 3-, 6-, and 12-week-old rat mandibles.

In 2-week-old rat mandible third molars (bell stage tooth germs), positive immunoreaction was found in a part of odontoblasts. At the stage of crown formation (2-week old first and second molars, 3-week-old third molars), the immunoreaction was found in odontoblasts and dental plus cells. At the stage of root formation (3-week-old first and second molars), the immunoreactivity was found in odontoblasts and dental pulp cells in the region of roots, but only in odontoblasts in the crown region. Thereafter, the immunoreaction in dental pulp cells declined. Some of the periodontal ligament cells, cementblasts, periodontal osteoblasts, and blood vessels were positive with the antibody.

4 Discussion

STRO-1 antigens were found in some cells of dental pulp and periodontium during tooth development and eruption. Further investigation will provide a better understanding of the relevance of the STRO-1 antigens to mesenchymal stem cells which reside in dental pulp and periodontium.

References

1. Simmons PJ, Torok-Storb B (1991) Identification of stromal cell precursors in human bone marrow by a novel monoclonal antibody, STRO-1. Blood 78(1):55–62
2. Shi S, Gronthos S (2003) Perivascular niche of postnatal mesenchymal stem cells in human bone marrow and dental pulp. J Bone Miner Res 18(4):696–704
3. Seo BM, Miura M, Gronthos S, et al (2004) Investigation of multipotent postnatal stem cells from human periodontal ligament. Lancet 364(9429):149–155

Interface Oral Health Science 2007: 375–376

Stealth authentication by communication with radio-frequency transponder embedded in a tooth

Hiroshi Ishihata*, Shigeru Shoji, and Hidetoshi Shimauchi
Division of Periodontology and Endodontology, Department of Oral Biology, Tohoku University Graduate School of Dentistry, Sendai 980-8575, Japan
*hiro@ddh.tohoku.ac.jp

Abstract. A radio frequency identification (RFID) transponder covering the 13.56 MHz band was adapted to be placed in the pulp chamber of an endodontically treated molar human tooth. In an animal experiment, the transponder could be fixed in the cavity of a mandibular canine of a dog. An RFID reader positioned close to the dog's face could communicate with the transponder in the dog's tooth.

Key words. radio frequency, endodontics, personal identification, mobile phone

1 Introduction

In recent years, automatic identification procedure (Auto-ID) exists to provide information about people, animal, goods, and procedure in transit. One of the technically optimal solution of Auto-ID, using a silicon chip is the radio frequency identification (RFID). The RFID technology shows remarkable development, and the miniature RFID transponders of less than 1 cm have been used [1]. RFID installation into the endodontic space of the tooth is not difficult. We have developed a method to prevent the medical incidents in the hospital by adopting the RFID embedded in patient's tooth that was preserved by dental treatment. In this study, we investigated the ability of the communication of RFID implanted inside the tooth and the feasibility of a novel authentication method using RFID installed in the oral cavity.

2 Materials and methods

A customized RFID transponder was embedded into the canine of a dog (male beagle, weighing 20 kg). This animal experiment was conducted according to the Committee of the Institute of Tohoku University Animal Experimentation. The RFID transponder, fabricated from an ISO/IEC15693 chip (13.56 MHz: SRF55V10S, Infineon Technologies) of integrated circuit including a contactless communication feature with security algorithm on mutual authentication, was customized into a rectangular bar for embedding in the cylindrical endodontic space of a tooth, and spatial directionality was established in communication with a reader positioned beside the dog's face. Pulp of the mandibular right canine was extirpated under general and local anesthesia. The endodontic treatment conformed to the same kind of treatment procedure as the case of human canines. The apex portion of the root canal of the canine was filled with endodontic sealer. The transponder was sealed and fixed with epoxy resin in the residual cavity of the canine after endodontic treatment.

3 Result and discussion

The communication range of the customized RFID transponder before installation was 30 mm. Although the transponder implanted to the canine was surrounded by hard tissues composed of dentin and alveolar bone, it could be operated when a conventional reader (FPRH100, 500 mW, Feig, Weilburg, Germany) was brought close to the cheek. The communication distance between the reader and the embedded transponder in the canine was estimated to be 25 mm.

We have developed a method to prevent the medical incidents in the hospital by adopting the RFID embedded in patient's tooth that was preserved by endodontic and restorative treatment. Hospitalized patients tend to wear personal identification (ID) wristbands to prevent medical incidents. An RFID transponder, holding ID information attached to the wristband, had been suggested for a swift identification of inpatients [2]. However, continuous wearing of a wristband may induce physical and mental stress in some patients. Embedded ID into oral cavity could be applied to a mobile communicator in combination with RFID reader. This engineering system is called 'stealth identification' that utilizes communication equipment such as mobile phones. In most of the voice communications, the face positions are close to the handset, so when the handset equips an RFID reader, the identification can be established by detecting the RFID during the phone usage. It would be difficult to detect the existence of an RFID transponder in a human tooth. Furthermore, it seems almost impossible to distinguish voice communication from the identification process using RFID in a tooth by the third party. This invisible method is relatively safe to hide the existence of an imbedded RFID from malicious intent and contribute to the protection of individual privacy. There are little physical and mental stresses for the owner.

References

1. Robertson ID, Jalaly I (2003) RFID tagging explained. IEE Commun Eng 1:20–23
2. Dzik WH (2003) Transfusion safety in the hospital. Transfusion 43:1190–1199

Author index

Subject index

AUTHOR QUERY-FORM

Dear Author:

During the preparation of your manuscript for publication, the questions listed below have arisen. Please attend to these matters and return this form with your proof.

Many thanks for your assistance.

Query References	Query	Remarks
1	Au: Please provide Department name and email, Tel or Fax for the corresponding author.	
2.	Au: Please provide upto 5 key words.	
3.	Au: Please confirm that this notation "at%" is correctly given.	
4.	Au: Your original, "sutento", is not found in dictionaries or on line. Do you mean "stent"? Please confirm.	
5.	Au: "stent"?	
6.	Au: "stents"?	
7.	Au: Please confirm this rephrasing.	
8.	Au: Unpublished material cannot be included in the reference list. It can be cited only parenthetically in the text. It has now been deleted from the reference list. Please renumber all subsequent references.	
9.	Au: Please confirm this phrase "with centimeter critical size". The intended meaning does not seem clear.	
10.	Au: From here forward, ref. cites have been renumbered because original #11 has been deleted.	
11.	Au: Unpublished material can be cited only in the text, parenthetically, and cannot be listed in References. It has been deleted from the reference list.	

12.	Au: Reference [32] was deleted from the text as no details regarding the reference were provided.	
13.	Au: Please provide year for this reference.	
14.	Au: Original ref. #11 was unpublished material. It has been deleted from the list. All subsequent numbers, both the citations in the text and on this reference list, have been renumbered accordingly.	
15.	Au: Please confirm whether the page range starts from 163 or 1163.	
16.	Au: This unpublished material has been deleted from the reference list.	
17.	Au: Please check the discrepancy in the legend between the word file and PPT file for Tables 3 and 4. We have retained the legend in the word file.	
18.	Au: Please explain this phrase. Where is it intended to be placed?	
19.	Au: Please confirm "stent," not "sutento".	
20.	Au: Please confirm "stent".	
21.	Au: The Chapter one is too much to fit in 10 pages, pls check and advise.	

If you have any concerns about our products,
you can contact us on
ProductSafety@springernature.com

[illegible] Publishing is established as [illegible] the EU
[illegible]

Springer Nature Customer Service Center GmbH
Europaplatz 3, 69115 Heidelberg, Germany

Printed by Libri Plureos GmbH
in Hamburg, Germany

MIX
Papier aus verantwortungsvollen Quellen
Paper from responsible sources
FSC® C105338

If you have any concerns about our products,
you can contact us on
ProductSafety@springernature.com

In case Publisher is established outside the EU,
the EU authorized representative is:
Springer Nature Customer Service Center GmbH
Europaplatz 3, 69115 Heidelberg, Germany

Printed by Libri Plureos GmbH
in Hamburg, Germany